Photographer's Guide to the Nikon Coolpix B700

Photographer's Guide to the Nikon Coolpix B700

Getting the Most from Nikon's Superzoom Camera

Alexander S. White

WHITE KNIGHT PRESS
HENRICO, VIRGINIA

Copyright © 2017 by Alexander S. White.

All rights reserved.

No part of this publication may be reproduced, stored in a retrieval system or transmitted in any form or by any means, electronic, mechanical, photocopying, recording or otherwise, without the prior written permission of the copyright holder, except for brief quotations used in a review.

The publisher does not assume responsibility for any damage or injury to property or person that results from the use of any of the advice, information, or suggestions contained in this book. Although the information in this book has been checked carefully for errors, the information is not guaranteed. Corrections and updates will be posted as needed at whiteknightpress.com.

Product names, brand names, and company names mentioned in this book are protected by trademarks, which are acknowledged.

Published by
White Knight Press
9704 Old Club Trace
Henrico, Virginia 23238
www.whiteknightpress.com
contact@whiteknightpress.com

ISBN: 978-1-937986-56-8 (paperback)
       978-1-937986-57-5 (ebook)

Printed in the United States of America

To my wife, Clenise.

Contents

Introduction — 1

Chapter 1: Preliminary Setup — 3

    Setting Up the Camera 3
        Charging and Inserting the Battery 3
        Inserting the Memory Card. 4
    Introduction to Main Controls 6
        Top of Camera 6
        Back of Camera 7
        Front and Sides of Camera 7
        Bottom of Camera 8
    Setting the Date, Time, and Language 9

Chapter 2: Basic Operations — 11

    Fully Automatic—Auto Mode 11
        Basic Variations from Fully Automatic 12
            Focus 13
                Manual Focus 15
            Exposure 16
                Exposure Compensation 16
            Flash 17
    Movie Recording. 18
    Viewing Pictures and Movies 20
        Review While in Shooting Mode 20
        Reviewing Images in Playback Mode 20
        Playing Movies 20

Chapter 3: The Shooting Modes — 22

    Auto Mode 22
    Program Mode 23
    Shutter Priority Mode 23
    Aperture Priority Mode 25

Contents | vii

Manual Exposure Mode	27
Scene Modes	30
Landscape	30
Night Portrait	31
Night Landscape	32
The SCENE Setting on the Mode Dial	33
Scene Auto Selector	34
Portrait	34
Sports	35
Party/Indoor	35
Beach	36
Snow	36
Sunset	36
Dusk/Dawn	36
Close-up	37
Food	37
Fireworks Show	38
Backlighting/HDR	38
Easy Panorama	40
Pet Portrait	41
Moon	42
Bird-watching	42
Soft	43
Selective Color	44
Multiple Exposure Lighten	44
Time-lapse Movie	45
Superlapse Movie	46
Creative Mode	47
User Settings Mode	49

Chapter 4: The Shooting Menu — 51

Image Quality	53
Image Size	55
Picture Control	56
Standard	57
Neutral	57
Vivid	57
Monochrome	58
Adjustments to Picture Control Settings	58
Custom Picture Control	59
White Balance	60
Metering	63
Continuous shooting	64
ISO Sensitivity	68
Minimum Shutter Speed	70
Exposure Bracketing	70
AF Area Mode	71
Face Priority	71
Manual (Spot, Normal or Wide)	72
Subject Tracking	72
Target Finding AF	73
Autofocus Mode	73
Flash Exposure Compensation	74
Noise Reduction Filter	75
Active D-Lighting	75
Multiple Exposure	76

 Save User Settings 77
 Reset User Settings 78
 Zoom Memory and Startup Zoom Position 78
 Manual Exposure Preview 79

Chapter 5: Physical Controls   81

 Power Switch 81
 Shutter Release Button 81
 Mode Dial 82
 Zoom Lever 82
 Function Button 1 (Fn1) 82
 Flash Pop-up Button 83
 Side Zoom Control 84
 Snap-back Zoom Button 84
 Fn2 Button 85
 Playback Button 85
 Diopter Adjustment Dial 85
 Monitor Button and Eye Sensor 85
 Display Button 85
 Movie Button 86
 Command Dial 86
 Menu Button 87
 Delete/Trash Button 87
 Multi Selector and its Buttons and Dial 88
  Multi Selector Dial 88
  OK Button 88
  Direction Buttons 89
   Up Button: Flash Settings 89
   Right Button: Exposure Compensation 90
   Down Button: Focus Mode 90
   Left Button: Self-timer; Smile Timer; Pet Portrait Release 91
 AF Assist/Red-eye Reduction/Self-timer Lamp 92
 Ports on Right Side of Camera 93
 Tilting and Swiveling LCD Screen 93

Chapter 6: Playback   95

 Normal Playback 95
 Index Views, Calendar View, and Enlarging Images 95
 Various Playback Screens 97
 Viewing Shots Taken in a Sequence 98
 The Playback Menu 100
  Mark for Upload 100
  Quick Retouch 100
  D-Lighting 101
  Skin Softening 101
  Filter Effects 102
  Slide Show 105
  Protect 105
  Rotate Image 106
  Small Picture 106
  Sequence Display Options 107
  Choose Key Picture 107
 Printing Images 108
  Printing Directly from the Camera 108

Contents | ix

Chapter 7: The Setup Menu — 110

- Time Zone and Date — 110
- Slot Empty Release Lock — 111
- Monitor Settings — 112
  - Image Review — 112
  - Monitor Options — 112
  - EVF Options — 112
  - View/Hide Framing Grid — 113
  - View/Hide Histograms — 113
- EVF Auto Toggle — 114
- Date Stamp — 114
- Vibration Reduction — 115
- AF Assist — 116
- Digital Zoom — 116
- Assign Side Zoom Control — 118
- Sound Settings — 118
- Auto Off — 119
- Format Card — 119
- Language — 120
- Charge by Computer — 120
- Image Comment — 121
- Copyright Information — 121
- Location Data — 122
- Toggle Av/Tv Selection — 122
- Reset File Numbering — 123
- Peaking — 123
- Reset All — 123
- Firmware Version — 124

Chapter 8: Motion Pictures — 125

- Movie-making Overview — 125
  - Quick Guide to Recording a Movie Clip — 125
- Other Settings for Movies — 126
  - Still Photo Settings Available for Movies — 126
    - Focus Mode — 126
    - Exposure, Exposure Compensation, and Exposure Lock — 127
    - Self-timer — 128
    - Picture Control — 128
    - White Balance — 128
    - Metering — 128
    - Vibration Reduction — 128
    - Zoom — 129
  - Taking Still Images During Movie Recording — 129
  - Pausing Recording with the OK Button — 130
  - Fn Buttons: Do Not Operate During Video Recording — 130
  - Settings That Are Not Adjustable for Video Recording — 130
- The Movie Menu — 130
  - Frame Rate — 131
  - Movie Options — 132
    - HS (High Speed) Movie Options — 133
  - Autofocus Mode — 134
  - Electronic VR (Vibration Reduction) — 135
  - Wind Noise Reduction — 136
  - Zoom Microphone — 136
  - Frame Rate — 136

  Movie Playback and Editing. 136
    Playback. 136
    Editing. 137

Chapter 9: SnapBridge App, Superzoom Lens, and Other Topics   139

  SnapBridge App 139
    Initial Connection and General Overview 139
    Summary of Options for Transferring Images and Movies to a Smart Device Wirelessly . . . 143
  The Network Menu 143
    Airplane Mode 144
    Connect to Smart Device. 144
    Send While Shooting 144
    Wi-Fi 144
    Bluetooth 145
    Restore Default Settings 145
  Using the Superzoom Lens 145
  Macro (Close-up) Photography 149
  Using Flash 151
  Infrared Photography 153
  Street Photography 154
  Connecting to a Television Set 156

Appendix A: Accessories   157

  Cases 157
  Batteries and Chargers 158
  External Flash 159
  Filters and Filter Adapters 161
  Pole for Extended-reach Shots 162
  Tripods 162

Appendix B: Quick Tips   163

Appendix C: Resources for Further Information   165

  Photography Books 165
  Digital Photography Review 165
  Reviews of the Coolpix B700 165
  The Official Nikon Site 165
  Photography Information 165

Index   167

Introduction

In 2015, I published *Photographer's Guide to the Nikon Coolpix P610*, a guide to the features and operation of an earlier Coolpix "superzoom" camera. The P610 was popular because of its extremely long zoom range and other advanced features, which make it an ideal camera to take along on a trip, especially for photographing birds or other wildlife, or any subject that lends itself to telephoto shots. With the Coolpix B700, Nikon has issued a new model with the same zoom range and general attributes as the P610, but it has redesignated the camera in a new "B" series, perhaps because of the addition of a few major new features.

In this model, Nikon has added several important refinements without dramatically changing the features that made the P610 so successful. The B700 includes an upgraded digital sensor and the ability to use the Raw format for still images, which adds considerably to the options for post-processing images. In addition, the B700 can record 4K video, providing more resolution than ordinary HD (high-definition) video formats. The camera also takes advantage of Nikon's new SnapBridge app for smartphones and tablets, which lets you upload images and movies from the camera to a smart device and control the camera remotely from the device.

The Coolpix B700 produces excellent image quality with its 21-megapixel digital sensor. That sensor is of the BSI CMOS variety (backside-illuminated, complementary metal oxide semiconductor), which gathers light more efficiently than earlier types, resulting in excellent performance in low light. The B700 lets you make manual settings for focus and exposure if you want to have maximum control over image-making. The camera provides several modes of rapid continuous shooting, exposure bracketing, and numerous special features, including a variety of ways to manipulate colors, a built-in HDR (high dynamic range) shooting option, high-speed video shooting, and time-lapse photography. It has special settings for typical superzoom subjects such as birds and the moon.

The B700 has an electronic viewfinder that provides a clear view of your image even in bright sunlight, when the LCD screen could be washed out by the glare. The LCD display has high resolution, with 921,000 dots, providing fine detail when viewing your images. Moreover, the screen swivels to positions that allow you to take low-level shots near ground level or to hold the camera over your head to shoot over crowds of people or other obstacles.

The B700 is not the perfect camera, of course; no camera can serve as the ideal tool for all situations. One drawback is that the camera lacks an accessory shoe, which could be used to attach items such as an external flash unit. (There is a way to use external flash units with the B700, as discussed in Appendix A.) Also, the camera has a limited range of aperture settings available: from f/3.3 to f/7.6 at the widest focal length, and only from f/6.5 to f/8.2 when the lens is zoomed in fully. This narrow range can limit your ability to make certain kinds of shots, such as those requiring slow shutter speeds, although you can compensate to some extent by using other techniques for such shots. In addition, the camera lacks a built-in GPS functionality for adding location information to images, although you can use the SnapBridge app for geotagging in connection with a smartphone or tablet.

This discussion of the camera's features is not complete, but it serves to illustrate that the Coolpix B700 has capabilities that should be attractive to serious amateur photographers—those who want a camera that has many options for creative control of images without needing to change lenses, and that is compact enough to be carried around at all times, so it will be available when a good picture-taking opportunity arises.

My goal with this guide is to provide a thorough introduction to the camera's features, explaining how they work and when you might want to use them. The book is intended largely for beginning and intermediate photographers who are not satisfied with the documentation provided with the camera and who need a more user-friendly explanation of its many controls and menus. For those seeking more advanced information, I discuss some topics that go beyond the basics, and I include information in the appendices to help you uncover additional resources. This book is not a replacement for the official Nikon Coolpix B700 Reference Manual, which contains a great deal of useful information; my book should be viewed as a supplementary resource to illustrate and explain the use of the camera's features.

One note on the scope of this guide: I live in the United States, and I bought my camera in the U.S. market. I am not familiar with any variations for cameras sold in Europe, the United Kingdom, or elsewhere, such as different batteries or chargers. The photographic functions are not different, though, so this guide should be useful to photographers in all locations. I have stated measurements in both the Imperial and metric systems for the benefit of readers in various countries around the world.

If you find any problems in this book, including typographical errors or information that appears to be confusing or incorrect, please let me know through the contact form at whiteknightpress.com or by e-mail to contact@whiteknightpress.com. If the images do not look good on a particular device, such as an iPad or Kindle, let me know that as well so I can take steps to remedy the situation. Feedback from readers is the best source of information for improving books such as this one. If you have general comments or feedback to provide, you also may want to post a review of the book at Amazon.com or another site that sells the book.

Chapter 1: Preliminary Setup

Setting Up the Camera

If you purchase your Nikon Coolpix B700 new, the box should contain the camera itself, Nikon EN-EL23 lithium-ion battery, charging adapter, neck strap, USB cable, lens cap with cord for attaching it to the camera (or to the neck strap), and a brief Quick Start Guide pamphlet. There also should be a registration card and one or two other items, such as an advertising sheet or safety notice. Nikon does not include an HDMI cable for connecting the camera to a TV set, and it does not include the full instruction manual or software on a disc.

The full Nikon reference manual is available for download as a PDF document from the following website: http://nikonimglib.com/manual/. The Nikon software for viewing and editing images, ViewNX-i, is available at http://downloadcenter.nikonimglib.com. At that site, you also can download Capture NX-D, the Nikon software that lets you process and manipulate the Raw images taken by this camera. To find each of those programs at the site, type its name in the Search box that appears at the site.

It is a good idea to attach the neck strap to the camera right away, and in the same procedure to attach the lens cap to its cord. In this process, you loop the other end of the lens cap cord over the neck strap before it is attached to the camera. When you're finished, the lens cap should be tethered by its cord to the strap. The lens cap cord should be attached to the neck strap where it joins the camera, on the left side as you hold the camera in position for shooting, as shown in Figure 1-1.

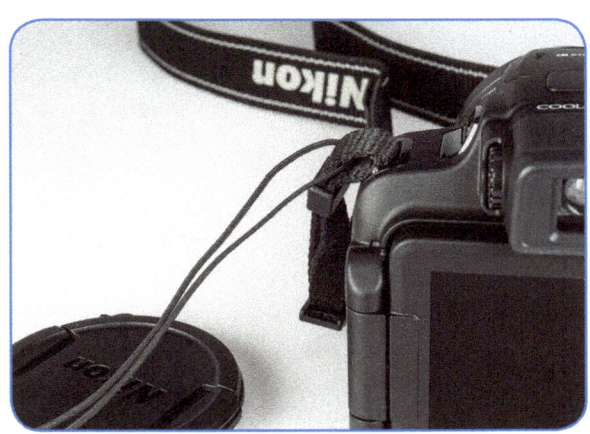

Figure 1-1. Lens Cap Attached to Neck Strap

Charging and Inserting the Battery

The Nikon battery for the Coolpix B700 is the EN-EL23. With this camera, the standard procedure is to charge the battery while it's inside the camera. To do this, you use the supplied USB cable to connect the camera to an AC outlet using the supplied charger, or to a USB port on a computer or other device.

There are pluses and minuses to this approach to battery-charging. On the positive side, the battery can charge automatically when the camera is connected to your computer to upload images, and you need only one cable for both charging the battery and connecting the camera to the computer. Also, you don't have to remove the battery from the camera to recharge it. You can refresh its charge by just plugging the camera into a power source.

If you use the Nikon charger and cable that are supplied with the camera, you can use the camera while the battery is charging, although you can't record movies, and charging will take longer in that case. It also is possible to recharge the battery using a generic USB charger such as those made by Anker and other companies. However, with those chargers, I have found that you cannot operate the camera while charging the battery. Also, Nikon's documentation warns that using a generic USB charger can lead to overheating and possible damage to the camera, so I do not recommend using such a charger.

The main drawback with the system of charging the battery inside the camera is that you cannot charge a spare battery outside the camera. Once the battery dies, you cannot readily replace the battery; you have to stop and recharge the battery. The solution to this situation is to purchase extra batteries and a device that will charge those batteries outside the camera. I'll discuss batteries and other accessories in Appendix A.

To charge the battery, open the battery compartment door on the bottom of the camera and put in the battery. You can only insert it fully into the camera one way. Look for the three gold-colored contacts at one edge of the battery, and insert the battery so those contacts are next to the outside edge of the camera, under the trash-can icon on the camera's back, as it goes into the compartment. Figure 1-2 shows the battery lined up to go into the camera.

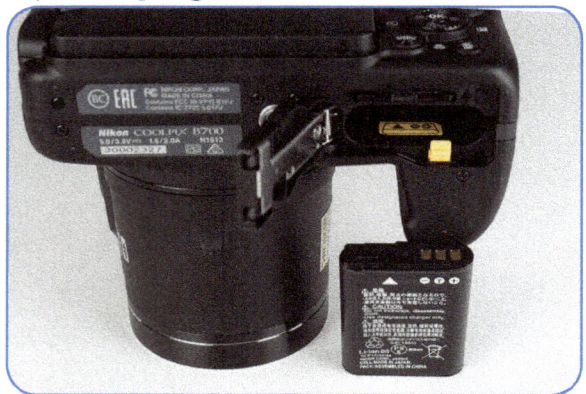

Figure 1-2. Battery Lined Up to Go Into Camera

If the battery will not go all the way down into the compartment, don't force it; check its orientation and make sure it is being inserted the correct way. You may have to push the orange plastic retaining latch to one side to allow the battery to slip all the way into its slot; the latch will then anchor the battery in place, as shown in Figure 1-3.

With the battery inserted into the camera, plug the small end of the USB cable into the USB port under the flap marked HDMI and with a USB symbol on the right side of the camera, as shown in Figure 1-4, and plug the other end of the USB cable into the charging adapter that ships with the camera.

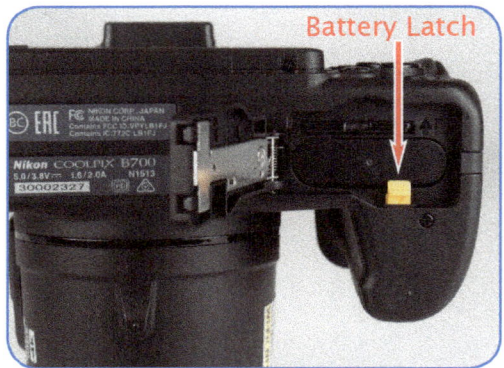

Figure 1-3. Battery Secured by Latch

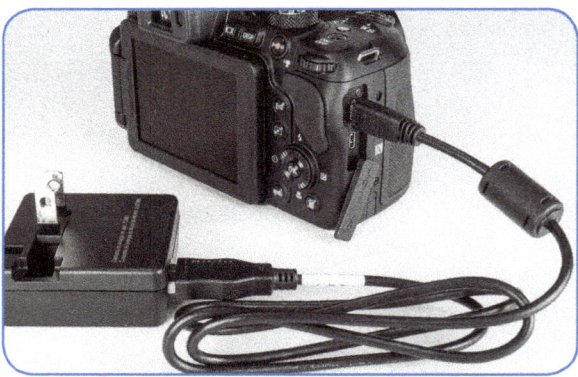

Figure 1-4. Charging Adapter Plugged into Camera

Then plug that charging adapter into a standard electrical outlet or surge protector. A green light around the power switch on top of the camera will blink about twice per second to show that the battery is charging. When the light goes off, the battery is fully charged. It takes about three hours to charge a fully depleted battery using this system. (This length of time is another factor that makes it a good idea to obtain other batteries and an external charger, as discussed in Appendix A.)

You also can charge the battery in the camera by connecting the USB cable to a compatible USB port on a computer, if the Charge by Computer option is turned on through the camera's menu system. I'll discuss that process in Chapter 7.

Inserting the Memory Card

The Coolpix B700, like most cameras these days, does not ship with a memory card included. If you turn the camera on with no card inserted, you will see the message "No card present" in the center of the screen. With default settings, if you ignore this message and press the shutter button to take a picture, nothing will happen; the camera will beep and will not allow you to operate the shutter with no memory card installed.

However, you can change this behavior using the Slot Empty Release Lock item on the Setup menu. If you set that menu item to Enable Release instead of Release Locked, the shutter will operate and allow you to take a few still photos and play them back. However, each of those photos will have "Demo Mode" displayed on it and will not be permanently saved. So, in practical terms, in order to use the camera effectively, you need to obtain and insert an appropriate card.

The B700 uses SD cards, which are quite small, about the size of a postage stamp. They come in several varieties, a few of which are shown in Figure 1-5.

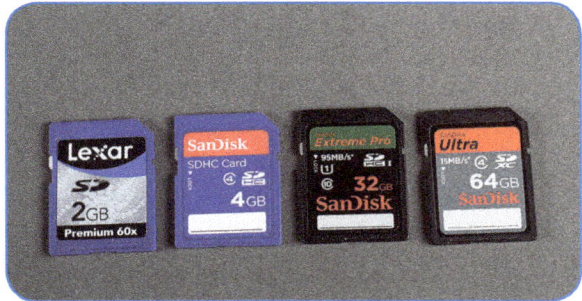

Figure 1-5. SD Cards of Various Capacities

The standard card, called simply SD, comes in capacities from eight MB to two GB. The next higher-capacity card, SDHC, comes in sizes from four GB to 32 GB. The newest, and highest-capacity card, SDXC (for extended capacity), comes in sizes of 48 GB, 64 GB, 128 GB, 256 GB, and up to 512 GB at this writing; this version of the card can have a capacity up to two terabytes (TB), theoretically, and SDXC cards have faster transfer speeds than the smaller-capacity cards. I have used a SanDisk 512 GB Extreme Pro SDXC card, rated in UHS Speed Class 3, in the B700 with no problems.

The B700 also can use micro-SD cards, which are smaller cards, often used in smartphones and other small devices. These cards operate in the same way as SD cards, but you have to use an adapter that is the size of an SD card, as shown in Figure 1-6, to insert this tiny card into the Coolpix B700 camera. You might want to use one of these cards so you can transfer images and videos to a smartphone or other device that accepts that size of card.

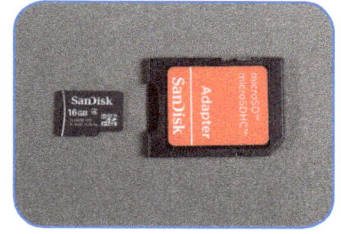

Figure 1-6. Micro-SD Card with Adapter

The type and size of memory card you use depends on your needs and intentions. If you're planning to record a good deal of high-definition (HD) video or large numbers of high-resolution still photos, you should get a card with a large capacity. There are several variables to take into account in computing how many images or videos you can store on a particular size of card, such as which aspect ratio you're using (16:9, 4:3, 3:2, or 1:1), picture size, and quality.

Here are a few examples of how many images (approximately) can be stored on a 64 GB SDXC card: Using Raw+Fine quality, with the Fine images set to the largest size of 20 megapixels, a 64 GB card can hold about 1458 still photos; using Raw quality with no Fine image, the same card can hold about 1924 images. With the image quality set to Fine at the largest size of 20 megapixels, the card can store about 5935 images. With the image quality set to Normal and image size set to eight megapixels, the card can hold more than 10,000 images. With the image quality set to Fine and the image size set to 15 megapixels (and aspect ratio to 1:1), the card can hold about 7854 images. With settings of Fine and eight megapixels, it can hold more than 10,000 images.

Another consideration is the speed of the card. If you plan to record high-quality video or do a lot of continuous (burst) shooting, you should get a card that is rated in Class 6 or higher for its speed. If you plan to record movies using the highest quality setting of 4K UHD at the 2160 setting, you should use a card rated in UHS Speed Class 3 or faster. (As of this writing, that is the fastest speed class available.)

As I write this, good-quality 64 GB SDXC cards cost about $20 and up, depending on speed. If you don't mind the risk of losing a great many images or videos if you lose the card, you might want to choose an SDXC card with a capacity of 64 GB, or even 128 GB. At this writing, 256 GB SDXC cards are selling for about $100 and up, so you might want to consider that option as well. Cards with a capacity of 512 GB are selling for about $220 and up, but the prices of those cards should continue to decline, making them a reasonable choice in the near future, I expect.

You have quite a few options for choosing a memory card. I like to use a high-speed 64 GB SDXC card, just to have extra capacity and speed in case they are needed.

Once you have selected your memory card, open the door on the bottom of the camera that covers the battery compartment, and slide the card into the card slot until it catches, with the label facing the back of the camera. Once the card has been pushed down until it catches, close the compartment door and push the latch back to the locking position. To remove the card, push down on it until it releases and springs up so you can grab it. Figure 1-7 shows a card being inserted into the B700.

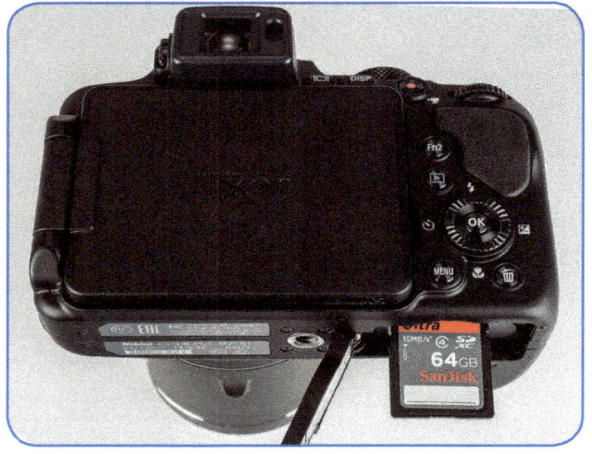

Figure 1-7. Memory Card Being Inserted Into Camera

One note for shooting continuous pictures with the B700: When the camera is writing its image data to the memory card, the camera displays a circulating series of greenish blocks in the center of the display screen, as shown in Figure 1-8. While that indicator is displayed, it's important not to turn off the camera or otherwise interrupt its functioning, such as by taking out the battery or disconnecting an AC power adapter. You need to let the card complete its recording process before taking further steps.

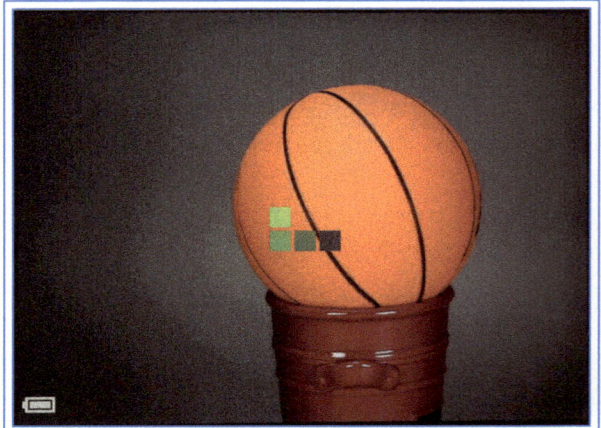

Figure 1-8. Blocks on Screen During Memory Card Access

Introduction to Main Controls

Before I discuss options for setting up the camera to take images and videos, I will introduce the main controls so you'll have a better idea of which button or dial is which as I discuss them later in this and other chapters. I will briefly mention the functions of the controls here; I will cover them in more detail in Chapter 5. The following series of images shows the major controls. As I discuss each item, I will describe its position and function; you may want to refer to these images later for a reminder about each control.

Top of Camera

On top of the camera are several important controls and dials, shown in Figure 1-9.

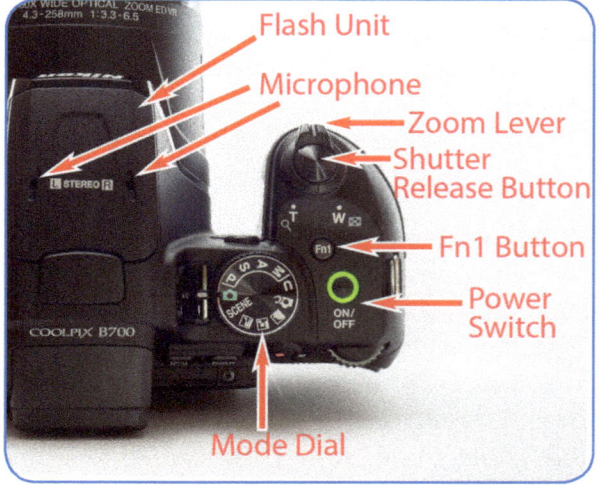

Figure 1-9. Controls on Top of Camera

The mode dial is used to select the shooting mode for still images. For basic shooting with the camera making most of the decisions, turn this dial so the green camera icon is next to the white selection marker. The shutter release button is used to take pictures; press it halfway down to lock focus and exposure, and press it all the way down to record the image. The zoom lever that surrounds the shutter release button is used to zoom the lens between its wide-angle and telephoto focal lengths. It also can be used to enlarge images on the camera's display and show index screens of images in playback mode. The Fn1 button is a versatile control that provides quick access to a single menu option of your choice. The power switch is used to turn the camera on and off. The microphone picks up sounds when you are recording movies. The pop-up flash unit is stored inside the camera until you release it with the pop-up switch, discussed later in this section.

Back of Camera

The controls on the camera's back are seen in Figure 1-10.

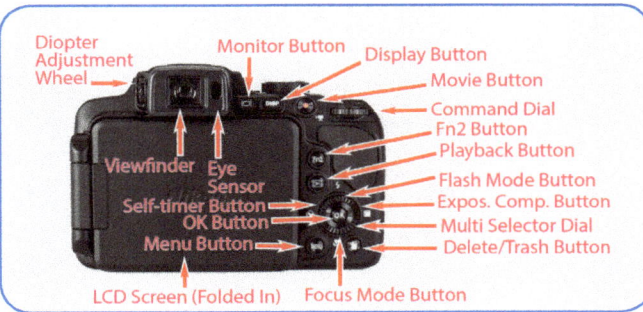

Figure 1-10. Controls on Back of Camera

The viewfinder is where you can look to compose your shots when you have the LCD screen folded in against the camera or you have selected the viewfinder using the Monitor button.

The diopter adjustment wheel directly to the left of the viewfinder is used to adjust the view according to your vision. The eye sensor, to the right of the viewfinder window, senses the presence of your head near the viewfinder and switches the display from the LCD screen to the viewfinder. (That behavior can be changed using the EVF Auto Toggle option on the Setup menu, as discussed in Chapter 7.)

The Monitor button switches between using the LCD screen and the viewfinder. The Display button selects screens for viewing information about shooting settings when images are being recorded and about the images themselves when they are being played back. The red Movie button starts and stops the recording of a video sequence. The Delete button (also called the Trash button) is used to delete recorded images.

The command dial is used to adjust settings such as shutter speed, manual focus, and a few others. The Playback button places the camera into playback mode so you can view your recorded images. The Menu button calls up the camera's system of menu screens with various settings for shooting and other values, such as control button functions, audio features, and others. The Fn2 button is a second programmable button that you can assign to carry out a function of your choice. The multi selector dial acts as a wheel for setting values such as aperture and for navigating through menu screens. In addition, each of its four edges acts as a button when you press it in, for selecting items including flash mode, exposure compensation, focus mode, and the self-timer. (Those buttons are called the Up, Down, Left, and Right buttons in this book.) The OK button in the center of the dial is used to confirm selections and for some miscellaneous operations. The LCD screen, also called the monitor, displays the live view of the scene the camera is viewing as well as shooting settings when you are not using the viewfinder, and it displays your recorded images and videos when the camera is in playback mode. It can tilt and swivel to help in viewing scenes from various angles.

Front and Sides of Camera

On the sides and front of the camera, there are several controls and other items, as shown in Figures 1-11 through 1-13.

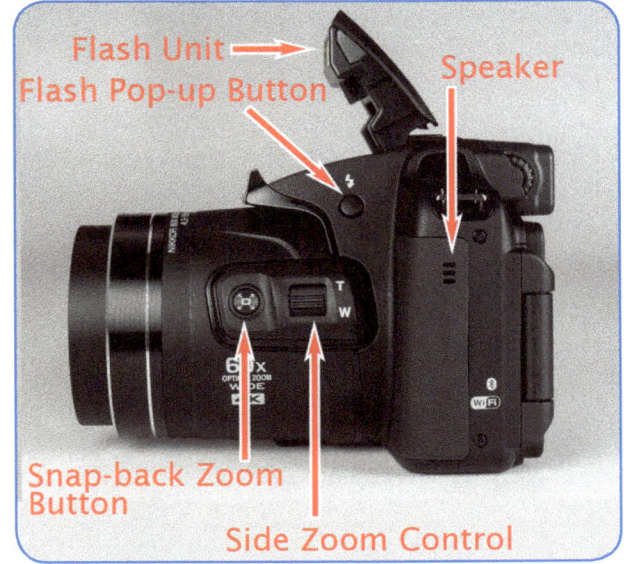

Figure 1-11. Items on Left Side of Camera

The round flash pop-up button, located on the upper part of the left side of the B700, is used to release the camera's built-in flash unit so it can fire if it is needed, as seen in Figure 1-11. The unit pops up at an angle, as shown in that figure. The single speaker, which emits audio for movies and operational sounds, is also on the left side of the B700. The side zoom control is used to zoom the lens, or you can assign it to control manual focus instead, using the Setup menu. The snap-back zoom button is used to move the zoom lens quickly to a wider view while you hold the button down, and back to its zoomed-in position when you release it.

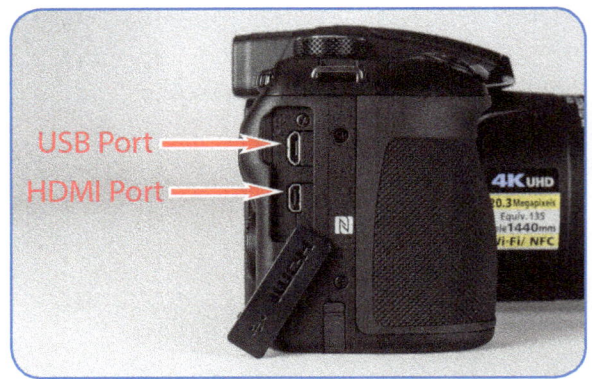

Figure 1-13. Ports on Right Side of Camera

The ports on the right side of the camera, located under a flap that can be rotated out of the way, as shown in Figure 1-13, are the micro-USB port, top, and the micro-HDMI port, bottom. The micro-USB port is for connecting the camera to a computer, charger, or printer, using the cable provided with the camera. The micro-HDMI port is for connecting the camera to an HDTV set for playback of images and videos using an optional HDMI cable. There is no audio-video port for a cable to connect the camera to a standard (non-HD) TV set. The fancy letter N marks the location of the camera's NFC (near field communication) antenna. You can touch this spot against the corresponding spot on an Android smartphone or tablet to initiate a wireless connection without having to use the camera's menu system.

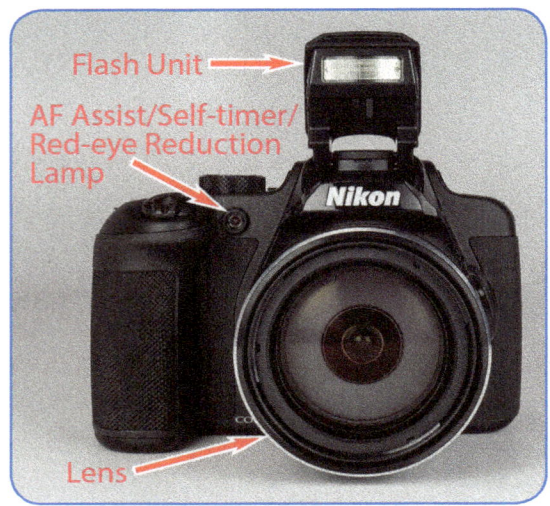

Figure 1-12. Items on Front of Camera

The AF Assist/Self-timer/Red-eye Reduction Lamp, shown in Figure 1-12, helps the camera use its autofocus technology in dimly lighted areas, and it lights up to indicate the functioning of the self-timer. In addition, when the flash mode is set to Auto with Red-eye Reduction, this bright lamp lights up before the flash fires, to constrict the pupils of a human subject's eyes, so the flash will not bounce off the retinas to cause the unpleasant "red-eye" effect. The camera's lens is a variable focal length, or zoom lens, with an optical zoom range from 24mm to 1440mm. It has aperture settings ranging from f/3.3 to f/8.2, as discussed in Chapter 3.

Bottom of Camera

Finally, as shown in Figure 1-14, there are two main items on the bottom of the camera: the tripod socket, where the camera can be attached to a tripod with a standard screw, and the latching door that covers the compartment where the memory card and battery are located.

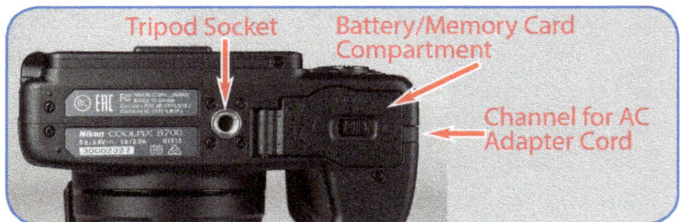

Figure 1-14. Items on Bottom of Camera

When the camera is placed on a tripod, you cannot open this door to get access to the battery or memory card. At the outer edge of the door is a flap that must be opened up when you install the optional AC adapter in the camera, so the door can close with the cord running

Chapter 1: Preliminary Setup | 9

through the channel occupied by the flap. I discuss that AC adapter in Appendix A.

Setting the Date, Time, and Language

You need to set the date and time correctly before you start taking pictures, because the camera records that information (sometimes known as "metadata," meaning data beyond the information in the picture itself) invisibly with each image, and displays it later if you want. Someday you may be very glad to have the date (and even the time of day) correctly recorded with your archives of digital images. If you purchase the camera new, it will prompt you to set the date and time when you first power it on.

If you later need to set the time and date, you can use the Setup menu. To do this, press down on the camera's power switch, marked On/Off, on top of the camera, to turn the camera on. Then press the Menu button at the lower left of the OK button on the camera's back. Next, press the Left button, which is marked with a timer icon. When you press the Left button, the yellow selection highlight will move to the far left of the screen, to the list of icons, with an icon or letter representing the current shooting mode, such as P for Program or an icon for Landscape or Night Portrait, at the top of the list.

Use the Down button, marked with a flower icon, to move the selection highlight down to the wrench icon that represents the Setup menu, as shown in Figure 1-15.

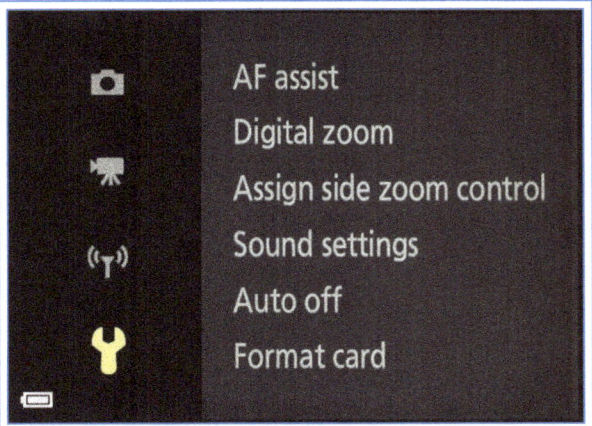

Figure 1-15. Wrench Icon for Setup Menu Highlighted

Press the Right button, marked with a plus and minus sign, to move the highlight back to the right, where it will become a yellow rectangle highlighting a menu item.

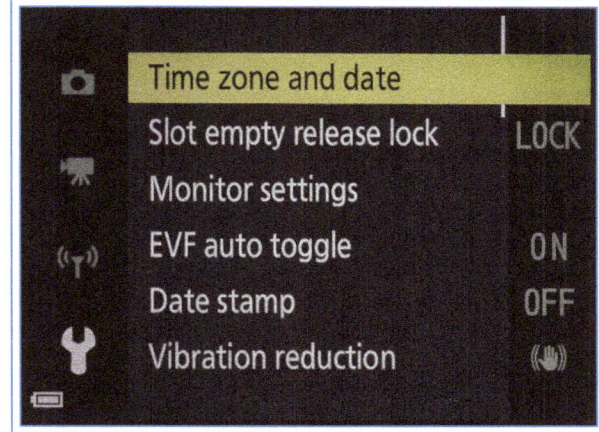

Figure 1-16. Time Zone and Date Highlighted on Setup Menu

Then use the Up and Down buttons (or turn the multi selector dial) to move the yellow selection bar to the Time Zone and Date line on the menu, as shown in Figure 1-16, and press the OK button to move to a screen with choices of Sync with Smart Device, Date and Time, Date Format, and Time Zone.

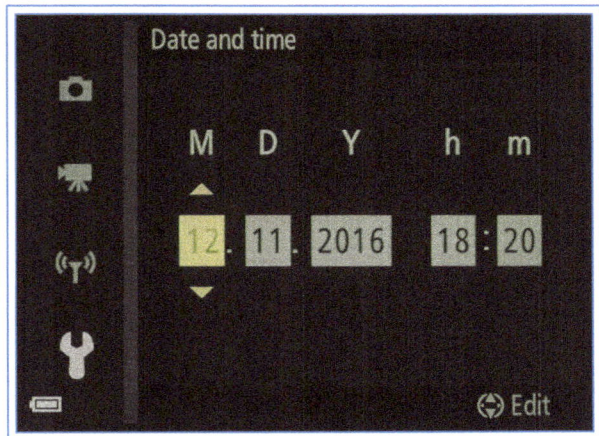

Figure 1-17. Time and Date Settings Screen

Highlight Date and Time, then press the OK button to move to the screen with a selection of settings, shown in Figure 1-17. You can press the Right button instead of the OK button to move to these menu screens, if you prefer.

On the settings screen, use the Left and Right buttons to move through the date, year, and time settings, and change the settings by pressing the Up and Down buttons. (You also can turn the multi selector dial or the command dial to change the settings.) When everything is set correctly, press the OK button to confirm and press the Menu button to exit the menu system.

As I will discuss in Chapter 7, you can also choose the Sync with Smart Device option, which will set the camera's clock to agree with the date and time on a smartphone or tablet that is connected to the camera using the SnapBridge app. (That app is discussed in Chapter 9.) For general use, though, it is easier to use the Time Zone and Date menu option, as discussed above.

If you need to change the language that the camera uses for the menus and other messages, navigate on the Setup menu to the line that says Language, and press the OK button or the Right button to select the Language menu item. Then navigate with the Up and Down buttons (or the multi selector dial) to the language of your choice, as shown in Figure 1-18, and press the OK button to select it.

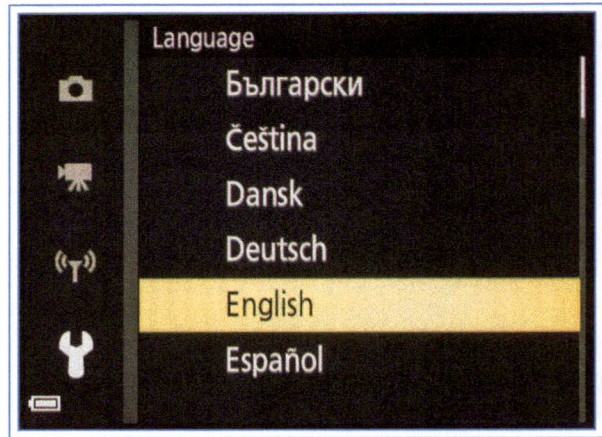

Figure 1-18. Language Selection Screen on Setup Menu

Then press the Menu button to exit from the menu system.

Chapter 2: Basic Operations

Once the Coolpix B700 has the correct time and date set and has a fully charged battery inserted along with a memory card, it is ready to take pictures and capture videos. For now, I won't discuss all of the available options and why you might choose one over another. I'll just outline a set of steps that will get the camera into action and will record a decent image or video sequence on your memory card.

Fully Automatic—Auto Mode

Here's a procedure to use if you want to let the camera make (almost) all of the decisions for you. This is a good system to use if you need to grab a quick shot without fiddling with settings, or if you're new at this and would rather let the camera take control without having to provide much input.

1. Remove the lens cap from the lens and let it dangle by its cord. (If you forget to remove the cap before turning on the camera, that's okay; Nikon has engineered the B700 to have the lens cap attached to the moving part of the lens, so the cap will not block the motion of the lens in extending out from the camera. But you will notice that the camera's display is black, because the lens cap will be blocking the view.)

2. Press the On/Off button. The LCD screen will illuminate to show that the camera has turned on. (If the LCD screen is folded in the closed position, the viewfinder will operate instead of the screen, as discussed in Step 5, below.)

3. Find the mode dial on top of the camera and turn the dial until the green camera icon is next to the white indicator line. This selects Auto shooting mode, as shown in Figure 2-1.

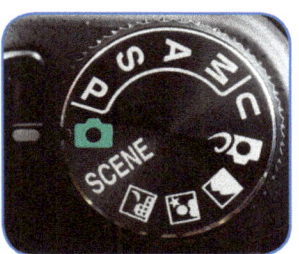

Figure 2-1. Mode Dial at Auto

4. Press the Menu button at the bottom left of the control area on the back of the camera. Use the direction buttons (four edges of the ridged multi selector dial) to navigate to the entry for Image Quality, select that line with the OK button or the Right button, highlight the Fine setting, and press OK. Then navigate down to the Image Size setting, select it, and choose 5184 x 3888 pixels, the top option, as shown in Figure 2-2. Press the Menu button to return to shooting mode.

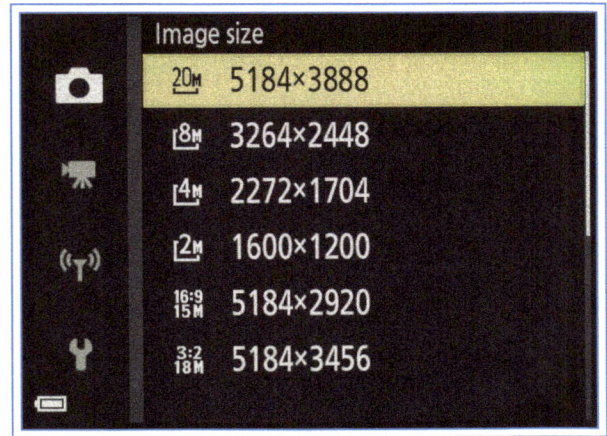

Figure 2-2. Image Size Menu Item Highlighted

(You can choose other settings for either or both of these options if you wish, but the ones I mentioned provide the highest quality, apart from using Raw for Image Quality, which I will discuss later.)

5. If you want to compose your image on the LCD screen on the back of the camera, no action is needed if the LCD screen has its active surface

exposed and is turned on. If the screen is folded against the camera, pull it out to the left and swivel it so the screen is visible. If necessary, press the Monitor button to activate it. To use the viewfinder instead, press the Monitor button again.

6. Look at the screen or into the viewfinder to compose the shot and view the camera's settings. You can adjust the viewfinder for your eyesight by turning the diopter adjustment wheel on the left side of the viewfinder's housing. Toggle between the LCD and the viewfinder by pressing the Monitor button. If the EVF Auto Toggle menu option is turned on in the Setup menu, the view will switch from the LCD to the viewfinder when your head approaches the viewfinder. (That option is discussed in Chapter 7.)

7. If you are indoors or otherwise in conditions that might call for the use of flash, press the button on the left side of the camera's built-in flash unit, marked with a lightning bolt, to pop up the flash.

8. If you have popped up the flash, press the Up button on the multi selector, marked with another lightning bolt, to bring up the flash mode menu, shown in Figure 2-3. Make sure the Auto setting, at the top of this menu, is highlighted. (Later in this chapter and in Chapter 9, I'll discuss the other flash options.)

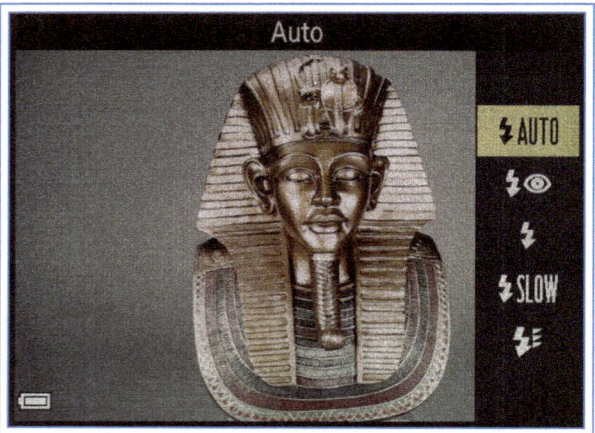

Figure 2-3. Flash Mode Menu

9. Aim the camera toward the subject and look at the LCD screen (or into the viewfinder window, depending on your choice in Step 5) to compose the scene as you want it. Locate the zoom lever on the ring that surrounds the shutter button on the top right at the front of the camera. Push that lever to the left, moving its indicator toward the letter W, to get a wider-angle shot (including more of the scene in the picture), or to the right, moving the indicator toward the letter T, to get a telephoto, zoomed-in shot. Or, if you prefer, use the equivalent zoom switch on the left side of the lens, moving it up for telephoto or down for wide-angle.

10. Once the picture is composed as you want it, press the shutter release button halfway down. You should hear a beep and see one or more green rectangles on the display, meaning the picture is in focus. A flashing red rectangle means the camera is having difficulty achieving focus. In that case, try moving the camera to a different angle before pressing the shutter button halfway again.

11. Press the shutter button all the way down to take the picture.

Basic Variations from Fully Automatic

I won't discuss all of the shooting modes now, except to name them. Besides Auto, which I just discussed, there are Program, Shutter Priority, Aperture Priority, Manual, User Settings, Creative, Landscape, Night Portrait, Night Landscape, and Scene. I'll discuss all of those modes in Chapter 3, and movie shooting in Chapter 8. In this section, I will discuss some functions and features of the Coolpix B700 you can adjust for various picture-taking situations you may be faced with. Not all of the settings can be adjusted in Auto mode, so you should set the camera to a lower level of automation for now, to Program mode. In that mode, you'll be able to control most of the camera's functions for taking still pictures.

I'm not going to repeat the preliminary steps for taking a picture, because those are pretty basic. If you need a refresher on those items, see the list in the above discussion of Auto mode.

Start by setting the mode dial on top of the camera to P, for Program, as shown in Figure 2-4. When you press the Menu button there will be more options available on the Shooting menu. Instead of the two menu lines available in Auto mode, you are presented with four menu screens containing 20 settings you can adjust, including white balance, ISO, metering mode, exposure bracketing, and others.

Chapter 2: Basic Operations | 13

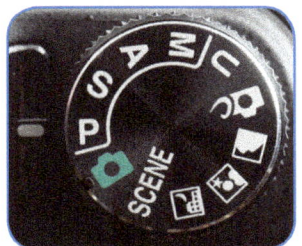

Figure 2-4. Mode Dial at P for Program

In Program mode, the camera will determine the proper exposure, both the aperture (size of opening to let in light) and the shutter speed (how long the shutter stays open to let in light). In this mode you can't make many decisions about those two settings; you can have more control over your settings in other modes, which I'll discuss in Chapter 3. That still leaves lots of decisions you can make, though, so I'll discuss a few settings you can adjust in Program mode.

Focus

Now that the camera is not in Auto mode, you have more control over focus. Specifically, you can set the camera to the MF option, for manual focus, which is not available in Auto mode. You also can select which of several types of autofocus operation you want the camera to use, if you opt for autofocus.

I'll discuss focus in more detail in Chapters 4 and 5. For now, here is how to select a standard focus mode. First, press the Down button (marked by a flower icon). This action puts a small menu with four options on the display, as shown in Figure 2-5. Starting at the top, they are the letters AF, for normal autofocus; the flower icon, for macro (close-up) focus; the mountain icon, for focus on infinity; and the letters MF, for manual focus.

For now, use the direction buttons to select the top icon, for normal autofocus. (You have to be quick; the four choices disappear within a few seconds.) Press the OK button to select and confirm your choice. The letters or icon for your choice will appear at the upper left of the display, as shown in Figure 2-6, unless the choice is AF; if you select AF, those letters will appear for a few seconds and then disappear, because that is the default setting. The icon shown in Figure 2-6 represents the infinity setting.

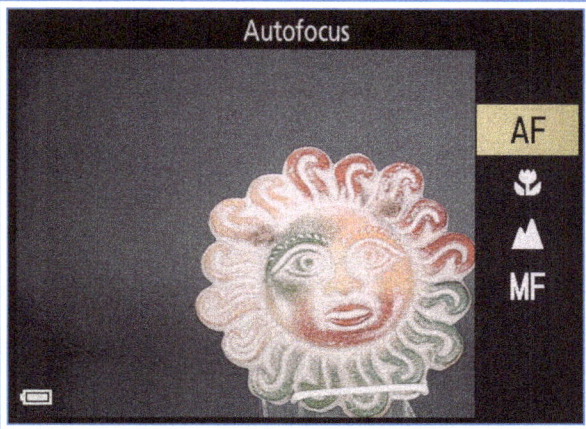

Figure 2-5. Focus Mode Menu

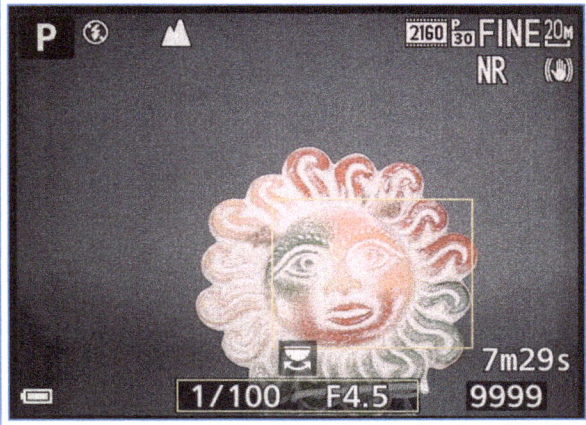

Figure 2-6. Infinity Focus Mode Icon on Shooting Screen

There are several other focus-related options you can set, but for now I will discuss only one of them. Press the Menu button at the bottom of the camera's back to enter the shooting menu. With the selection block in the list of menu items, scroll with the Up and Down buttons until the yellow selection rectangle highlights the line for AF Area Mode on the second screen of the menu, as shown in Figure 2-7.

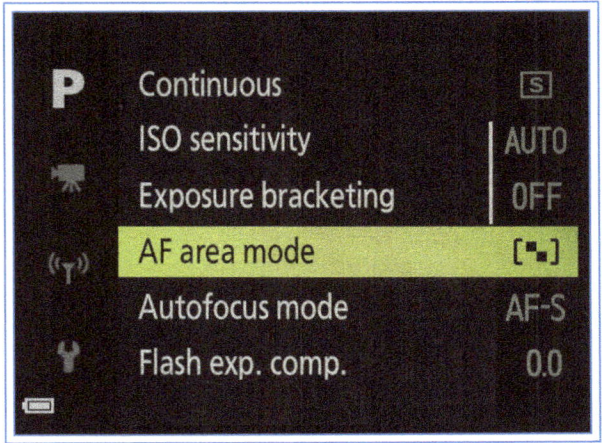

Figure 2-7. AF Area Mode Menu Options Screen

Press OK to select that item, then use the Up and Down buttons to change the setting to Manual (Normal), as shown in Figure 2-8.

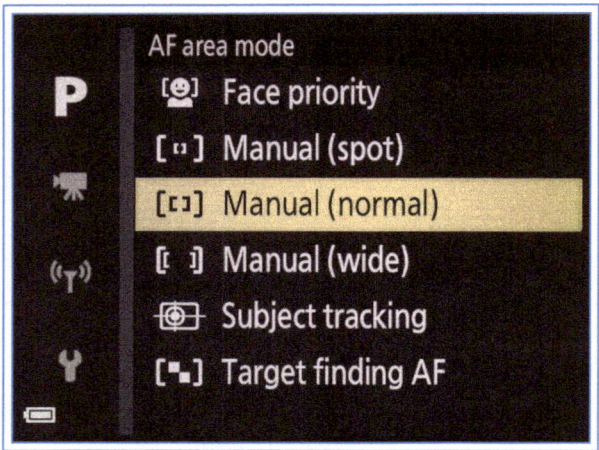

Figure 2-8. Manual (Normal) Option for AF Area Mode

Press the OK button to confirm the selection, then press the shutter button halfway to go to the shooting screen. You should see a focus frame with four arrows, as in Figure 2-9.

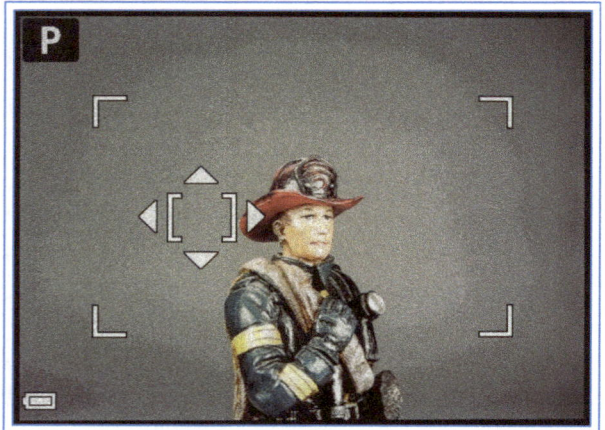

Figure 2-9. Movable Focus Frame on Shooting Screen

(If you don't see the arrows, press the OK button to make them appear.)

With this setting, the camera places a focus frame in the center of the screen. You can move the frame around the screen using the direction buttons or the multi selector dial; I'll discuss that process in Chapter 4. For now, press OK to confirm, and the focus frame will be locked at its current position.

When you aim the camera, center the subject between the white focus brackets, as shown in Figure 2-10.

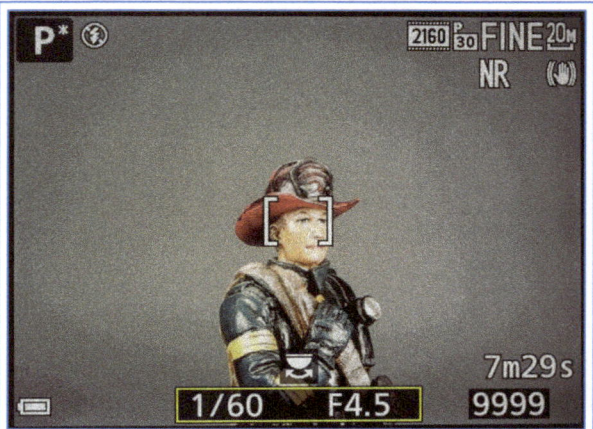

Figure 2-10. Subject Centered in White Focus Frame

Press the shutter button halfway so the camera will evaluate exposure and focus. You should hear a beep and the focus brackets should turn green to confirm the focus, as shown in Figure 2-11.

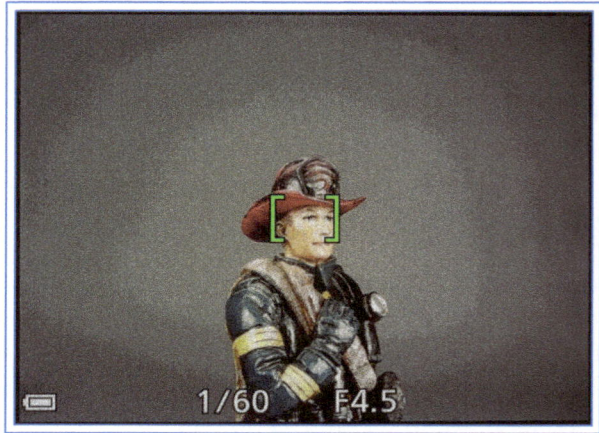

Figure 2-11. Focus Frame Turned Green When Focus Achieved

If everything looks good to you, press the shutter button all the way down to take the picture.

Suppose you want to take a picture in which your main subject is not in the center of the screen. Maybe your shot is set up so that a person is standing off to the right of center, and there is some attractive scenery to the left in the scene. Place the focus frame over the part of the picture that needs to be in focus—in this example, the person to the right. Then press the shutter button halfway down until the camera focuses and beeps. Keep the button pressed halfway to lock in the focus (and exposure) while you move the camera back to the left to create your desired composition, with the person off to the right. Then take the picture, and the area you originally focused on will be in focus.

There is another way to handle this sort of situation, by moving the focus frame using the direction buttons or multi selector dial so it covers the subject, as mentioned above; I'll discuss that option, as well as several other options for autofocus, in Chapter 4.

Manual Focus

Manual focus is the other major option for focusing. Many photographers like the amount of control that comes from being able to set the focus exactly how they want it. And, in some situations, such as focusing in dark areas or areas behind glass or wire fences, taking extreme close-ups, or cases where there are objects at various distances from the camera, it may be useful to control exactly where the point of sharpest focus lies.

For example, one day I went into the back yard to experiment with using the superzoom lens to capture images of birds in our small fountain. It was a cloudy day, and the autofocus mechanism was having problems settling on a sharp focus. I finally decided to switch to manual focus, and the results improved. Of course, manual focus was useful on that occasion partly because I was sitting in one place, the birds tended to stay in one place to take a bath, and the fountain wasn't going anywhere; with a moving subject, manual focus is not likely to be as useful.

To use manual focus, with the camera in Program mode, press the Down button, with the flower icon; on the menu that appears, navigate to the MF icon and select it by pressing the OK button, as seen in Figure 2-12.

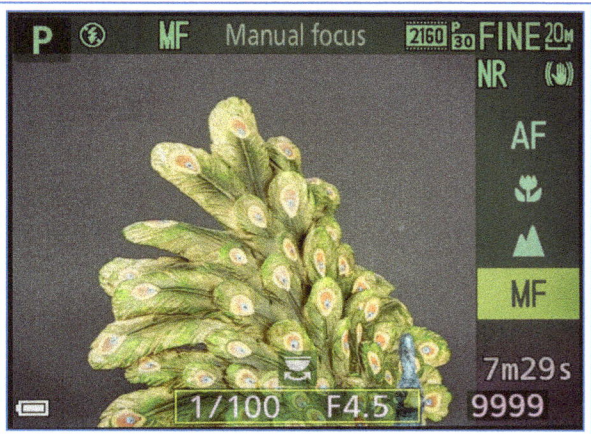

Figure 2-12. MF Icon Highlighted on Focus Mode Menu

Now the camera is set for manual focusing. When the camera is first set to manual focus mode, the screen is displayed at its normal magnification, or at two or four times its normal magnification, as shown in Figure 2-13, where 2x magnification is in effect.

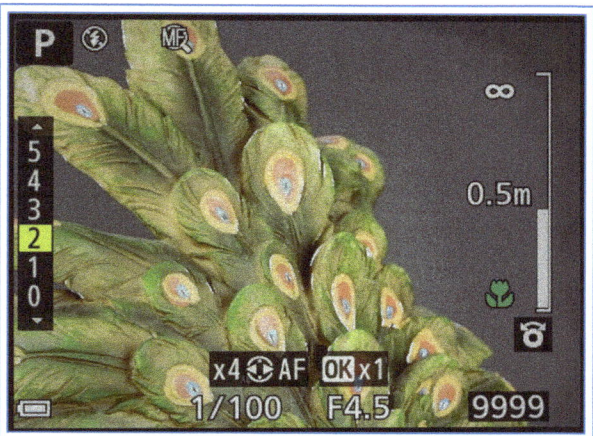

Figure 2-13. Manual Focus with 2x Magnification in Effect

The amount of magnification is saved from the last time manual focus was used. Note, though, that the system Nikon uses for displaying the current magnification can be somewhat confusing. If you see the number 4 next to an icon for the Left button, as in Figure 2-13, that means 2x magnification is in effect, and you can press the Left button to switch to 4x magnification. If you see the number 2 in that position, that means 1x magnification is in effect, and you can press the Left button to switch to 2x magnification. If you see the number 1, that means 4x magnification is in effect, and you can press the Left button to switch to 1x magnification.

To adjust focus using the manual focus setting, turn the multi selector dial or the command dial. Look at the focusing scale on the right side of the screen and turn either dial until the focus appears as sharp as possible. As shown by the prompts in the lower left corner of the screen, you can press the Left button to switch the magnification amount or the OK button to set the screen to normal size (shown as x1 on the display) with focus locked at the current setting. If you return the screen to its normal magnification using the OK button, you can press the OK button again to go to the magnified screen to continue adjusting focus.

As you move the focus point using the multi selector dial or the command dial, you will see a white bar go up and down inside the focus scale. (The bar turns green when focus is in the macro range.) Continue adjusting until the focus is as sharp as you can get it, and then take the picture.

If you need to use the multi selector dial for another function, such as adjusting aperture in Aperture Priority mode, press the OK button to return the screen to its normal size and lock focus. You can then use the dial for its other operations. To return to adjusting manual focus, press the OK button again.

If you want the camera to assist you with its autofocus capability, you can press the Right button, as prompted on the screen, and the camera will autofocus on the subject in the center of the screen. You can then continue adjusting focus manually using the multi selector dial or the command dial.

If the Peaking feature is turned on through the Setup menu, you can press the Up and Down buttons to change the intensity of the Peaking effect on the scale numbered from 1 to 5 at the left of the display. This feature displays an increasing number of white pixels at the areas of sharp focus as the focus gets sharper. I'll discuss that option in Chapter 7.

Exposure

Next, I'll discuss some possibilities for controlling exposure, beyond just letting the camera make the decisions. The Coolpix B700's Auto mode does a good job of choosing the right exposure, and so does the Program mode. But there are some situations in which you may want to override the camera's automation.

Exposure Compensation

First, the camera provides a control for adjusting exposure to account for an unusual, or non-optimal, lighting situation.

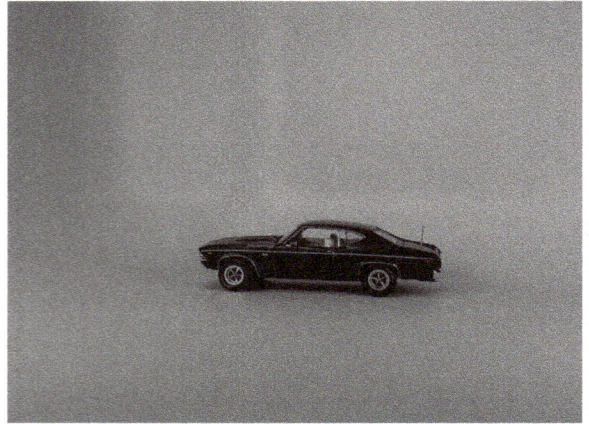

Figure 2-14. Subject in Need of Exposure Compensation

For example, consider Figure 2-14, which shows the B700's view of a dark green model car. Because the model is in front of a white background, the camera's autoexposure system makes the exposure too dark, to account for the large expanse of white.

One solution to this problem is to use the camera's exposure compensation control. Look closely at the Right button on the multi selector. That button is labeled with little plus and minus signs, with the plus on a black background and the minus on white. This control activates the exposure compensation system, which will override the automatic exposure as much as you tell it to, within limits.

Select Program mode and aim at your subject. (You also can select Auto mode, because exposure compensation is available in that mode.) Press the Right button, and a vertical scale will appear on the right side of the display, with a plus sign at the top and a minus sign at the bottom, as shown in Figure 2-15.

Once the exposure compensation scale has appeared, press the Up and Down buttons or turn the multi selector dial or the command dial to move the values higher or lower. The changes in exposure are indicated by yellow tick marks that appear on the scale to show the value that is being set. If you move the yellow marks all the way to the bottom of the scale, the picture will be considerably darker than the automatic exposure would produce. If you move the yellow tick marks to the top, the picture will be noticeably brighter.

The camera's screen brightens and darkens to show you how the exposure is changing, before you take the picture. The camera also displays a histogram—a chart showing peaks and valleys of brightness values—on the left side of the screen. In this case, you would adjust the exposure to be brighter, so the camera will expose for the model car properly, and let the background show up as a brighter (and more accurate) white than in the first image.

I'll discuss the histogram in more detail in Chapter 6. Basically, brighter values in your image skew the peaks in the chart to the right, and darker ones skew them to the left. In this case, I adjusted exposure upward by 1.7 EV (exposure value), moving the histogram to the right and resulting in a more normally exposed image, as shown in Figure 2-15.

Chapter 2: Basic Operations

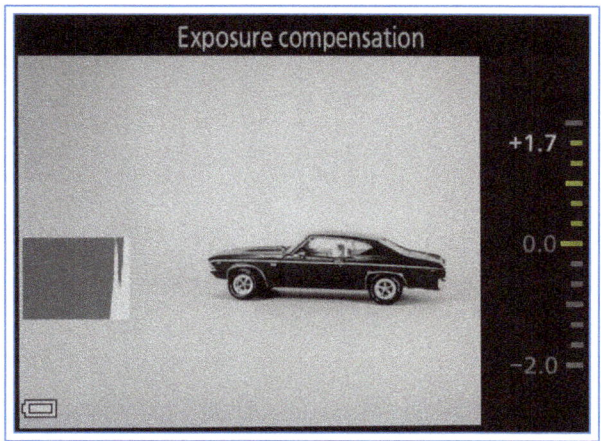

Figure 2-15. Exposure Compensation Adjustment of +1.7 EV

After taking the picture, you should reset the exposure compensation back to zero, in the middle of the scale, so you don't unintentionally affect the pictures you take later. You need to be careful about this, because the camera will retain any exposure compensation value you set, even when it's turned off and then on again. If a positive or negative value for exposure compensation is in effect, the camera will display that value, along with the exposure compensation icon, in the lower right corner of the display, as shown in Figure 2-16.

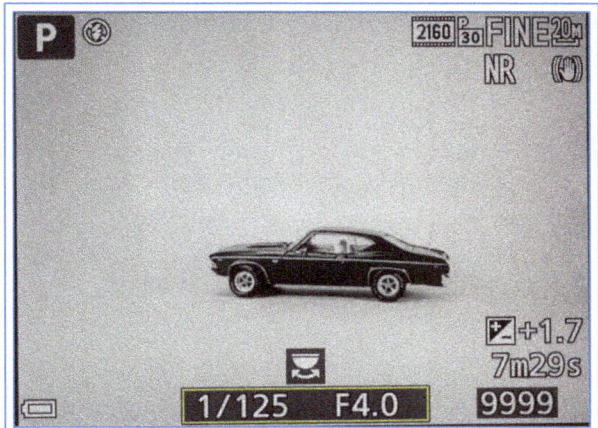

Figure 2-16. Exposure Compensation Icon on Shooting Screen

Flash

In Chapters 3 and 4, I'll discuss several other topics dealing with exposure, such as the Manual exposure, Aperture Priority and Shutter Priority shooting modes, exposure bracketing, Active D-Lighting, and others. Now I will discuss the basics of using the Coolpix B700's built-in flash unit. In Chapter 9, I'll discuss other options for using the flash, such as the Slow Sync mode and correcting "red-eye." In Appendix A, I'll discuss using other flash units.

The built-in flash can provide enough illumination to let you take pictures in dark areas and to brighten up areas otherwise lost in shadows, even outdoors on a sunny day. Here is a fundamental point to be aware of: The built-in flash will not pop up by itself. If you are in a situation in which you think flash may be needed or desirable, you need to take the first step of popping up the flash unit. To do so, find the small, round button on the left side of the flash unit, marked by a lightning bolt. Press in on this button, and the flash springs up into place. (When you're done with the flash, just push down on the unit until it catches again.)

Even though you have popped up the flash unit, in some shooting situations it will never fire. In the situations in which you're likely to want it to, though, it will be ready to illuminate your subject as well as it can.

Here is a common scenario to show how the flash works. Make sure the camera is turned on and the flash has been released by pressing the flash pop-up button. Turn the mode dial to select the Night Landscape shooting mode, represented by the icon just below the Scene mode icon, as shown in Figure 2-17.

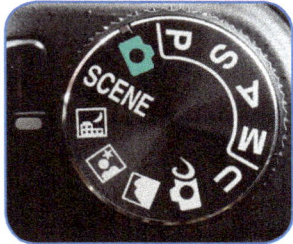

Figure 2-17. Mode Dial at Night Landscape

Next, press the Up button, with the lightning bolt icon on it. Nothing will happen. In the upper left corner of the display you will see the universal negative symbol—a circle with a line through it—over a lightning bolt, indicating that the flash is turned off, as shown in Figure 2-18.

(If you don't see this symbol, press the Display button, to the right of the viewfinder, to switch to the more detailed shooting screen.) In this situation, because you have chosen a shooting mode (Night Landscape) that will not use flash under any circumstances, you cannot turn the flash on.

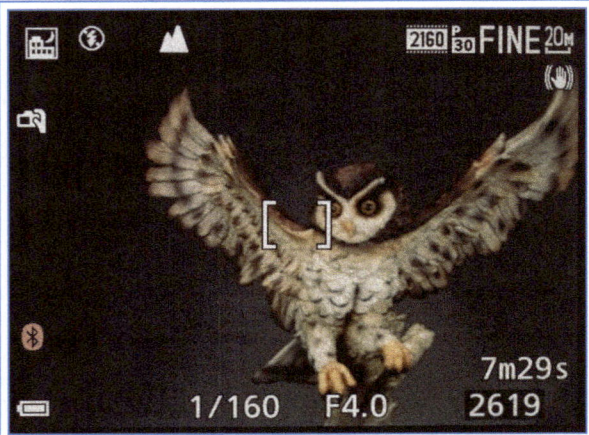

Figure 2-18. Flash Off Icon on Shooting Screen

Next, with the flash still popped up, set the mode dial to the green camera icon, for Auto mode, and then press the Up button on the multi selector. You will see a menu on the screen with five options available: Auto, Auto with Red-eye Reduction, Fill Flash, Slow Sync, and Rear-curtain Sync, as shown in Figure 2-19.

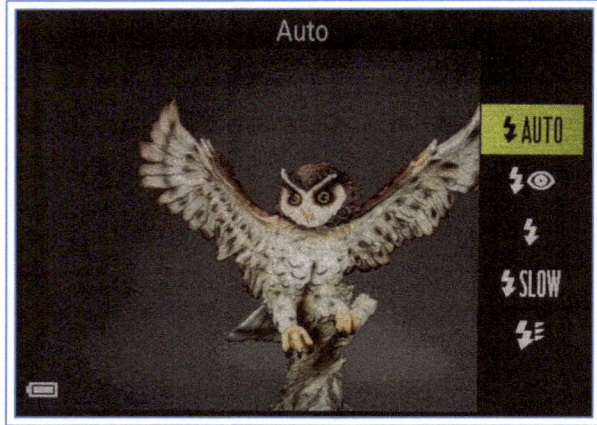

Figure 2-19. Flash Mode Menu

Move through this list by pressing the Up and Down buttons or by turning the command dial or multi selector dial to select the flash mode. Then set up your shot in the current lighting conditions and press the shutter button halfway.

If the lightning bolt icon in the lower right corner of the screen lights up in solid orange when you press the shutter button halfway, as shown in Figure 2-20, that means the flash is ready to fire if it is needed. If the icon is blinking orange, that means the flash unit is not available because it is charging. If you have selected Auto for the flash mode, the flash will not fire unless the camera determines that it is needed, even if the orange flash icon is lit steadily. The orange icon will not appear if you have selected a mode, such as Night Landscape, that will not use flash.

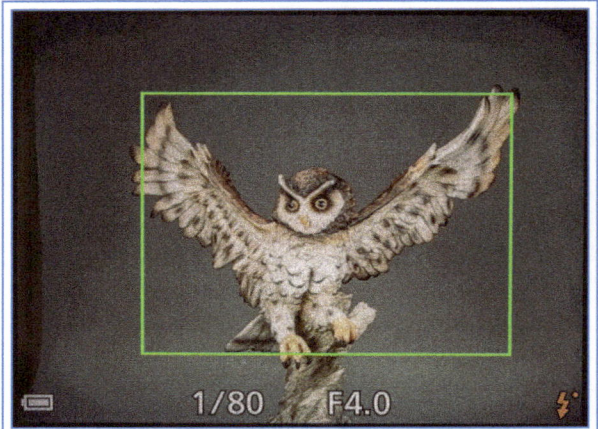

Figure 2-20. Orange Icon Meaning Flash Ready to Fire

In Chapter 9 I'll discuss other flash options, such as Slow Sync and Rear-curtain. For now, you know how to choose one of the flash modes, and you know that they will not all be available at all times.

Movie Recording

Next, I'll describe steps for recording a video sequence with the Coolpix B700. In Chapter 8, I'll discuss other options for video recording, but for now I will stick with the basics.

1. With the camera in Auto mode, make sure the flash unit is pressed down in the off position, because it will not be needed. Once the camera is turned on, press the Menu button and then the Left button to place the yellow selection block in the line of icons at the far left of the screen. Navigate down to the second icon, which looks like a movie camera, then press the Right button to move the selection rectangle back into the list of menu items. You will see six items on this Movie menu screen, which provides the options for recording video footage.

2. Navigate to the last item on the menu screen, called Frame Rate, and press the OK or Right button to move to the screen with the two options for this setting, as shown in Figure 2-21: 30 fps (30p/60p) and 25 fps (25p/50p).

Chapter 2: Basic Operations | 19

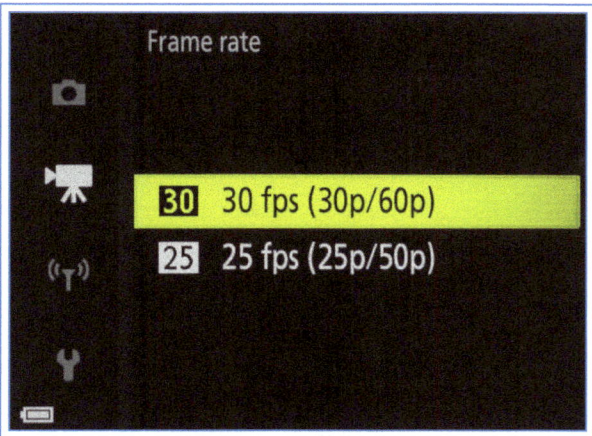

Figure 2-21. Frame Rate Menu Options Screen

This setting determines the frames per second at which the camera will record video in the highest-quality (HD and 4K) formats. The 30 fps setting is the standard for the NTSC video system, which is used in the United States, Canada, Mexico, Japan, and some other areas. The 25 fps setting is the standard for the PAL video system, in use in Europe and other locations. If you are in the United States, select the 30 fps option unless you have a specific need to do otherwise. This setting will affect the choices available for the Movie Options item, discussed next.

3. Navigate to the top option on the menu screen, which is Movie Options, press the OK button or the Right button to move to the next screen, and look at the list of options. If you selected 30 fps for Frame Rate, as discussed above, the first five options on this screen will include the number 30 or 60. If you selected 25 fps, the first five options will include the number 25 or 50.

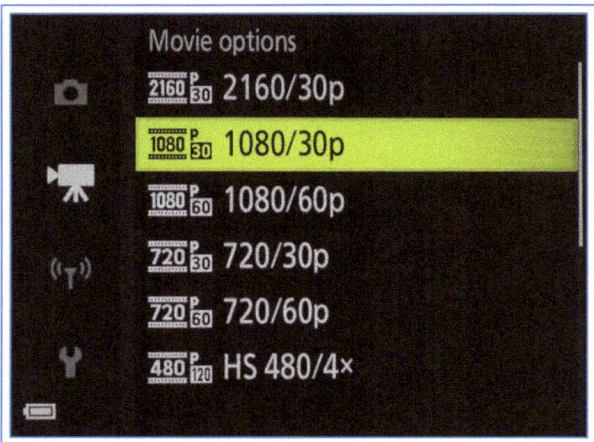

Figure 2-22. 1080/30p Option for Movie Options

4. Choose the second selection for Movie Options, which, if you chose 30 fps for Frame Rate, will be 1080/30p, as shown in Figure 2-22. (If that item is 1080/25p, that means the Frame Rate menu item is set to 25 fps. Unless you have a reason to make that choice, go back and change it to 30 fps.) This setting provides very high quality for shooting HD movies with the B700.

5. Press the Left button to move back one screen, and navigate down to the second option on the screen, Autofocus Mode. Move to the next screen by pressing the OK button or the Right button, and select the second option, Full-time AF, as shown in Figure 2-23.

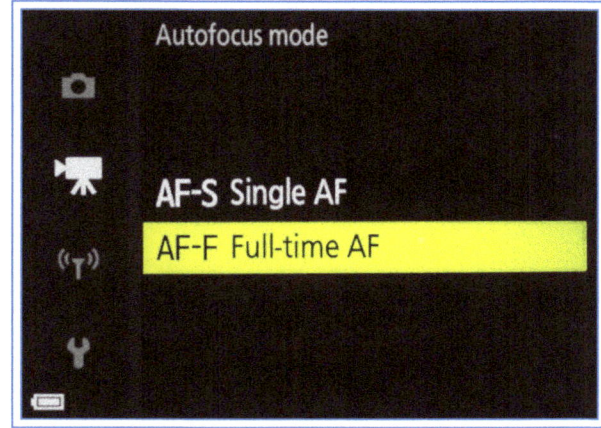

Figure 2-23. Full-time AF Option for Autofocus Mode

6. Press the OK button to confirm this selection. This option will cause the camera to adjust its focus continuously as the distance from the camera to your main subject changes. Then, if you want, make sure Electronic VR and Zoom Microphone are turned on. (If you are using a tripod, turn Electronic VR off.) Exit from the menu system by pressing the Menu button.

7. Compose the shot the way you want it, and when you're ready, press the red Movie button once. (That button is at the top of the camera's back, just below the mode dial.) You don't need to hold the button down; just press and release. The camera's display will blank out briefly, then it will show a red REC indicator at the upper left and the minutes and seconds remaining for your recording at the lower right, as shown in Figure 2-24.

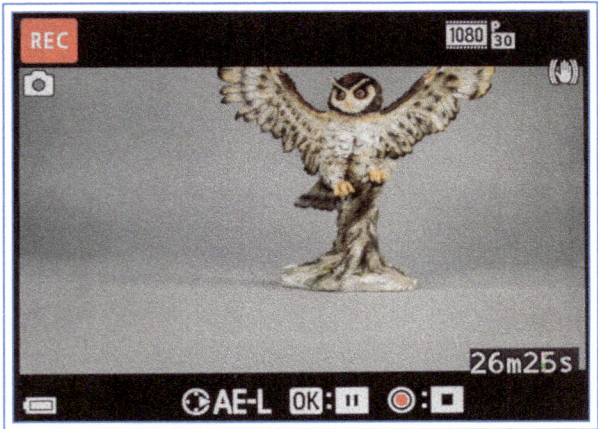

Figure 2-24. REC indicator on Screen During Movie Recording

The camera will keep recording until it reaches a recording time limit or until you press the red button again to stop the recording. Don't be concerned about the level of the sound that is being recorded, because you have no control over the audio volume while recording.

The camera will automatically adjust the exposure as lighting conditions change. As noted earlier, with the Full-time Autofocus option turned on, the camera will continue to adjust its focus as needed, when the distance to the main subject changes.

8. One other point that's not specific to the Coolpix B700: Unless you have a good reason to do otherwise, try to hold the camera as steady as possible (use a tripod or monopod if possible), and don't zoom unnecessarily or move the camera except in very smooth, slow motions, such as a pan (side-to-side motion) to take in a wide scene gradually. Video from a jerkily moving camera can be very disconcerting to the viewer.

Viewing Pictures and Movies

Before I delve into more advanced settings for taking still pictures and movies, as well as other matters of interest, I will discuss the basics of viewing your images in the camera.

Review While in Shooting Mode

Every time you take a still picture, the recorded image will show up on the screen (or in the viewfinder if it's in use) for a brief amount of time, if you have the Setup menu's Monitor Settings option set to turn on the Image Review function. I'll discuss the details of that setting in Chapter 7. By default, your image will stay on the display for about one second after you take a new picture.

Reviewing Images in Playback Mode

To review images that were taken previously, enter playback mode by pressing the Playback button, marked with a small triangle icon, to the right side of the LCD screen. You can then scroll through the recorded images using the Left and Right or Up and Down buttons on the multi selector or by turning the multi selector dial (the dial that surrounds the OK button). You can enlarge any image using the zoom lever on top of the camera, and you can scroll around in the enlarged image using the direction buttons. You can speed through the images by holding down any one of the four direction buttons.

If you have used the continuous-shooting features of the camera, you may see some images labeled with an OK followed by a colon and a triangle at the bottom center, as seen in Figure 2-25.

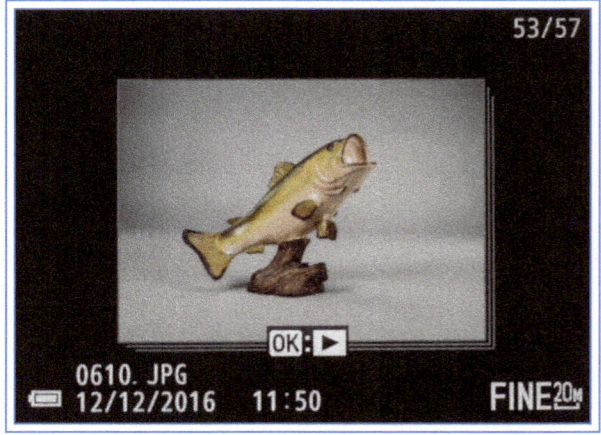

Figure 2-25. Burst of Shots Ready to Play in Camera

In those cases, you can press the OK button to "open" a series of continuous shots, and use the direction buttons to move among the shots in that series. To exit to the main viewing screen so you can see other images and series of images, press the Up button. (I'll discuss playback options in more detail in Chapter 6.)

Playing Movies

To play back motion pictures, move through the recorded images by the methods described above until you find an image with an icon at the lower right of the screen showing a movie format such as 720p, and a file name that ends in .mp4, as shown in Figure 2-26.

Chapter 2: Basic Operations | 21

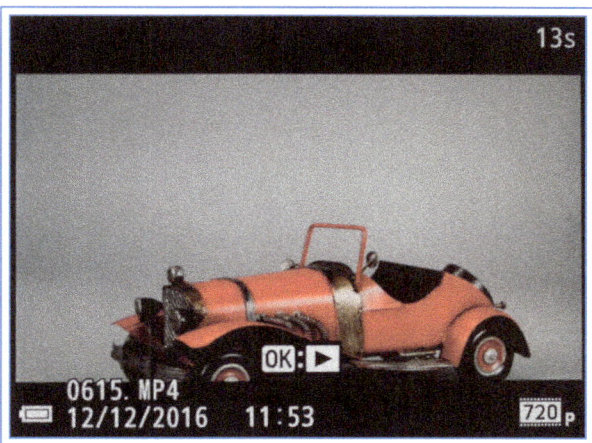

Figure 2-26. Movie Ready to Play in Camera

While the frame from the movie is displayed on the screen, press the OK button and the movie will start playing on the LCD, or in the electronic viewfinder if that display option is active instead of the LCD.

At the bottom left of the display there will be a line of DVR-like controls, as seen in Figure 2-27.

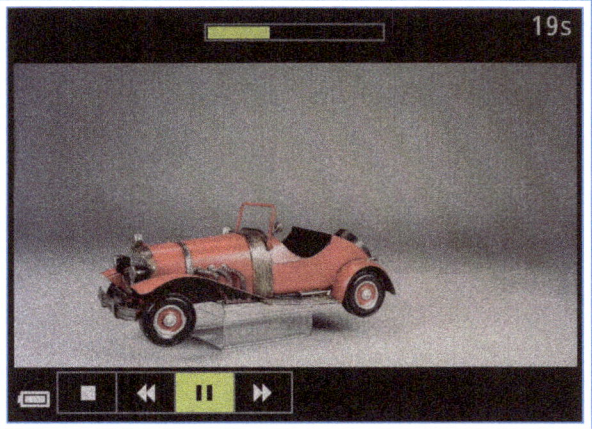

Figure 2-27. Initial Movie Playback Controls

Scroll through the line of controls using the direction buttons on the multi selector and press the OK button to activate one. You also can turn the multi selector dial or the command dial to the right to fast-forward or to the left to rewind. You can raise or lower the volume of the audio by turning the zoom lever (surrounding the shutter button) toward the T position (louder) or the W position (softer). You will see a little set of volume "waves" increase or decrease next to a speaker icon at the lower right of the screen when you adjust the sound in this way. Note, though, that you cannot adjust the sound with this control if the camera is connected to a TV set; in that case, you must use the TV's volume control to change the sound level.

If you want to play the movies on a computer or edit them with video-editing software, they will import nicely into software such as iMovie for the Macintosh or any other program for Mac or Windows that can deal with video files with the extension .mp4. For some Windows-based video editing software, you may need to convert the B700's movie files to the .avi format before importing them into the software. You can do so with a program such as mp4cam2avi, which can be found on the internet at http://mp4cam2avi.sourceforge.net/. You also can use the Nikon Movie Editor software that comes with the camera as part of the ViewNX-i package.

I will discuss more options for playing movies and editing them in the camera in Chapter 8.

Chapter 3: The Shooting Modes

So far, I have discussed setting up the camera for quick shots, relying on features such as Auto mode for taking pictures with settings controlled mostly by the camera's automation. As with other sophisticated digital cameras, though, the Coolpix B700 has a wide range of settings available, particularly for shooting still images. One of the goals of this book is to provide clear guidance about this range of features. To get started, I will turn my attention to the B700's several shooting modes, which provide you with many options for your photography.

To record still images, you need to select one of the available shooting modes: Auto, Program, Shutter Priority, Aperture Priority, Manual exposure, User Settings, Creative, Landscape, Night Portrait, Night Landscape, or Scene. So far, I have discussed the use of the Auto and Program modes. Now I will describe the others, after some review of the first two.

Auto Mode

The Auto shooting mode is a good choice if you need to have the camera ready for a quick shot, maybe in an environment with fast-paced events when you won't have much time to fuss with settings.

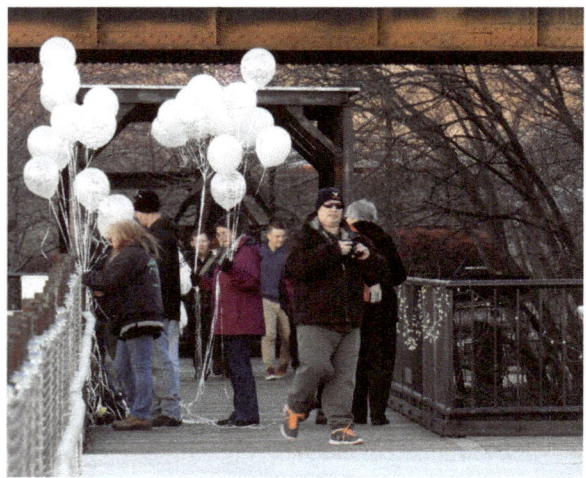

Figure 3-1. Auto Mode Example

For example, in Figure 3-1, I used this mode to grab a quick shot of a group of people holding a celebration on a pedestrian bridge over the river. In this mode, the camera does not try to figure out what kind of scene it is photographing, though it will detect human faces and focus on them if possible.

To set this mode, turn the mode dial, on top of the camera to the right of the viewfinder, to the green camera icon, as shown in Figure 3-2.

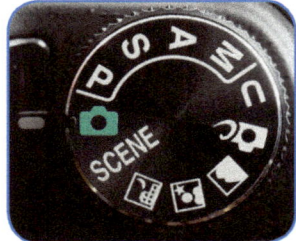

Figure 3-2. Mode Dial at Auto Mode

In this mode, the camera makes several decisions for you and limits your options in some ways. For example, you can't set ISO or white balance to any value other than Auto, and you can't choose the metering method, use exposure bracketing, or use the Picture Control settings to alter the appearance of your images. In addition, you cannot select continuous shooting.

There are still a few settings you can control, however. For instance, you can choose any options for Image Size and Image Quality, including the Raw format, you can use exposure compensation, and you can select any of five available modes for the built-in flash (if you have raised the flash unit). You can select normal autofocus, macro (close-up) focus, or infinity focus (but not manual focus), and you can use the self-timer, including its Smile Timer option. My recommendation is that, for general shooting, you set Image Size to the maximum value of 5184 x 3888 pixels and Image Quality to Fine, and use the other available settings (such as exposure compensation and flash mode) as needed.

Program Mode

Choose this option by turning the mode dial to the P slot, as shown in Figure 3-3.

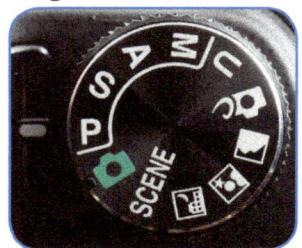

Figure 3-3. Mode Dial at Program Mode

In this mode, the camera evaluates the light and selects both shutter speed and aperture so as to produce an exposure that the camera's programming considers to be normal. The Program shooting mode lets you control many of the settings available with the camera, but not shutter speed and aperture. However, even though you can't directly set those two values, you can override the camera's automatic exposure to a fair extent by using exposure compensation, the Flexible Program feature, and exposure bracketing.

I discussed exposure compensation in Chapter 2, and I'll explain exposure bracketing in Chapter 4. Flexible Program is the name Nikon uses for what is often called "Program Shift" for some other cameras. This option lets you adjust the values the camera selects in Program mode for shutter speed and aperture. For example, if the camera selects, say, 1/60 second at f/4.5, the Flexible Program feature will find equivalent combinations that result in the same exposure, such as 1/50 second at f/5.0, 1/40 second at f/5.6, or 1/30 second at f/6.3. To use this feature, when the camera is in Program mode, aim at your subject and turn the command dial (the wheel at the top right corner of the camera's back) to find an equivalent pair of shutter speed and aperture values.

When the camera is using one of these equivalent match-ups of settings rather than the originally chosen settings, it displays an asterisk at the upper right of the letter P that signifies Program mode in the upper left of the display, as seen in Figure 3-4.

To cancel Flexible Program, turn the command dial back to reset the original shutter speed and aperture, select a different shooting mode, or turn off the camera.

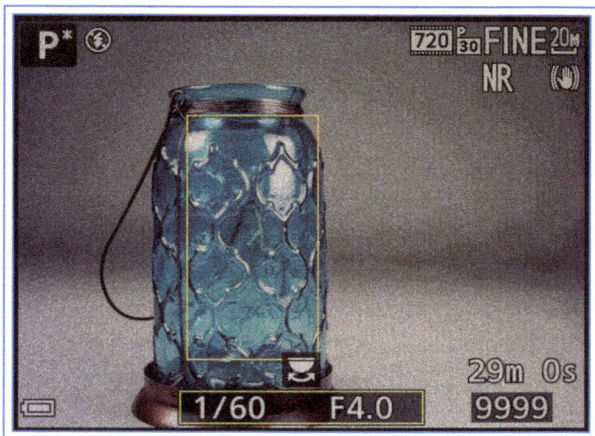

Figure 3-4. Asterisk on Display for Flexible Program

The Flexible Program feature is useful in several situations. For example, you may want to see what the "normal" settings are and then see if you can use a wider aperture to achieve a blurred background, or a faster shutter speed to stop the action or prevent blur from camera motion. And, when you're experimenting with the camera to see what it is capable of, it can be helpful to try various combinations of aperture and shutter speed to find out which combination gives you the best results in different situations. With a digital camera, there's no added cost for trying these different approaches, and Flexible Program is a useful way to experiment.

One way to look at Program mode is that it expands the choices available through the Shooting menu. You can make choices for image size and quality, white balance, ISO sensitivity, metering method, autofocus mode, and others. I won't discuss all of those choices here; if you want to explore that topic, go to the discussion of the Shooting menu in Chapter 4 and check out all of the different selections that are available to you.

Shutter Priority Mode

Select Shutter Priority mode by setting the mode dial to the S indicator, as shown in Figure 3-5.

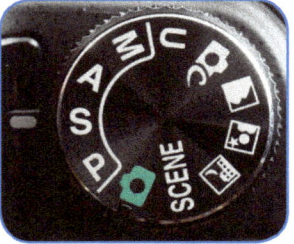

Figure 3-5. Mode Dial at Shutter Priority Mode

In this shooting mode, you set the shutter speed and the camera will set the corresponding aperture in order to achieve a proper exposure. In Shutter Priority mode, you can set the shutter for intervals ranging from eight full seconds to 1/4000 of a second, although the camera has built-in limitations on the use of the fastest and slowest shutter speeds. For example, if the ISO is set to 800, the slowest shutter speed available is two seconds, and if the aperture is set to f/3.3 or f/3.8, the fastest shutter speed available is 1/2000 second. In addition, the zoom range of the lens has a limiting effect on the availability of the fastest shutter speeds. For example, the fastest shutter speed available when the lens is zoomed fully in to the telephoto position is 1/2500 second. The chart in Table 3-1 sets forth some of these limitations.

Table 3-1. Limits on Shutter Speed Settings in S Mode

Slowest Shutter Speed	ISO Value
8 seconds	100
4 seconds	200 or 400
2 seconds	800
1 second	1600
0.5 second	3200
Fastest Shutter Speed	**Aperture Value**
1/4000 second	f/7.6*
1/2500 second	f/8.2**

 * At wide-angle zoom setting
 ** At telephoto zoom setting

If you are photographing fast action like a baseball swing or a race at a track meet and you want to stop the action with a minimum of blur, you will need a fast shutter speed, such as 1/1000 of a second. In other cases, for creative purposes, you may want to use a slow shutter speed of one second or more to achieve a certain effect, such as leaving the shutter open to capture a trail of automobiles' taillights at night.

Figures 3-6 and 3-7 illustrate the effects of different shutter speeds on the same action. For both shots, I poured a cup of uncooked rice into a clear plastic pitcher. In Figure 3-6 I set the shutter speed to 1/2500 second. With that setting, the fast shutter speed froze the motion of the rice, so you can see many of the grains individually.

In Figure 3-7 I set the speed to the much slower value of 1/30 second, which made the grains of rice appear to flow together in a stream.

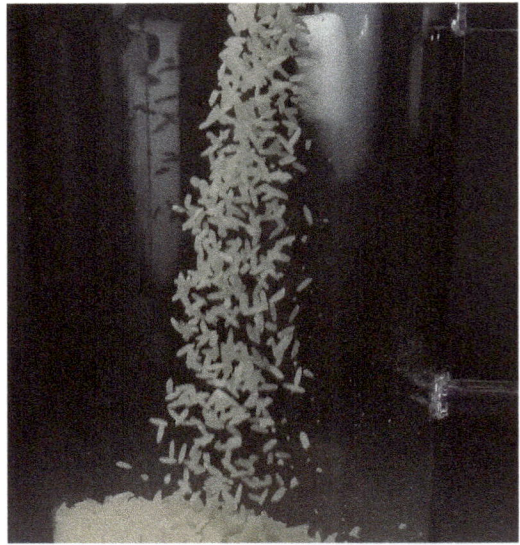

Figure 3-6. Shutter Speed Example: 1/2500 Second

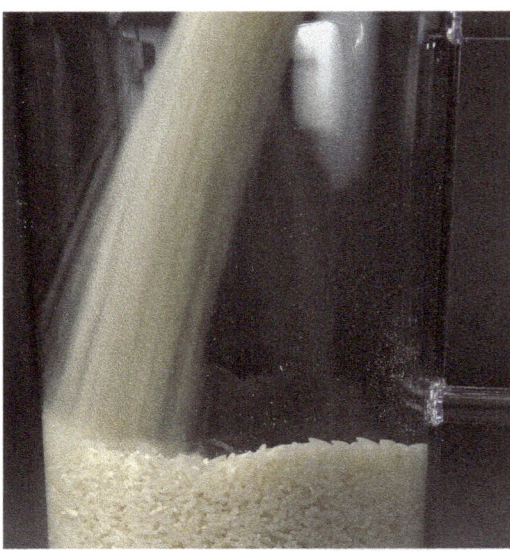

Figure 3-7. Shutter Speed Example: 1/30 Second

To set the shutter speed on the Coolpix B700, turn the command dial—the ridged dial at the top right of the camera's back, below the power switch. (As discussed in Chapter 7, you can switch this function to the multi selector dial with the Toggle Av/Tv Selection option on the Setup menu.) The LCD (or viewfinder, if selected) will display the selected shutter speed inside a yellow rectangle at the bottom center of the screen, as shown in Figure 3-8, where the shutter speed was set to 1/3 second.

Chapter 3: The Shooting Modes

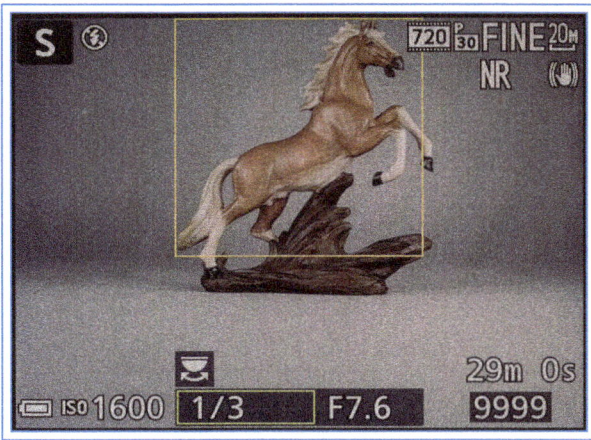

Figure 3-8. Shutter Speed Value on Display Screen

As you point the camera at scenes with varying lighting, the camera will select and display the appropriate aperture (such as f/7.6 in this example) to achieve a proper exposure.

Once you've pressed the shutter button halfway, watch the shutter speed number on the screen. If that number blinks, that means proper exposure at that shutter speed is not possible at any available aperture, according to the camera's calculations. For example, with a shutter speed of two seconds in a well-lighted room, the shutter speed number may begin to blink, indicating that proper exposure is not possible.

The camera will still let you take the picture, despite having blinked the number to warn you. The camera is saying, in effect, "Look, maybe you shouldn't do this, but that's your business. If you want an overly bright picture for some reason, help yourself." (This situation is less likely to take place when the camera is in Aperture Priority mode, because in that mode, there is a wide range of shutter speeds for the camera to choose from—a range from eight seconds to 1/4000 second in some situations, depending on factors such as ISO and aperture.)

When you are setting shutter speed, the fractions of a second are easy to read because they are displayed as standard fractions, such as 1/5 or 1/200. Some of the longer times are a bit harder to read; the camera displays them using quotation marks. So, for example, two seconds is displayed as 2", and 1.3 second is displayed as 1.3."

One feature of the shutter speed display on the Coolpix B700 is a bit confusing, at least to me. Some of the camera's shutter speeds are displayed as fractions whose denominators are decimal numbers, such as 1/1.3. I would have trouble understanding that number without doing some arithmetic, so Table 3-2 provides a brief chart that converts these few values into terms that may be easier to comprehend:

Table 3-2. Shutter Speed Equivalents

1/2.5	= 0.4 = 2/5 second
1/1.6	= 0.625 = 5/8 second
1/1.3	= 0.77 = 10/13 second (0.8 sec)

Aperture Priority Mode

Aperture Priority mode, represented by the A setting on the mode dial as shown in Figure 3-9, is the inverse of Shutter Priority.

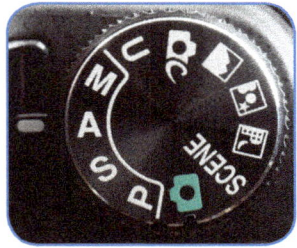

Figure 3-9. Mode Dial at Aperture Priority Mode

In this mode, you select an aperture value and the camera selects a corresponding shutter speed to achieve a proper exposure. The camera's aperture is a measure of the current width of its opening that lets in light to create the image. This width is stated numerically in f-stops. For the Coolpix B700, the range of f-stops is from f/3.3 (wide open) to f/8.2 (most narrow), though this range is limited in some circumstances, as discussed below. The amount of light that is let into the camera to create an image is controlled by the combination of aperture (how wide open the lens is) and shutter speed (how long the shutter remains open to let in the light).

For some purposes, you may want to control the width of the aperture, but let the camera choose the corresponding shutter speed, so you can control the depth of field. Depth of field is a measure of how well a camera is able to keep multiple objects or subjects in focus at different distances. For example, say you have three friends lined up so you can see all of them, but they are standing at different distances—five, seven, and nine feet (1.5, 2.1, and 2.7 meters) from the camera. If the camera's depth of field is shallow at a particular focal length, such as five feet (1.5 meters), then, if you focus on the friend at that distance, the other two will be out of focus and blurry. But if the camera's depth of field when focused at five feet is broad, then it may be possible for all three friends to be in sharp focus in your photograph, even if the focus is set for the friend at five feet.

The wider the camera's aperture is, the more shallow its depth of field is at a given focal length. So in the example discussed above, if you have the camera's aperture set to its widest opening, f/3.3, the depth of field will be relatively shallow, and it will be possible to keep fewer items in focus at varying distances from the camera. If the aperture is set to the narrowest, f/8.2, the depth of field will be greater, and it will be possible to have more items in focus at varying distances.

It is hard to illustrate this effect with a camera like the B700, for a couple of reasons. First, the image sensor, where the light is gathered to form the image, is relatively small, which results in the depth of field being relatively deep at all apertures. Second, the largest aperture available is f/3.3, whereas some compact cameras have lenses that open as wide as f/2.0, or even f/1.4. With such cameras it is easier to achieve a blurred background, because the depth of field can be quite shallow at such a wide aperture. With the B700, the widest aperture you can shoot with is f/3.3, and that aperture is available only when the lens is zoomed back to its extreme wide-angle setting, where depth of field is greater. If you zoom the lens in to a telephoto setting, the maximum aperture decreases steadily. At the maximum zoom range, the widest aperture available is only f/6.5, which is not far from the narrowest aperture of f/8.2.

Despite the difficulty of demonstrating the effects of using different apertures, the images in Figures 3-10 and 3-11 illustrate these effects to some extent. For both images, the lens was zoomed out to 24mm. The first image was taken at f/3.3, the widest aperture setting available; the second one was taken at f/7.6, the narrowest aperture setting available on the camera at that focal length.

As you should be able to see, in Figure 3-10, with the wider aperture, the background is noticeably blurred because the depth of field is relatively shallow at that setting. In Figure 3-11, on the other hand, the background is in somewhat sharper focus because the depth of field is greater at the narrower aperture setting.

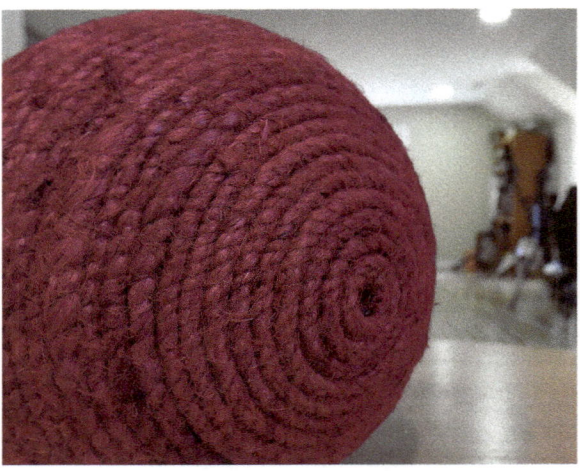

Figure 3-10. Aperture Set to f/3.3

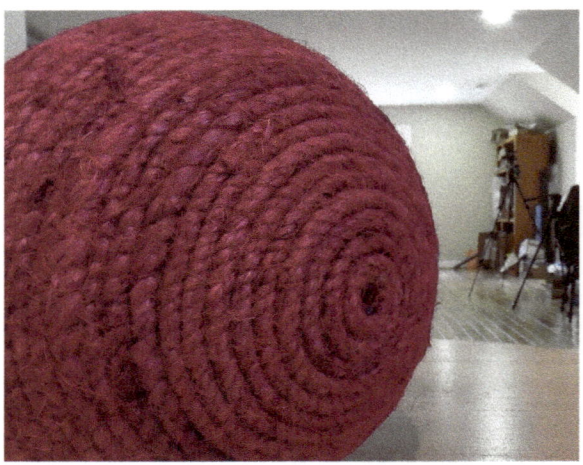

Figure 3-11. Aperture Set to f/7.6

If you want to have the sharpest picture possible, especially when you have subjects at varying distances from the lens and you want them all to be in focus, then you may want to control the aperture and make sure it is set to the highest number (narrowest opening) possible. It also helps to have the lens zoomed back toward its wide-angle setting and to be somewhat distant from the subject.

On the other hand, there are times when photographers prize a shallow depth of field. This situation arises often in the case of outdoor portraits. For example, you may want to take a photo of a person standing outdoors with a background of trees and bushes, and possibly some other, more distracting objects, such as a swing set or a tool shed. If you can achieve a shallow depth of field, you can have the person's face in sharp focus, but leave the background quite blurry and indistinct. This effect is sometimes called "bokeh," a Japanese term describing an aesthetically pleasing blurriness of the background.

To achieve the greatest blurring of the background, you should try to use a wide aperture, zoom the lens in as much as possible, and get as close to the subject as possible. It also is helpful to have as much distance as possible between the main subject and the background.

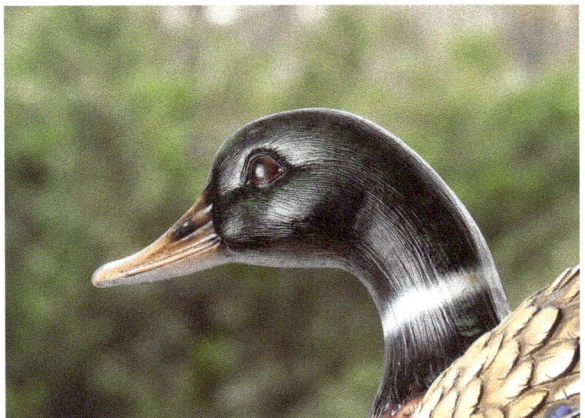
Figure 3-12. Example of Bokeh Effect

Figure 3-12 is an example using this effect. For this image, I located the camera fairly far from the subject and zoomed the lens in to a focal length of 300mm to decrease the depth of field. The blurriness of the background can reduce the distraction factor from unwanted objects and highlight the sharply focused portrait of your subject.

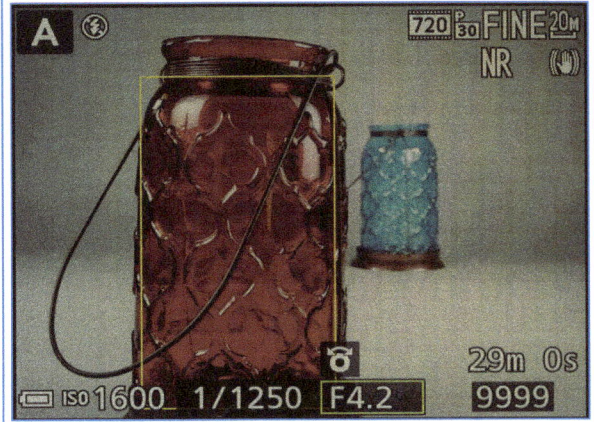
Figure 3-13. Aperture Value on Display Screen

To set the aperture, once you have moved the mode dial to the A setting, aim the camera at your subject and turn the multi selector dial to change the aperture. The number of the f-stop will appear inside a yellow rectangle at the bottom right of the screen. The shutter speed chosen by the camera will show up also, to the left of the aperture, as seen in Figure 3-13.

When you press the shutter button halfway, the camera will lock in the selected shutter speed.

As I noted briefly above, not all apertures are available at all times. The widest aperture, f/3.3, is available only when the lens is zoomed out to its wide-angle setting (zoom lever moved toward the letter W). At the highest zoom levels, the widest aperture available is f/6.5.

To see an illustration of this point, here is a quick test. Zoom the lens out by moving the zoom lever all the way to the left, toward the W setting. Then select Aperture Priority mode and choose an aperture of f/3.3 by turning the multi selector dial all the way to the left. Now zoom the lens in by moving the zoom lever to the right, toward the T setting. When you release the lever, the aperture displayed at the bottom of the screen will change to f/6.5. If you try to reset the aperture to f/3.3 after the zoom action is finished, you will see that the lowest aperture number you can set is f/6.5, because that is the widest aperture available on the B700 at the telephoto zoom level. (The aperture will change back to f/3.3 if you move the zoom back to the wide-angle setting.)

In addition, the narrowest apertures are not available in all cases. The overall aperture range for the B700 is from f/3.3 to f/7.6 when the lens is at the wide-angle setting and from f/6.5 to f/8.2 at the fully zoomed-in setting.

Manual Exposure Mode

The Coolpix B700 has a manual exposure mode for control of both aperture and shutter speed, which helps you enjoy full creative control over exposure decisions.

This mode is useful when you want to use settings that result in an unusual effect, such as an abnormally dark image. For example, I used Manual exposure mode for Figure 3-14 to produce an image of a mannequin against a window with a silhouette effect. I also often use Manual mode to take a series of photographs at different exposures to create HDR (high dynamic range) images using special software. I will discuss that process later in this chapter, in the discussion of the Backlighting/HDR setting.

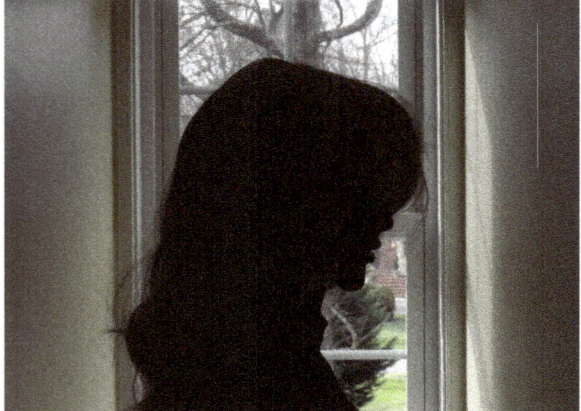

Figure 3-14. Manual Exposure Example

To use this mode, set the mode dial to the M indicator, as shown in Figure 3-15.

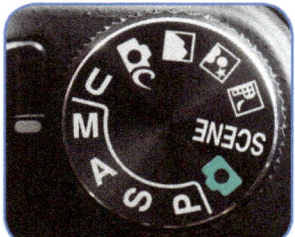

Figure 3-15. Mode Dial at Manual Exposure Mode

You now have to control both shutter speed and aperture by setting them yourself.

To set these values, first look at the camera's display and find where the shutter speed (such as 1/80) and aperture (such as F7.1) are displayed at the bottom of the screen, as shown in Figure 3-16.

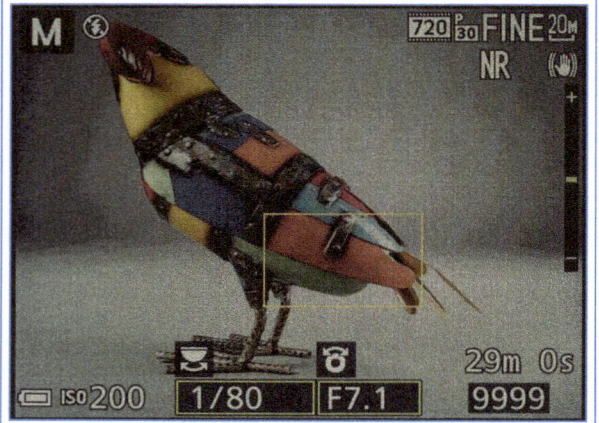

Figure 3-16. Shutter Speed and Aperture Values on Display Screen

At the bottom of the display is the shutter speed on the left, inside a yellow rectangle. Above that value is a curved arrow beneath an icon that represents the command dial. These icons mean that the shutter speed value is controlled by the command dial (the wheel at the top of the camera's back, just below the power button). To the right of that value is the value for the aperture, or f-stop, inside another rectangle; above that value is an icon showing a dial that is oriented vertically; that icon represents the multi selector dial, on the back of the camera surrounding the OK button.

To adjust the settings, turn the command dial until you have selected the shutter speed you want, and turn the multi selector dial to set your desired aperture. As you adjust these values, watch the vertical scale that appears at the right of the screen, as shown in Figure 3-17.

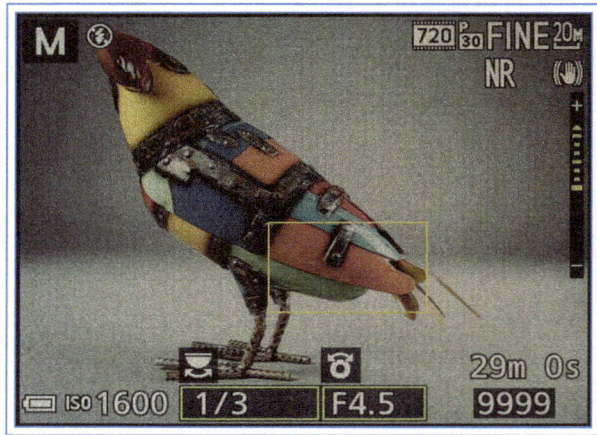

Figure 3-17. Exposure Scale for Brighter Image

You will see the tick marks turn yellow, either above or below the scale's center point, as the values change. When the exposure is set as the camera judges to be normal, there will be a lone tick mark in the center of the scale, as shown in Figure 3-16.

If the marks above the center of the scale turn yellow, as in Figure 3-17, the exposure is too bright; if they turn yellow below the center, it is too dark, as shown in Figure 3-18. If the setting becomes more extreme than the scale can indicate, a yellow triangle appears at the top or bottom of the scale, indicating that the scale's limit has been exceeded, as shown in both Figures 3-17 and 3-18.

Chapter 3: The Shooting Modes

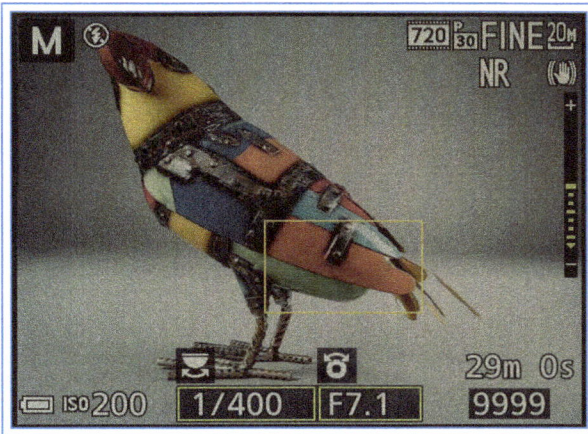

Figure 3-18. Exposure Scale for Darker Image

If you are shooting in dim light, such as indoors or in a shadowed area, you may find it impossible to center the yellow tick mark on the exposure scale by adjusting the shutter speed and aperture unless you use a very slow shutter speed, such as one second or longer. If you are handholding the camera, you won't be able to hold it steady for more than about 1/30 second, so it will be difficult to get a clear exposure.

In that situation, you can adjust exposure by changing the ISO setting. I will discuss ISO in more detail in Chapter 4, because it is an option found on the Shooting menu. Briefly, ISO is a setting that controls the sensitivity of the camera's digital sensor. The higher the ISO value, the more sensitive the sensor is to light. With higher ISO values, you can achieve a normal exposure with narrower apertures and faster shutter speeds.

With other shooting modes, the B700 can use the Auto ISO setting, which means the camera will set the ISO value as needed to reach a good exposure level. With Manual exposure mode, though, the camera will not set the ISO automatically. If you select Auto ISO from the ISO menu item, the camera will set ISO to 100, the lowest value possible.

If you find you need a higher ISO value in Manual exposure mode, you need to go to the ISO menu item and select a value such as ISO 400, 800, or even higher. The first screen of the ISO menu is shown in Figure 3-19.

Of course, you don't have to adjust the ISO or other values in an effort to center the indicator on the Manual exposure scale; that scale is there only to give you an idea of how the camera would meter the scene. You may want parts of the scene (or the whole image) to be darker or lighter than the metering system would indicate to be "correct."

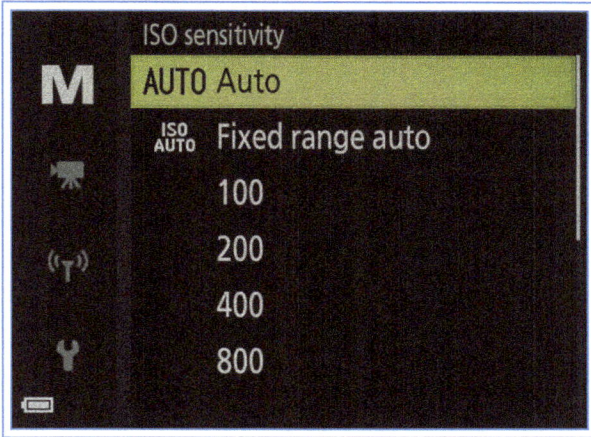

Figure 3-19. First Screen of ISO Menu

In Manual mode, the settings for aperture and shutter speed are independent of each other. When you change one, the other one stays unchanged until you change it manually. The camera is leaving the creative decision about exposure entirely up to you, even if the resulting photograph would be washed out by excessive exposure or underexposed to the point of near-blackness.

As with Aperture Priority mode, the range of available apertures in Manual exposure mode varies as the lens is zoomed to various focal lengths. Also, the range of shutter speeds has certain limits, although the camera's slowest shutter speed of 15 seconds is available only in this mode. (There is a slower shutter speed of 25 seconds available, but only with the Star Trails option for the Multiple Exposure Lighten setting in Scene mode, discussed later in this chapter.)

The slowest shutter speed available for setting in this mode at ISO 800 is two seconds; at ISO 3200, the slowest is 1/2 second. If ISO is set to 100, the shutter speed of 15 seconds is available. In addition, as discussed above, if ISO is set to Auto in Manual exposure mode, the camera actually uses an ISO setting of 100, so the shutter speed of 15 seconds is available when ISO is set to Auto, in this shooting mode only. Auto Flash mode is not available with Manual exposure mode. The fastest shutter speeds available in this mode have the same restrictions as in Shutter Priority mode, discussed earlier.

Scene Modes

The Coolpix B700 offers several of what I will call scene modes. The terminology can be a bit confusing, because the camera's menus and documentation use the word "scene" in several similar and overlapping contexts. First, there are three modes that occupy slots marked by icons on the mode dial: Landscape, Night Portrait, and Night Landscape. Next, there is another slot on the mode dial marked SCENE. When you select that setting, you can press the Menu button to the lower left of the multi selector and scroll through a list of 20 specific scene settings: Portrait, Sports, Party/Indoor, Beach, Snow, Sunset, Dusk/Dawn, Close-up, Food, Fireworks Show, Backlighting, Easy Panorama, Pet Portrait, Moon, Bird-watching, Soft, Selective Color, Multiple Exposure Lighten, Time-lapse Movie, and Superlapse Movie. Finally, there is a 21st entry, which actually is the first entry at the top of the list: Scene Auto Selector. I will discuss all of these scene settings individually, but first I will provide some general remarks about these shooting mode options.

Scene modes are different from the other shooting modes I have discussed up to this point. These modes do not have a single defining feature, such as permitting control over one or more aspects of exposure. Instead, when you select a scene shooting mode, you are in effect telling the camera what sort of environment the picture is being taken in, and what type of image you are looking for, and you are letting the camera decide what settings to use to produce that result.

Some photographers may not like scene modes because they take some creative decisions away from you and limit your options in some ways. For example, you will find that the Shooting menu options are severely limited when the mode dial is turned to the SCENE setting or any of the scene modes with slots on the mode dial, such as Night Landscape. For example, you cannot set the white balance, but must rely on the camera's Auto White Balance setting, which may not always properly evaluate the existing light source. In most cases, you cannot select features such as continuous shooting, and you can't choose a metering mode or an ISO setting. You can set the Image Quality to Fine or Normal, but you cannot set it to Raw, Raw+Fine, or Raw+Normal.

Despite the limitations, I have found the scene settings useful in certain situations. You don't have to use these settings only for their labeled purposes; you may find that some of them are well-suited for shooting scenarios you are regularly faced with. For example, you may find the Sports setting works well for shots of children at play, or that the Sunset setting, which emphasizes red hues, is great for images in a particular garden that is rich with reddish plants and flowers. The Bird-watching setting can work well for taking images of various sorts of wildlife, not just birds.

You need to know something about each of these options to decide whether it's one you would want to select. In general, a given scene setting carries with it a variety of values, including things like focus mode, flash status, range of shutter speeds, sensitivity to various colors, and others. I will discuss the complete list of scene settings so you can make informed choices. I will first discuss the settings that have their own slots on the mode dial, followed by the settings that are grouped under the SCENE setting on the dial.

LANDSCAPE

This is the setting with a mountain icon on the mode dial next to the Creative setting, as seen in Figure 3-20.

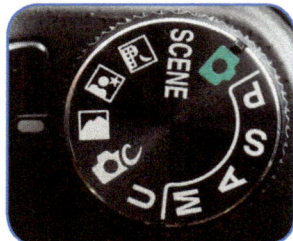

Figure 3-20. Mode Dial at Landscape Mode

With some other cameras, including the Nikon Coolpix P500, a predecessor to the B700, this shooting mode is one of those that are selected through the Shooting menu when the dial is set to the SCENE setting. With the B700, though, the Landscape setting occupies its own slot on the mode dial, perhaps in recognition of the usefulness of this setting.

I use Landscape mode often, and I appreciate that it is easy to twist the mode dial to choose it. When you do, you can then go to the Landscape item on the Shooting menu, shown in Figure 3-21, and select one of two options from the Shooting menu: in this case, Noise Reduction Burst or Single Shot, as seen in Figure 3-22.

Chapter 3: The Shooting Modes | 31

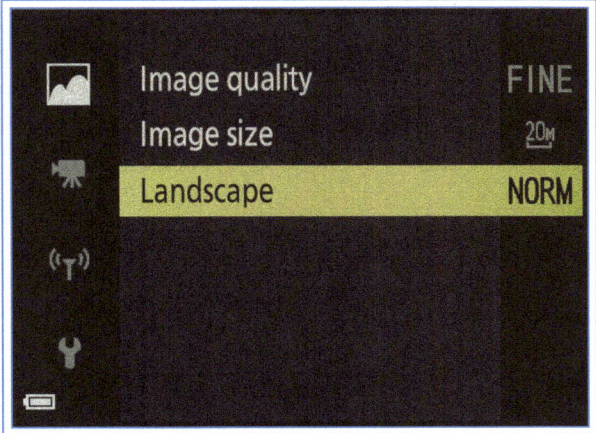

Figure 3-21. Landscape Item on Shooting Menu

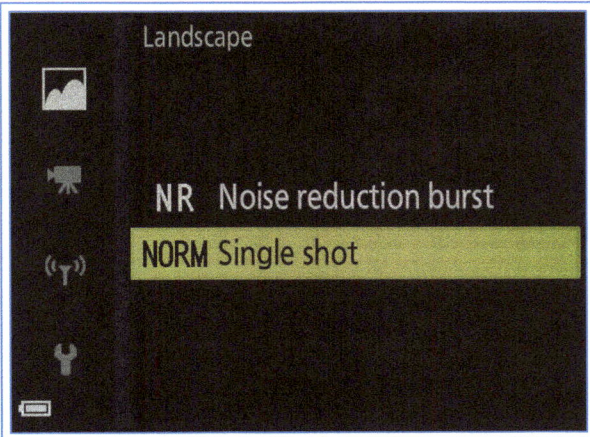

Figure 3-22. Options for Landscape Menu Item

If you select Noise Reduction Burst, the camera will take a rapid series of shots at a relatively high ISO setting and merge them in the camera into a final image. The camera's internal processing will combine the individual shots so as to reduce the visual "noise" that results from using higher ISO settings. The final image will be cropped slightly because of this processing, so less of the scene will be included in the image. This setting is useful when you are taking photographs in subdued light, to reduce the noise effects caused by a higher ISO setting.

If, instead, you select the Single Shot option, the camera will operate as you would expect for a normal Landscape setting—it will take just one shot at a lower ISO setting, which likely will result in a sharper picture than one taken with the Noise Reduction Burst setting. The example in Figure 3-23 was taken using the Single Shot setting.

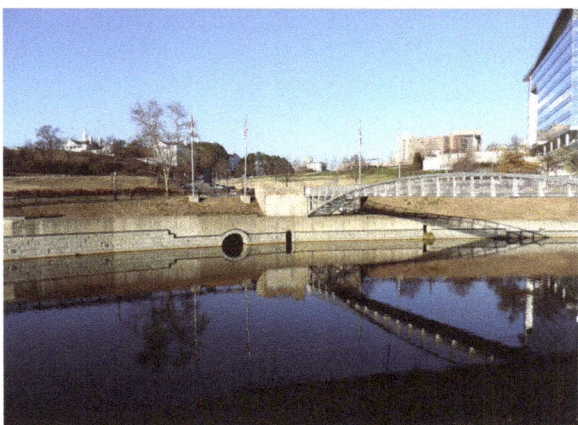

Figure 3-23. Landscape Example

Night Portrait

This mode is signified on the mode dial by an icon showing a human face under the stars, as seen in Figure 3-24.

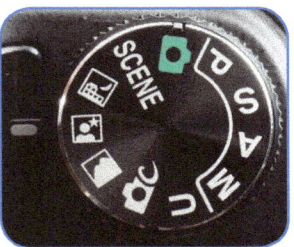

Figure 3-24. Mode Dial at Night Portrait Mode

With this setting, the camera will use the built-in flash. The subject presumably will be close to the camera and, unlike a landscape scene, can be illuminated by the flash. The camera will select Slow Sync for the flash mode, and will not let you change it. (If you haven't popped up the flash unit, the camera will display an error message until you do.)

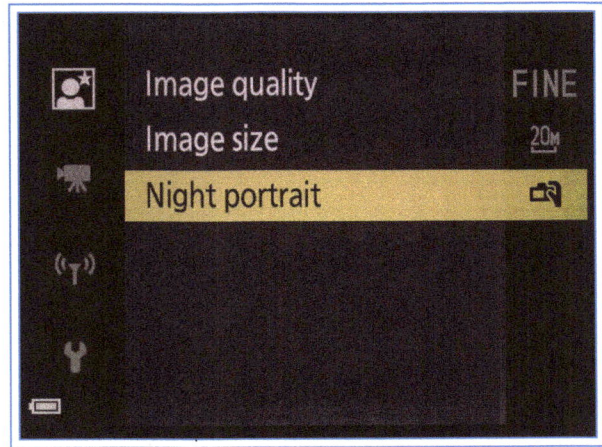

Figure 3-25. Night Portrait Item on Shooting Menu

From the Night Portrait item on the Shooting menu, shown in Figure 3-25, you can select either Hand-held or Tripod, as seen in Figure 3-26.

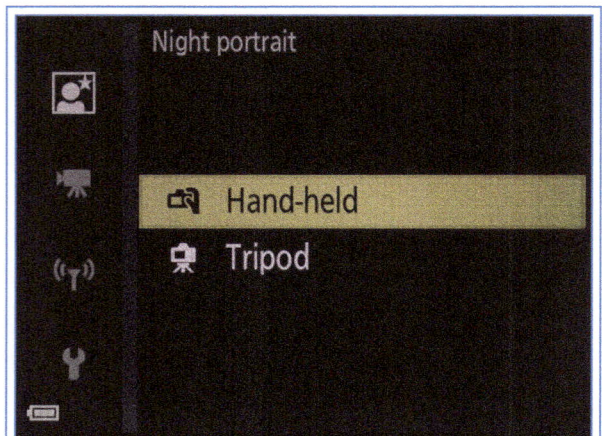

Figure 3-26. Options for Night Portrait Menu Item

If you select Hand-held, the camera may use a faster shutter speed than with the Tripod setting, to counteract camera shake. You can use the self-timer (including the Smile Timer) or exposure compensation, but you cannot change the focus method. If you choose Tripod, the camera will likely use a slower shutter speed. It still will use the flash.

The camera uses its face detection circuitry and attempts to find a face to focus on. If it finds a face, it automatically applies skin softening, which smooths out wrinkles and other harsh features on the skin. However, you cannot control the amount of skin softening or turn it off.

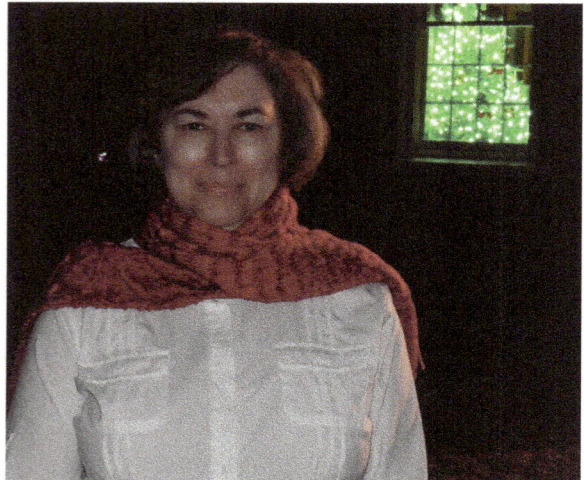

Figure 3-27. Night Portrait Example

In Figure 3-27, I used this setting for a portrait after dark in front of a window with holiday lights. The flash lit the subject appropriately and the camera used a shutter speed of one full second in order to allow the lights in the background to show up as well.

Night Landscape

This mode, symbolized by a crescent moon above a building, appears on the mode dial just below the SCENE setting, as seen in Figure 3-28.

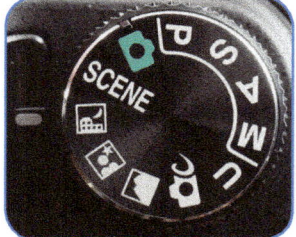

Figure 3-28. Mode Dial at Night Landscape

It is for use without flash at night in order to preserve the natural appearance of dark landscapes. When you select this shooting mode with the mode dial, the Shooting menu includes an option called Night Landscape, shown in Figure 3-29.

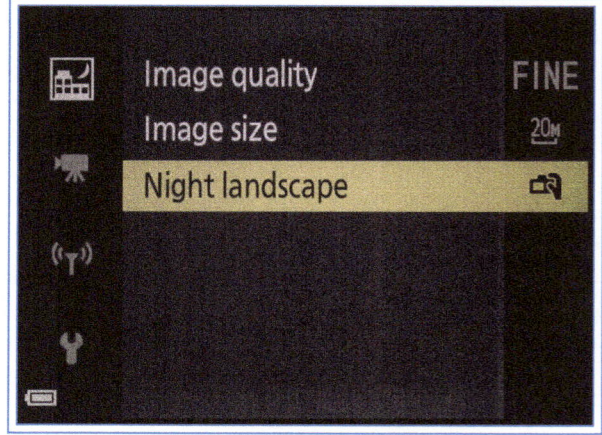

Figure 3-29. Night Landscape Item on Shooting Menu

That option has two sub-options, as with Night Portrait: Hand-held and Tripod. To choose one of these, press the Menu button, then select Night Landscape, the bottom item on the brief list that appears. Then press the OK button or the Right button, and choose either Hand-held or Tripod, as shown in Figure 3-30.

Chapter 3: The Shooting Modes | 33

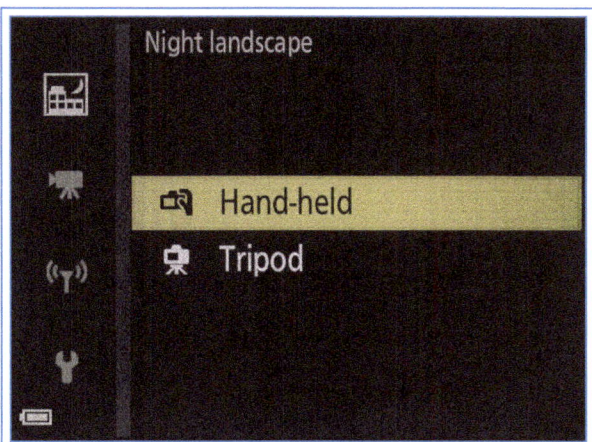

Figure 3-30. Options for Night Landscape Menu Item

With Hand-held, the camera will take a continuous group of pictures and combine them in the camera into a single image, to overcome the effects of the high ISO setting the camera uses to take a good exposure in dim light without flash. A single image could be degraded from the visual "noise" that results from the use of high ISO values; by combining several images, the camera can create a single image using the best aspects of each, and can digitally smooth away the noise. You should try to hold the camera as steady as possible when shooting, but it will use a relatively fast shutter speed if at all possible to avoid blur from camera shake.

If you choose the Tripod option from the menu, the camera will use a slower shutter speed and a lower ISO setting, to minimize noise. The use of the tripod will avoid the effects of camera shake. Of course, this setting is useful only if you actually attach the camera firmly to a tripod.

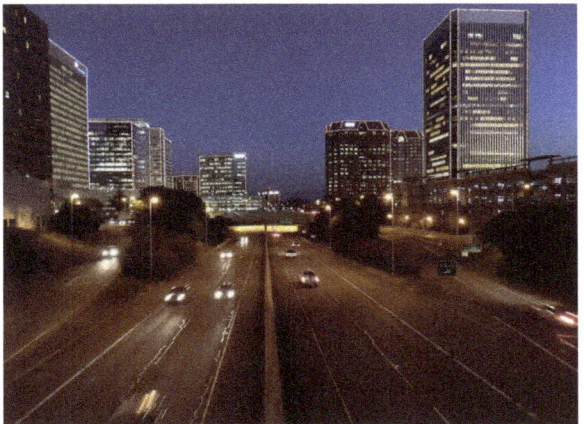

Figure 3-31. Night Landscape Example

I took the image in Figure 3-31 a few minutes after sunset with the B700 on a tripod. The camera exposed this image at f/3.3 for 1/15 second at ISO 1600.

The SCENE Setting on the Mode Dial

Turning the mode dial to SCENE, as in Figure 3-32, gives you access to 21 choices of settings, including Scene Auto Selector and 20 specific scene types.

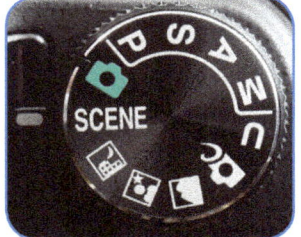

Figure 3-32. Mode Dial at Scene Mode

You can select any one of these choices by pressing the Menu button and selecting a scene setting from the menu that appears. The first screen of the Scene mode menu is shown in Figure 3-33.

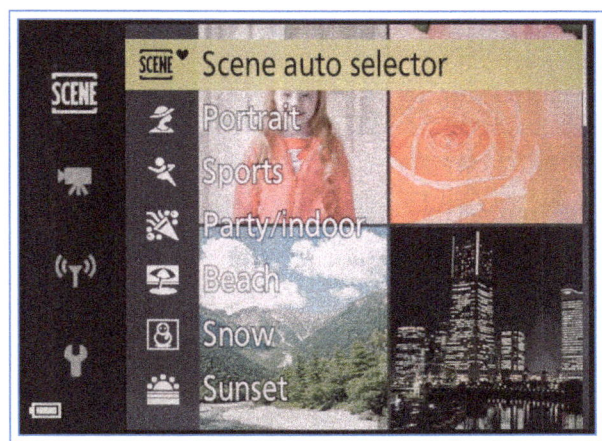

Figure 3-33. First Screen of Scene Menu

Scroll through the multiple screens of choices using the multi selector dial or the Up and Down buttons.

When you select any of the 21 Scene mode settings, the menu offers few other choices. For example, when you make a selection such as Sunset from the Scene menu, you cannot make any other menu choices except Image Size and Image Quality. The camera will make all other settings as it deems appropriate for that selection. These Scene mode settings are convenient if you are faced with a certain type of photographic situation and you want the camera to make reasonable choices for that situation, but you have very little control over the camera's other settings. Following are details about each of the types. I will include sample images for most of the selections.

Scene Auto Selector

With this first option, the camera analyzes the live view using its digital circuitry and tries to choose the most appropriate shooting mode to use from among these choices: Portrait, Landscape, Night Portrait, Night Landscape, Close-up, Backlighting, and Other Scenes. If the camera can identify what appears to be a scene calling for one of the listed settings, it displays an icon for that setting in the upper left corner of the screen.

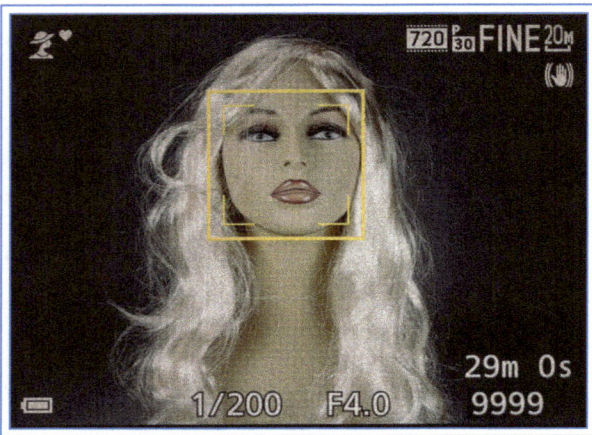

Figure 3-34. Scene Auto Selector - Portrait

For example, in Figure 3-34, the camera properly identified the view of a mannequin's head as calling for Portrait mode, represented by the icon in the upper left of the screen. The heart icon next to the Portrait icon indicates the use of Scene Auto Selector. The camera displays a slightly different icon with a number 1 if it detects more than two subjects.

With the Portrait settings, the camera applies skin softening if it detects faces, as it does with Night Portrait. With the Backlighting option, it will use an icon with the number 1 if it detects human subjects as opposed to non-human ones.

Figure 3-35 shows the landscape icon that the camera places in the upper left corner when it detects a landscape-oriented scene.

In Scene Auto Selector mode, you can use exposure compensation and the self-timer, but you cannot select an autofocus mode or a flash mode. The camera prompts you to raise the flash, and it uses Auto Flash mode, meaning the camera will decide whether or not to fire the flash.

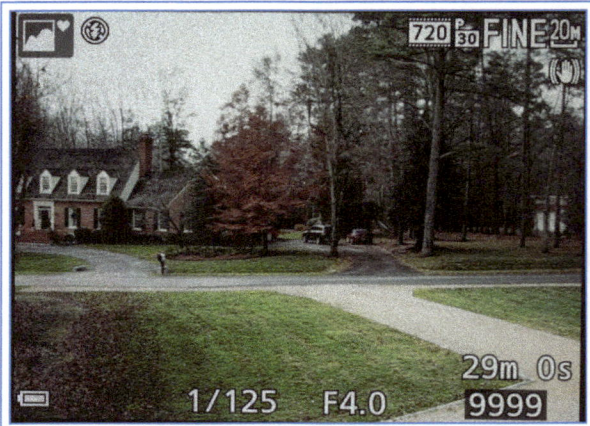

Figure 3-35. Scene Auto Selector - Landscape

If the camera chooses a scene type that you don't like, you of course have the option of choosing another option, such as Auto or Program mode, or a specific Scene mode setting. With all of the settings discussed below, the only Shooting menu settings available are Image Quality and Image Size, which appear at the bottom of the list of scene types on the scene-selection menu, as shown in Figure 3-36. However, some of the settings, such as Bird-watching, have sub-options that can be selected from a menu that is specific to that setting. I will discuss those options below.

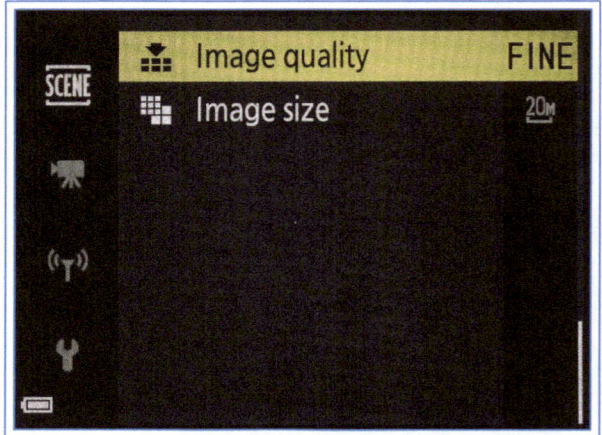

Figure 3-36. Image Size and Image Quality on Scene Menu

Portrait

With the Portrait setting, the camera automatically sets itself for face detection, which means it looks for human faces and focuses on the one closest to the camera. As with other portrait-oriented settings, the camera automatically applies skin softening to any faces it detects. You can pop up the flash and select a flash mode. You cannot make any changes to the focus mode, but you can use the self-timer and exposure compensation.

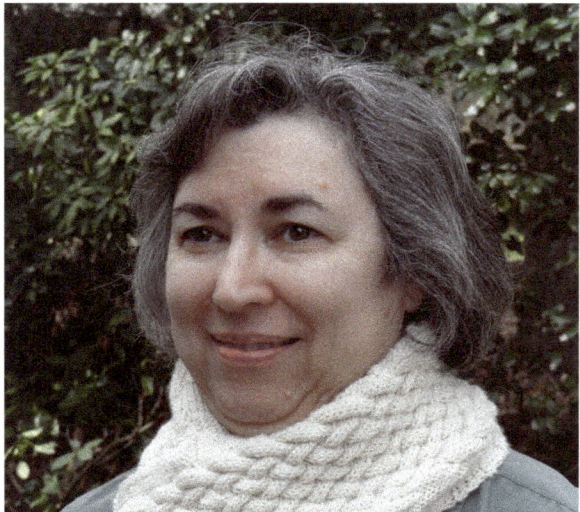

Figure 3-37. Portrait Example

Figure 3-37 is an example of an image I took hand-held without the flash. I zoomed the lens in to 85mm to try to blur the background somewhat.

Sports

The Sports setting is intended for fast-moving subjects. The camera sets itself for continuous shooting and takes a series of as many as five images at a rate of up to five frames per second when you hold down the shutter button, depending on conditions. The flash is forced off, and focus and exposure are locked when the first image is taken, to increase the speed of the sequence of shots. You can use manual focus and exposure compensation, but you cannot use the self-timer. This mode is useful when you need to stop action in relatively bright lighting conditions. In Figure 3-38, I used the Sports setting to capture an image of a kayaker going through the rapids on a cold day in December. The continuous-shooting function let me catch a variety of shots, of which this one was the best.

The Sports setting is great for stopping action and getting a fast series of shots, but this setting can limit your options because the camera takes time to recover after taking a rapid burst of shots. So, if another opportunity for a good image comes up while the camera is still writing the burst images to the memory card, you have to wait.

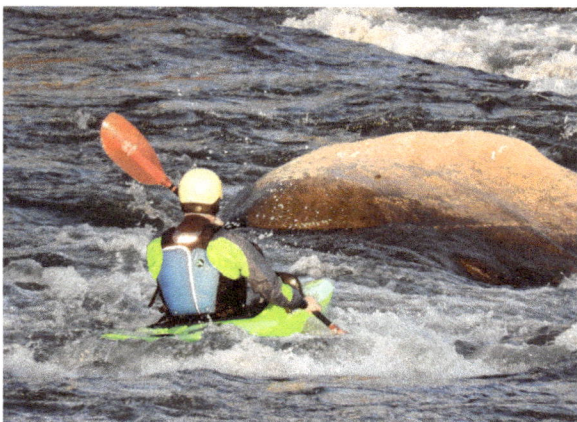

Figure 3-38. Sports Example

Be aware of this limitation and plan accordingly. If you expect to need a second burst of shots within about five seconds after the first burst, you should use a different setting, such as Program mode, and use single shooting or, possibly, low-speed burst shooting and take only one or two shots, so the camera will be ready for more action quickly.

Party/Indoor

This setting is meant for indoor photos of people and rooms. In most cases, you should pop up the flash; the flash mode is initially set to Auto with Red-eye Reduction, but you can change to another flash mode if you want to. If you don't want to use flash, you can leave the flash unit retracted, in which case the B700 will try to use a relatively slow shutter speed. In that situation, you should hold the camera very steady or place it on a tripod. (Realistically, though, you probably are not going to be setting up a tripod for candid or impromptu pictures at a party.) The camera will focus on the subject at the center of the frame.

I took the image in Figure 3-39 with the flash turned on. The camera set the shutter speed to 1/13 second and the aperture to f/4.2, with the ISO at 800. With the Party/Indoor setting, you can use exposure compensation or the self-timer, but you cannot change the focus method.

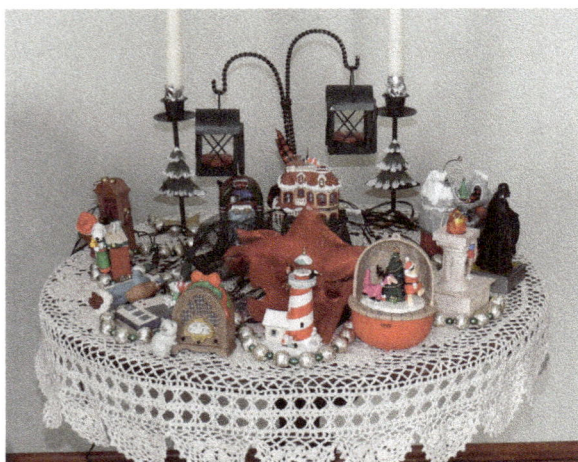

Figure 3-39. Party/Indoor Example

Beach

With this selection, the camera optimizes its settings for the beach, where there is likely to be bright sunlight reflected from the ground. In this environment, the camera will have a tendency to underexpose the subject because the exposure meter will be measuring the brightness of the beach. If you pop up the flash, the camera will set the flash mode to Auto in order to light the subject sufficiently, and it is quite likely the flash will fire so the subject will be clearly visible against the glare of the background. You can change the flash mode if you want to. However, you do not have to pop up the flash when using the Beach setting. You can use either macro focus or normal autofocus, but you cannot select manual focus or use infinity autofocus. You can use the self-timer or exposure compensation.

Snow

The Snow setting is similar to Beach. The camera may use flash to compensate for the brightness of the snowy background, if you have chosen to pop up the flash unit. The camera appears to use a greater amount of reddish hue than with the beach setting, as a balance against the bluish color temperature of a snowy scene. Other settings are similar to those for the Beach setting.

I took a few test shots of general scenes, not shown here, using both the Beach and the Snow setting. I found that the Beach setting resulted in slightly darker images, but otherwise the results were very similar.

Sunset

With this setting, the B700 disables the flash, but you can use the self-timer and exposure compensation. You cannot change the focus mode from infinity autofocus. The camera processes the shot to emphasize reddish tones in the rays of the late afternoon or early morning sun. Of course, you don't have to limit the use of this (or any other) scene setting by its label; if you are photographing autumn leaves, red-brick buildings, or other subjects with reds you want to emphasize, consider this setting as a tool that may be of use. For Figure 3-40, I used this setting for a view of a bridge over the river shortly before sunset.

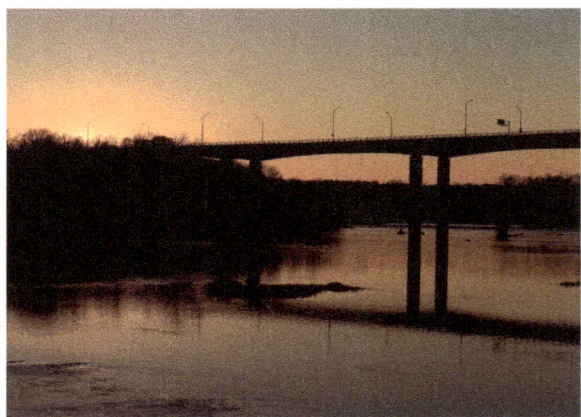

Figure 3-40. Sunset Example

Dusk/Dawn

If you take pictures before sunrise or after sunset, this is a setting to consider. The camera turns off flash and intensifies colors to add interest to images that otherwise might be flat or washed out because of dim light. You cannot adjust most settings, though you can use exposure compensation and the self-timer. This option emphasizes the purplish or bluish tones that may be present in the twilight or early morning hours.

In Figure 3-41, I used this setting to capture an image of the same general area of the river as in the previous shot, a few minutes later. You can see the dramatic difference in the colors used by the camera with this setting.

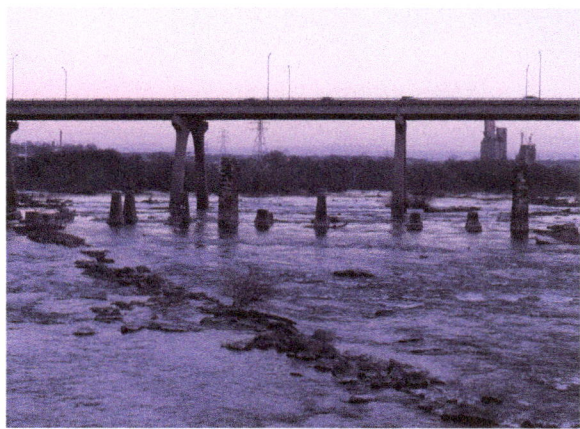

Figure 3-41. Dusk/Dawn Example

Close-up

With this setting, the camera switches into macro focus mode to capture images of items close to the lens. If the lens is zoomed in to a telephoto position, when you select the Close-up setting the lens will automatically zoom back out to a position that allows the camera to focus on the subject. (You can, however, zoom the lens back in if you want to.) You can use the self-timer or exposure compensation; you cannot, naturally enough, change the focus mode, which is set on macro.

The camera also sets AF Area Mode to Manual (Spot). This means you can control exactly where the focus point is placed. To do this, press the OK button in the center of the multi selector, then press the direction buttons on the multi selector to move the focus area around the screen so it covers the point where you want the camera to focus. You also can turn the multi selector dial to move the frame around the screen; it will move up and down as it reaches the right or left edge of the display.

If you need to use one of the direction buttons for its other function (self-timer, flash, or exposure compensation), press the OK button again, and those functions will be available. Press the OK button once more if you need to move the focus area another time. The camera also uses continuous autofocus with this setting, so it continues to adjust the focus until you press the shutter button halfway down to lock in the focus.

As with several other scene types, with Close-up the Scene menu gives you the option of choosing Noise Reduction Burst or Single Shot. You can choose one of these options by going to the Close-up item on the Scene menu. With Noise Reduction Burst, the camera takes a rapid set of shots to counter the effects of high ISO noise and forces the flash off. With the Single Shot setting, the camera takes just one image and lets you pop up the flash and select any flash mode. The camera also uses processing to sharpen the image's outline with added contrast.

You could, if you want, use another shooting mode, such as Program or Auto, and just select macro focusing using the focus button (Down button). But, if you want to quickly set up the camera for close-up shooting, it can be convenient to have this scene setting available. You should hold the camera very steady to avoid blurring the image. Use of a tripod or monopod is the best practice, but of course that is often impractical.

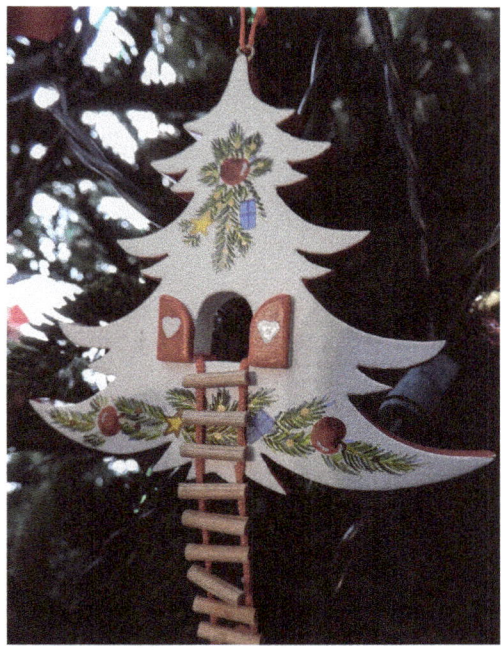

Figure 3-42. Close-up Example

In Figure 3-42, I hand-held the B700 with the lens set to nearly the minimum focusing distance to get a picture of an ornament hanging on the Christmas tree. I used the Noise Reduction Burst mode to improve quality. The camera used a fairly slow shutter speed of 1/20 second at f/3.3, with ISO set to 800.

Food

This setting is similar to the Close-up setting, discussed above, though it does not offer Noise Reduction Burst mode; only single shots are available. The camera switches to macro focus mode and zooms back if necessary so it can focus on a nearby subject. It turns on Manual (Spot) for the AF Area Mode, so you can move the focus point around, and it uses continuous autofocus. The flash is disabled, but you can use exposure compensation or the self-timer.

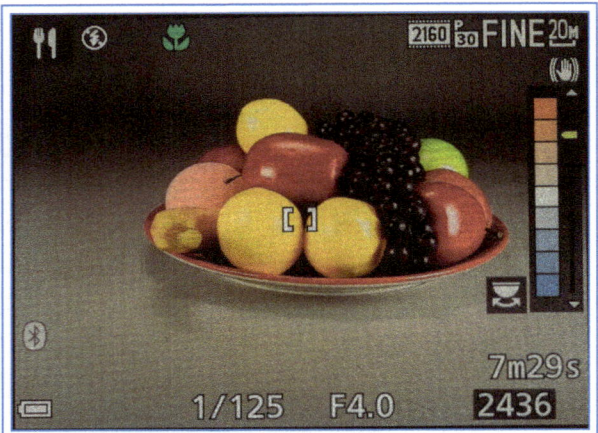

Figure 3-43. Shooting Screen for Food Mode

The one major difference from Close-up mode (apart from the lack of the burst option) is that, in Food mode, the camera places a scale of colors at the right side of the screen, as shown in Figure 3-43, and allows you to adjust the hues of your images by moving the pointer up and down along the scale using the command dial. Move the pointer toward the top for more reddish hues, and toward the bottom for more bluish ones.

In Figure 3-44, the final image resulting from setting the hue slider to the second position, I used the slider to emphasize the reddish hues of the artificial fruits on the plate.

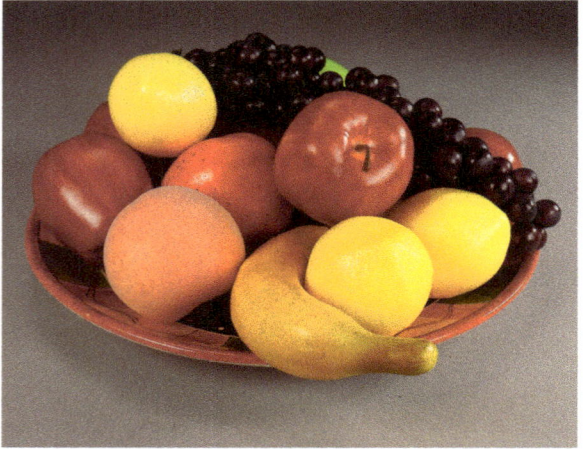

Figure 3-44. Food Example

This setting is suited for people who are in the habit of documenting their meals through food blogs or photographic diaries of their dining experiences. With the hue slider, you can experiment until you achieve the desired effect of emphasizing the colors of meats, vegetables, or other aspects of the meal.

Unless you want to leave a hue adjustment permanently in place, be sure to return the hue slider to the neutral position when you're done shooting, because the adjustments you make will remain in place the next time you choose the Food setting, even if the camera has been turned off in the meantime.

Fireworks Show

With this setting, the camera sets the focus to infinity and uses a shutter speed of four seconds so you can capture a long burst of color from a fireworks display. (You may notice that the camera uses a mathematical infinity symbol on the display for this focus mode. The camera has two focus modes called "infinity," oddly enough. It appears that this "infinity" focus mode causes the camera to focus at its farthest distance, while the other "infinity" mode, symbolized by a mountain icon, causes it to focus "near infinity.")

The camera also increases the vividness of the colors and uses a low ISO setting to maximize image quality. The flash is forced off and you cannot use exposure compensation or the self-timer. You can select manual focus if you want. You should set the camera on a tripod if possible, or hold it firmly on a fence post or other solid object as an alternative.

Backlighting/HDR

This mode is designed for difficult lighting situations, such as when there is bright light present, but it is behind the subject, or when the subject is partly in bright light and partly in shadows. Nikon calls this mode simply "Backlighting," but its single sub-mode is HDR, which, as discussed below, is an important feature for modern cameras, so I have added HDR to the heading here for easier identification.

When this mode is selected, you have two sub-options selectable by pressing the Menu button: HDR Off, or HDR On, as shown in Figure 3-45.

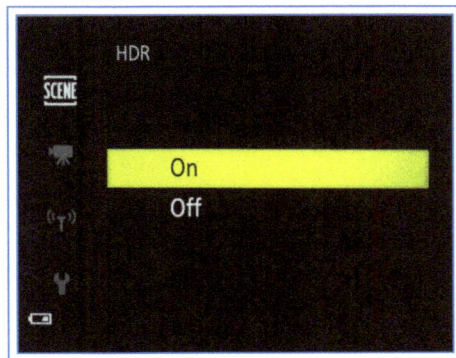

Figure 3-45. Backlighting Menu Options Screen

If you choose the default value, HDR Off, the camera requires that you raise the flash, and it forces the flash to fire to overcome the shadows caused by your subject's being lighted from behind or unevenly. If you choose HDR On, the camera uses a different approach. With this setting, the camera internally performs its own version of HDR, or high dynamic range, processing.

HDR photography uses special techniques to deal with scenes that include areas of extreme contrast between light and dark. For example, if a building is partly in sunshine and partly in deep shadow, the contrast is likely to be so great that a photograph cannot depict both parts of the building with normal exposure. Either one area of the image will be much too bright, so highlights are blown out, or one area will be much too dark, so details are swallowed in the shadows.

In the past, HDR was carried out in post-processing, using software such as Photoshop or special programs such as PhotoAcute or Photomatix Pro. The photographer takes multiple images of the scene at different brightness levels, some exposing dark parts of the scene properly and some exposing bright parts properly. When combined in HDR software, the images can be blended together to result in a final composite image that shows all parts of the image clearly exposed. These HDR composite images often have an unnatural or surrealistic appearance, because it is obvious that a "normal" photograph could not include such a wide range of well-exposed areas.

With many modern cameras, including the Coolpix B700, the manufacturer includes programming that lets you take multiple photographs that are combined inside the camera to result in an HDR-like image. With the B700, I would not say the result quite matches the "true" HDR you can obtain through software, but it does make a noticeable difference. Here is how it works.

When you use the HDR setting on the B700, you should hold the camera very steady, or, ideally, place it on a tripod. When you press the shutter button, you will hear multiple clicks as the camera takes several exposures. It will then create and save two final images. The first of these images will be one taken with HDR off and the Active D-Lighting feature turned on, to brighten shadowy areas of the image to bring out details. (Active D-Lighting is discussed in Chapter 4.) The second image will be an HDR composite that contains the best-exposed parts of multiple images, thereby expanding the dynamic range of the shot.

To illustrate the effects of this setting, I took several shots of a model truck in an area partly in bright sunlight and partly in shade. The first shot, Figure 3-46, was taken in Program mode with no special settings, to show the contrast between the bright and shadowed areas.

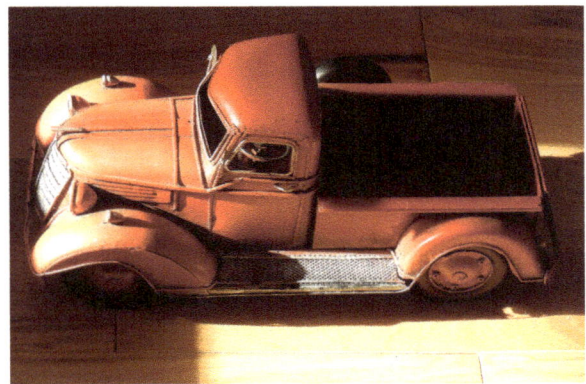

Figure 3-46. HDR Series: No Special Settings

Figure 3-47 was taken in Backlighting mode, with HDR turned off. In this mode, the camera always uses flash.

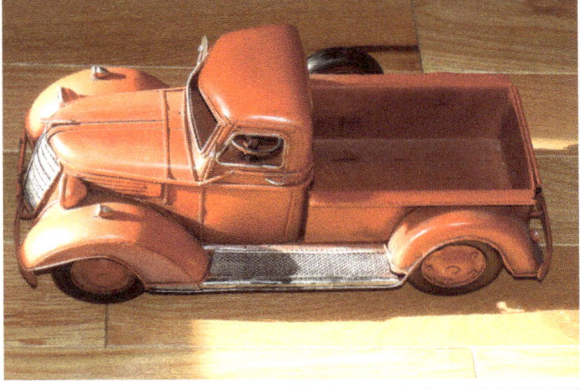

Figure 3-47. HDR Series: Backlighting Mode with HDR Off

Figure 3-48 was taken with the HDR setting turned on.

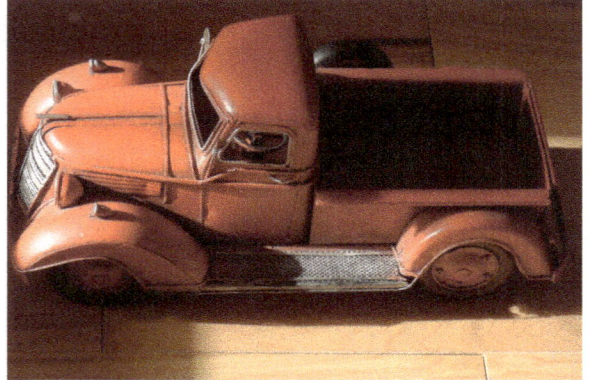

Figure 3-48. HDR Series: Backlighting Mode with HDR On

(This is the actual composite shot created in the camera from multiple exposures; as noted above, with this setting the camera also saves one image with HDR turned off.)

Finally, Figure 3-49 is a composite image created in Photomatix Pro, an HDR processing program. I took several images at different settings in Manual exposure mode to use as the basis for this composite.

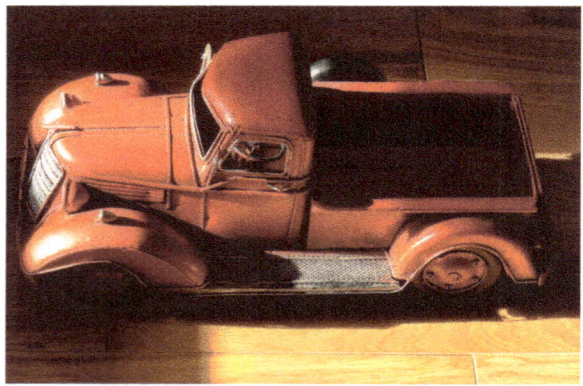

Figure 3-49. HDR Series: Photomatix Pro Composite

The B700's in-camera processing did a good job of reducing the contrast between light and dark areas, but it did not quite match the performance of Photomatix Pro, especially when, as here, I took several shots in Manual exposure mode, giving the software a wide variety of exposure values to work with. Another option for taking shots for this purpose is the exposure bracketing feature of the B700, which I will discuss in Chapter 4. With that option, the camera takes a series of three shots at different exposure levels; you can combine those shots using HDR software to create a composite.

The HDR setting is an excellent tool, because it is not always practical to take multiple shots to be combined using HDR software. If you take pictures in an area that is partly shaded and partly sunny, or otherwise has contrasting lighting, you can take advantage of this setting to improve the overall appearance of the image.

Easy Panorama

When you select this option from the Scene menu, you are presented with a sub-menu with two options—Normal (180°) or Wide (360°), as shown in Figure 3-50.

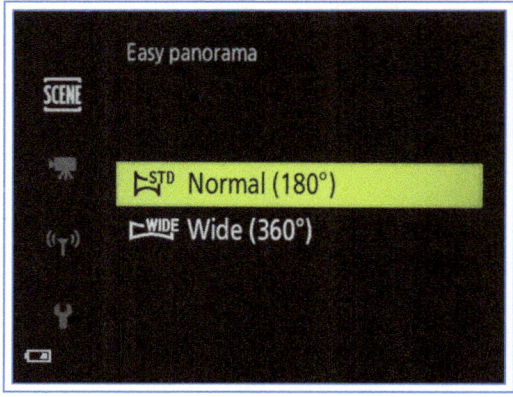

Figure 3-50. Easy Panorama Menu Options Screen

With this setting, you cannot select Image Quality or Image Size from the menu. If you select Normal, the final panorama will be 4800 x 920 pixels if shot horizontally, or 1536 x 4800 if shot vertically. If you select Wide, the dimensions with 4800 pixels are doubled to 9600 pixels. After you make your selection, the camera will display the message shown in Figure 3-51, telling you to press the shutter button and start moving the camera in your chosen direction.

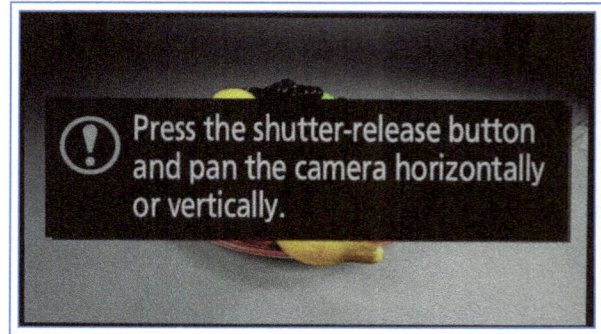

Figure 3-51. Message for Panorama Shooting

Aim the camera at the first part of your panoramic scene. For example, if you are shooting a panorama of a wide mountain range, you may want to aim at the far left side of the range.

Press the shutter button halfway down to lock in focus and exposure; the camera will automatically zoom back to the wide-angle position and will not let you zoom in. In addition, the flash is disabled. You can use exposure compensation, though. When you are satisfied with the initial view, press the shutter button all the way down and release it; you don't need to hold it down while the panorama shooting proceeds.

Hold the camera steady and level as you sweep it in the direction of your shot—in this case, from left to right— until you have covered the entire scene. The camera will detect the direction you are moving in, and it will automatically stop shooting when it detects the end point of the 180-degree shot. It will display a yellow progress bar at the top of the display. You should take about 15 seconds to complete this arc.

You can shoot the panorama moving either left to right or right to left, or you can shoot it vertically, moving from low to high or vice-versa. If you select the Wide option, you can move the camera through a complete circle to cover an entire scene. In that case, you should take about 30 seconds to complete the circuit.

When shooting panoramas, try to avoid including moving people, vehicles, or other objects, because they may end up appearing in multiple positions in the panorama.

When you are done, you can view the whole panorama on the screen in a small size by pressing the Playback button. To see the panorama scroll by on the screen in a larger size, press the OK button, and it will scroll in the same direction in which it was taken.

The sample panorama shown in Figure 3-52 was shot hand-held using the Easy Panorama setting.

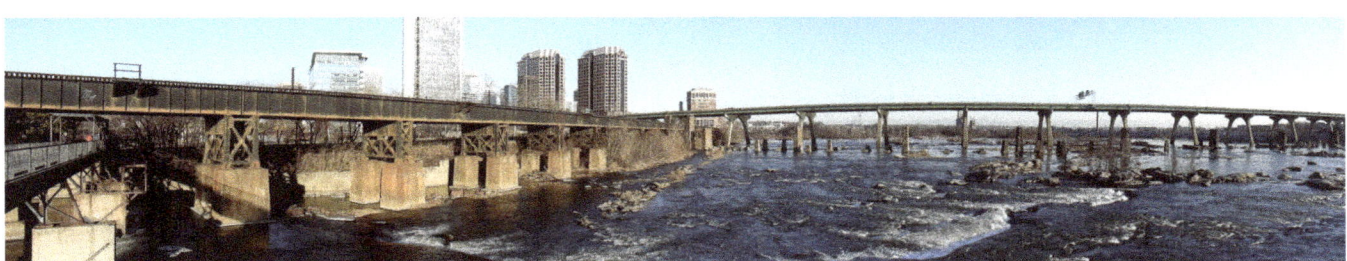

Figure 3-52. Easy Panorama Example: James River, Richmond, Virginia

Pet Portrait

This setting is for shooting pictures of the family dog or cat. With this option, the camera turns on continuous shooting and activates a feature called "Pet Portrait Auto Release." With this feature, the camera looks for the face of a pet, and, if it finds one, it automatically takes three quick pictures to try to capture a good expression. If the camera does not display the double-bordered yellow frame that indicates face detection, you can press the shutter button yourself to take the images when you're ready. The default setting with this option is continuous shooting, but you can change to single shots using the special menu that appears when you select this scene type.

You also can change the Pet Portrait Auto Release setting, which is turned on by default. To turn it off, press the Left button and select the Off setting, rather than the icon of a pet's face. If you turn this setting off but leave continuous shooting turned on, the camera will take up to three shots when you press and hold the shutter button. For Figure 3-53, I used this setting to capture an image of an active spaniel who paused long enough to let the camera capture a portrait.

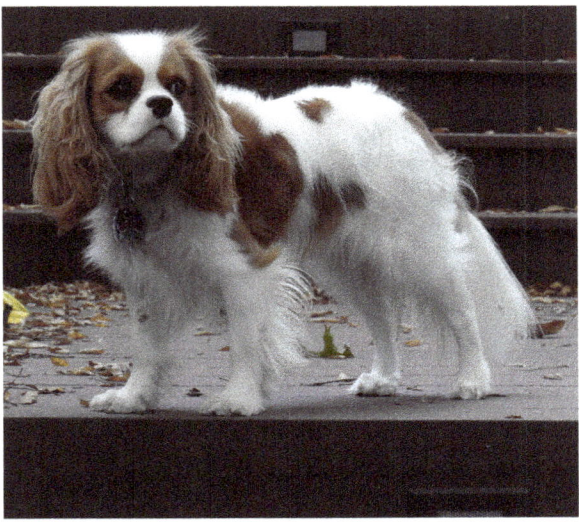

Figure 3-53. Pet Portrait Example

With the Pet Portrait setting, use of the flash is disabled, but you can use exposure compensation. You also can select macro focus instead of the default autofocus option.

Moon

The Coolpix B700, with its great optical zoom range, is a natural for shooting images of the moon without having to attach the camera to a telescope. With the Moon setting, Nikon has provided a shortcut to using appropriate settings for these shots.

With this option, the B700 disables the flash and turns on the self-timer to two seconds so the camera will have some time to settle down after you press the shutter button, to avoid camera shake. You can change this setting to 10 seconds or turn the self-timer off if you want. You can use exposure compensation, which is useful if the moon appears too bright or too dim because of its phase. Focus is fixed in the center of the frame, at the infinity setting.

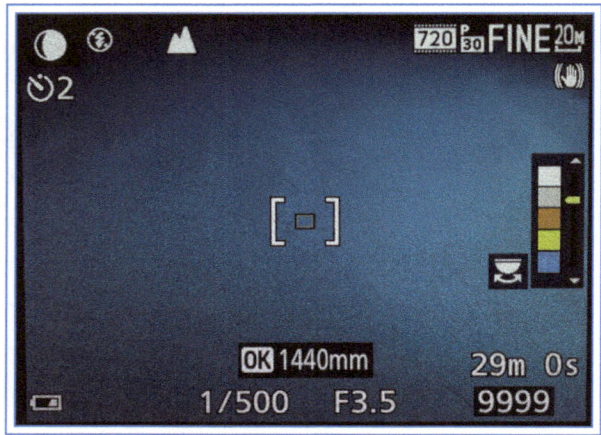

Figure 3-54. Shooting Screen for Moon Setting

The camera also uses two special settings to help with this type of shot. First, as shown in Figure 3-54, when the lens is zoomed out to its wide-angle setting, the camera places a small rectangle in the center of the focus frame. That very small frame represents the viewing angle at the full optical zoom range of 1440mm. You can place that frame over the moon with the lens zoomed back to its wide-angle setting, then press the OK button to cause the camera to zoom immediately to its full zoom length. In that way, you can easily locate the moon in the sky and zoom in on it with confidence that you will center it in the image, as shown in Figure 3-55.

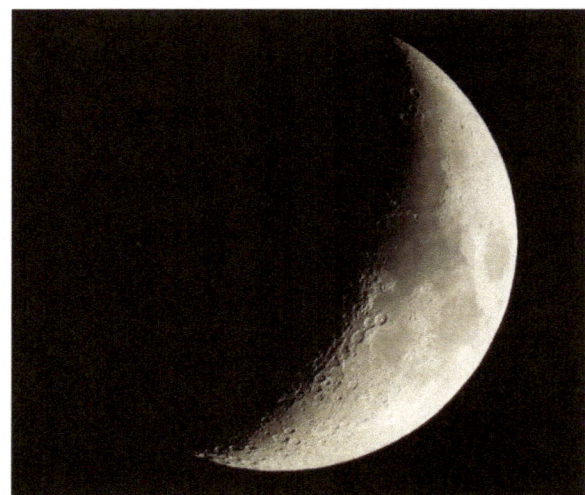

Figure 3-55. Moon Example

Also, the camera places on the right side of the screen a scale with a variety of hues. This is not a sliding scale, but, instead, is like a set of virtual color filters. You move through them by turning the command dial. If you select the top option, you will view the moon with no color change. You can try any of the other selections to enhance your view of the moon and its craters. The yellow filter is good for enhancing the overall contrast of the image, while the blue filter can be used to reduce the glare. My suggestion is to switch through all of the options while you are viewing the moon, to see which ones yield better results for the sort of image you are looking for. As with the Food setting, the hue setting you choose for the Moon setting will stay in place even after the camera is turned off and then on again.

Bird-watching

The next Scene mode option, Bird-watching, is, like the Moon option, an important and natural setting for the B700. With this setting, the camera disables the flash and turns on autofocus, but you can switch to infinity focus or manual focus. You also can use the self-timer or exposure compensation. You can turn on continuous shooting, so you can take a burst of shots with one shutter press, increasing the chances of capturing a good image as a bird moves around. To do that, press the Menu button, and select Bird-watching from the menu, as shown in Figure 3-56.

Chapter 3: The Shooting Modes | 43

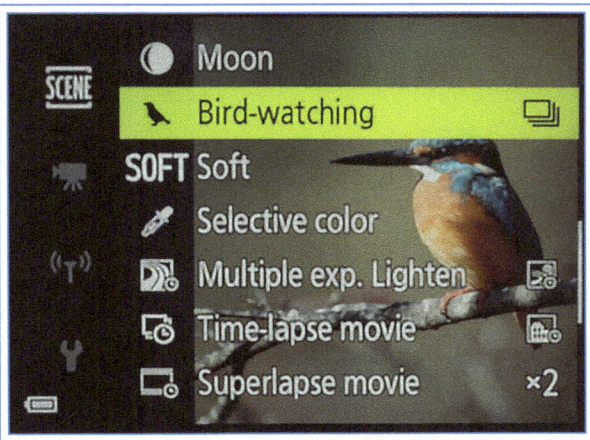

Figure 3-56. Bird-watching Item on Scene Menu

Press the OK or Right button to move to the next screen, shown in Figure 3-57, and select Continuous from that screen. Then, when you press the shutter button and hold it down, the camera will shoot a continuous burst of about five shots at a rate of about five frames per second.

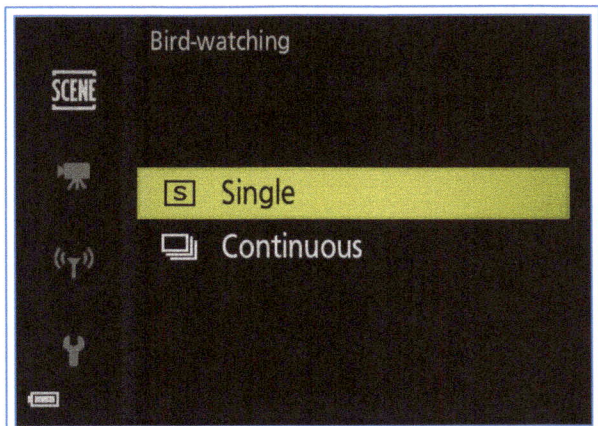

Figure 3-57. Bird-watching Menu Options Screen

One important feature is that, just as with the Moon option, when you are using a wide-angle setting, the camera places a special frame in the center of the image to indicate the angle of view at the telephoto end of the zoom scale. In this case, the frame shows the amount of the image that would be in view with the lens zoomed in to the 800mm point, rather than the full 1440mm that is used with the Moon setting.

So, when you are trying to capture a close-up shot of a distant bird, you can start with a wide-angle view so it is easy to find the bird, and center the subject in the small frame. Then, just press the OK button and the camera will automatically zoom the lens to the 800mm point, with the bird centered in the image. If you then need to zoom back to a wide-angle view, just press the zoom lever to the left to zoom out. If you want to zoom in beyond the 800mm point, you can use the zoom lever to do that, also.

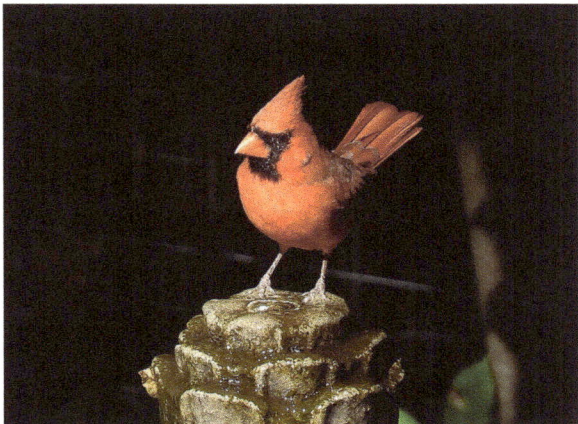

Figure 3-58. Bird-watching Example

For Figure 3-58, I used this setting to fire off several quick bursts of shots when I saw this cardinal land on top of a backyard fountain. It took several bursts before I captured this image of the bird with its head clearly visible and well lighted.

SOFT

This setting, as shown in Figure 3-59, applies a small amount of blur to the image. You can use the flash, self-timer, and exposure compensation, and you can select a focus mode.

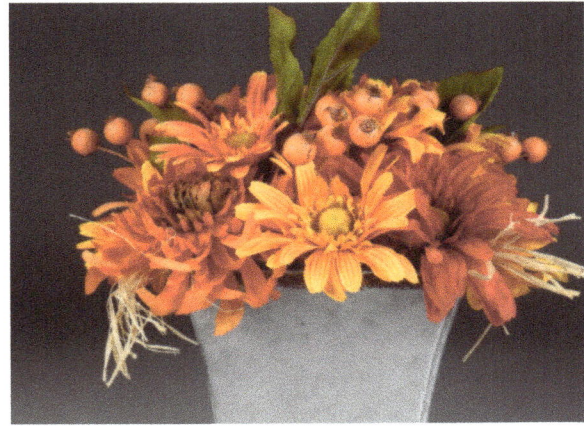

Figure 3-59. Soft Example

You could use this option when taking pictures of an aging movie star to soften facial wrinkles, or you might use it just to add a somewhat dreamlike or fantasy aura to the image, such as this shot of an artificial flower arrangement.

SELECTIVE COLOR

This setting is intended to capture a photograph with only a single color retained, leaving the rest of the image in black-and-white. When you choose this option from the menu screen, the camera displays a yellow indicator beside a vertical spectrum of 12 colors plus one entry for no color, as shown in Figure 3-60.

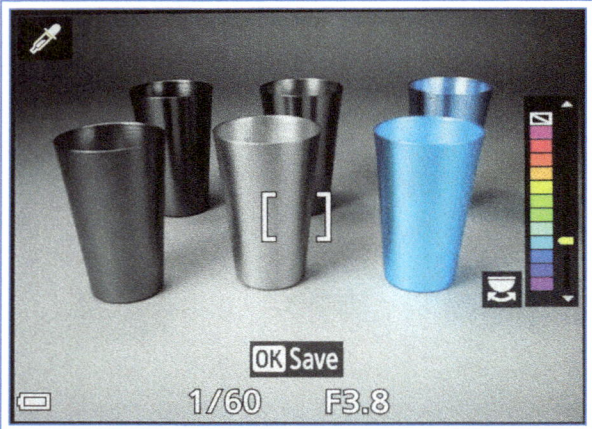

Figure 3-60. Shooting Screen for Selective Color Setting

Use the command dial or the Up and Down buttons to move the indicator to the block in the spectrum for the single color you want to retain in the image, and press the OK button to lock in that choice. The topmost block in the spectrum has a negative symbol in it; if you select that block, the camera retains all colors and does not use the Selective Color function. The result in that case is an ordinary image with no special processing at all. It will look as if you took the image using Program or Auto mode. To select a different color, press the OK button to activate the color scale again.

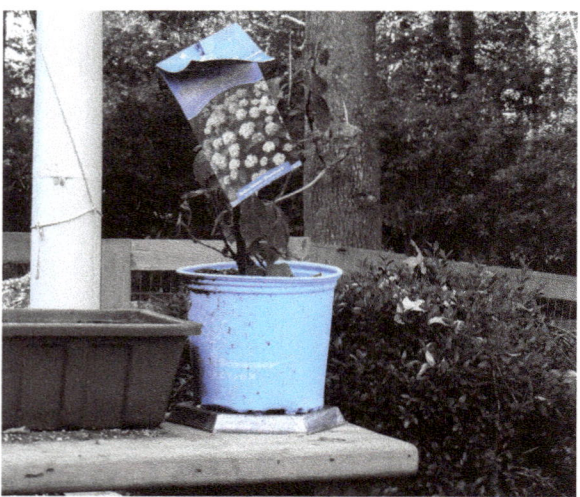

Figure 3-61. Selective Color Example

Figure 3-61 shows the image that resulted from setting the color to light blue for a shot I took that included planters of various colors outdoors.

MULTIPLE EXPOSURE LIGHTEN

This highly specialized setting is designed for taking multiple shots in two particular situations: capturing a trail of automobile headlights and taillights at night, or capturing star trails in the night sky. On the menu screen for this option, select either Nightscape+Light Trails or Star Trails, as shown in Figure 3-62.

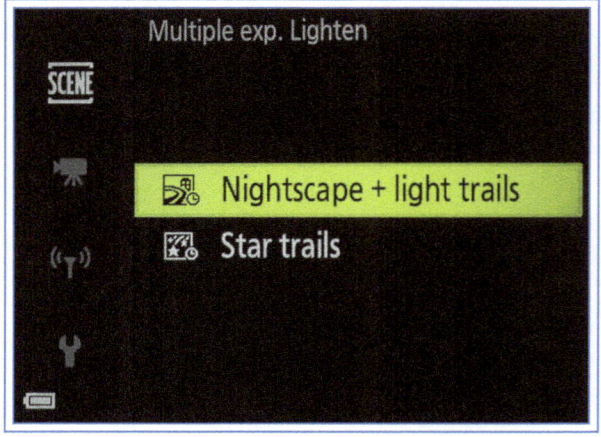

Figure 3-62. Multiple Exposure Lighten Menu Options Screen

If you choose the Nightscape+Light Trails setting, you can then turn the command dial to set the shutter speed to one, two, four, or eight seconds. Next, with the camera set firmly on a tripod or equivalent support, press the shutter button to begin the sequence of shots. The camera will automatically capture 50 images in a sequence, using the shutter speed you set, which will also act as the interval between shots.

For example, with the shutter speed set to four seconds, the camera will take 50 shots, one every four seconds, with a shutter speed of four seconds. The camera will save five images out of the 50, and each of those 5 images will be a composite that includes the brightest areas in 10 of the images. In this way, there is increased likelihood of ending up with several images that include a dramatic series of light trails.

Figure 3-63 is an example of the results I got when I set the shutter speed to one second and positioned the camera on a tripod after dark on an overpass over a busy highway. The image filled up rapidly with light trails, and I ended up pressing the OK button to stop the process several times, so the image would not be overwhelmed with excessive streaks of light.

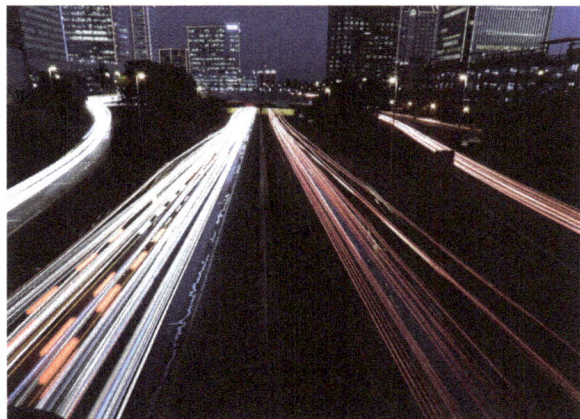

Figure 3-63. Nightscape+Light Trails Example

If you choose the Star Trails option, there are no further settings to make. Once you press the shutter button, the camera will take a series of 300 shots at 30-second intervals, using a shutter speed of 25 seconds for each shot. The camera will save 10 composite images, each one including the bright areas from 30 of the shots taken. Figure 3-64 is an example taken using this setting and letting the camera run for a period of about 90 minutes. The star trails are quite faint and hard to see, possibly because the night sky is not very bright in my area.

Figure 3-64. Star Trails Example

For either of these settings, you can interrupt the shooting sequence by pressing the OK button. Once you see that you have captured the views you were looking for, you should interrupt the sequence in this way, so the images are not washed out by adding excessive bright areas to the composite shots.

Time-lapse Movie

This next selection for Scene mode gives you a way to shoot a time-lapse sequence, in which the camera takes a series of images at a set interval. With this option, for example, you might capture a series of 300 images of a sunset taken at 10-second intervals. The result would show the progress of the sunset over a period of 3000 seconds, or 50 minutes. At the standard playback rate of 30 frames per second (in the United States), this sequence of 300 images would play back in 10 seconds, producing a dramatically speeded-up view of the sunset.

You can achieve a similar effect using the interval timer setting under the Continuous menu option on the Shooting menu. However, the shortest interval available with that option is 30 seconds. In addition, the Time-lapse Movie option provides you with several preset options for typical time-lapse subjects, so you do not have to choose the interval and number of shots, or make other settings; you can just choose the subject matter and let the camera make the appropriate settings.

The five available subjects, along with the elapsed time for each shooting sequence, are: Cityscape (10 minutes); Landscape (25 minutes); Sunset (50 minutes); Night Sky (150 minutes); and Star Trails (150 minutes). In each case, the resulting time-lapse movie is about 10 seconds long, when played back at 30 frames per second.

To use this option, with the mode dial at the SCENE setting, press the Menu button and select Time-lapse Movie from the menu, as shown in Figure 3-65.

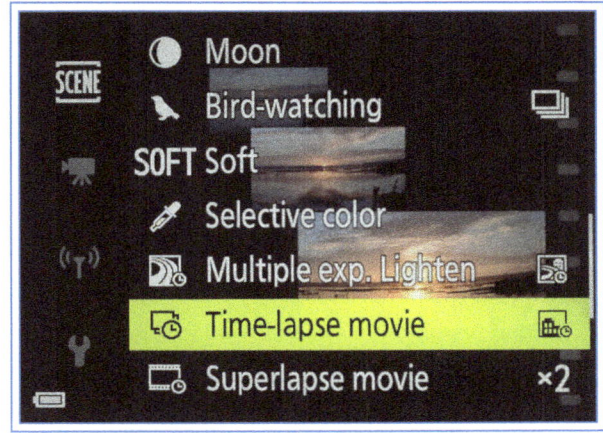

Figure 3-65. Time-Lapse Movie Item on Scene Menu

When you press the OK or Right button, you will see the screen shown in Figure 3-66, with five options for the subject matter.

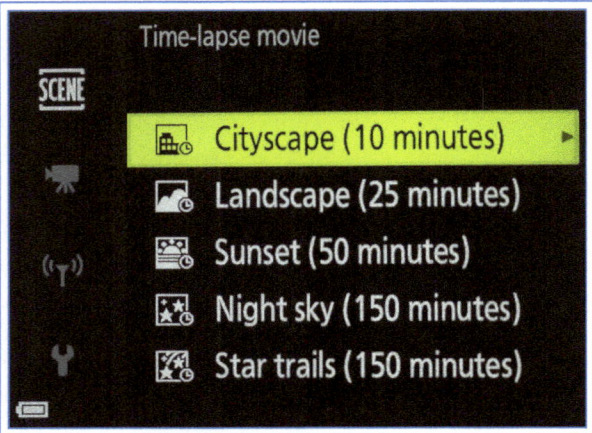

Figure 3-66. Time-Lapse Movie Menu Options Screen

Highlight the one you want, and press the OK or Right button to move to the screen shown in Figure 3-67, where you can select either AE-L On or AE-L Off.

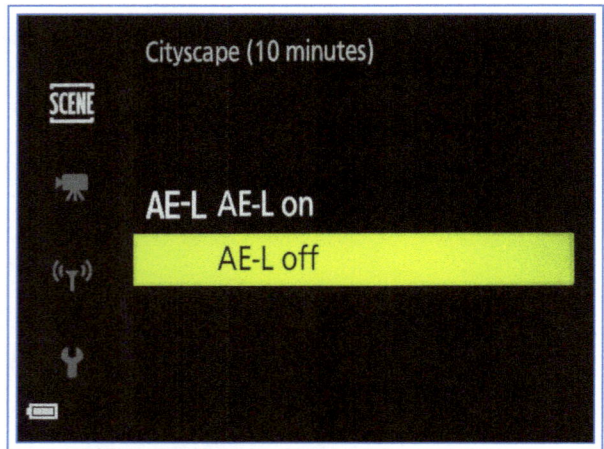

Figure 3-67. AEL Options Screen for Time-Lapse Movie

With the AE-L On setting, the camera will lock exposure with the first image, so the exposure will not be adjusted as the lighting conditions change. With the AE-L Off setting, the camera will automatically adjust exposure as appropriate. (The AE-L settings are not available with the Night Sky or Star Trails options.)

The choice of the AE-L setting depends on your preference. If you select AE-L Off, the exposure adjustments as clouds pass over the sun can be distracting. Also, for a sunset, the exposure adjustments are likely to make the scene appear somewhat unnatural, with the sky staying relatively uniform in brightness even as the sun sets. I generally prefer the AE-L Off setting to achieve a more natural effect, but the choice depends on the particular circumstances.

You should set the camera on a sturdy tripod and be careful not to touch it or disturb it while the sequence is being recorded. The camera does not record sound with the images. You can stop the sequence at any time by pressing the OK button. When the sequence has been recorded, it will appear as a movie file, with the file extension .mp4, as shown in Figure 3-68.

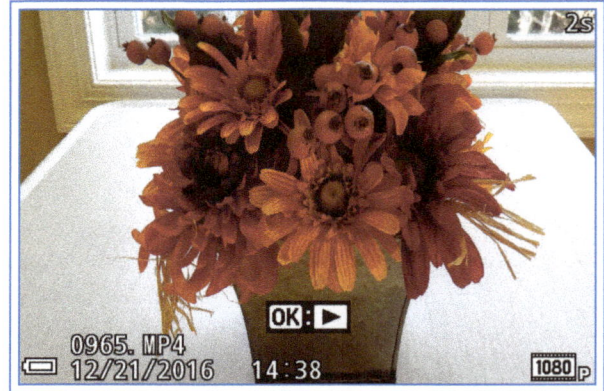

Figure 3-68. Time-Lapse Movie Ready to Play in Camera

To play it, press the OK button as with any other movie. You will see a greatly speeded-up version of the subject matter.

Superlapse Movie

This setting, despite its name, is not a "super" version of the time-lapse setting, though it has some similarities to that option. The Time-lapse Movie option is intended for making lengthy recordings that are compressed into a few seconds, with the camera on a tripod. The Superlapse Movie feature, by contrast, is intended for making fast-speed recordings with the camera handheld. You can choose to have the action's speed increased by a factor of two, four, 10, 20, or 30 times.

To use this setting, select Superlapse Movie from the Scene mode menu, and you will see the screen shown in Figure 3-69, with the five speed options noted above. That is the only setting to make for this feature. Once the speed is set, just press the red Movie button to start recording. You cannot zoom the lens or make any other control adjustments during the recording. Press the Movie button again to stop the recording and save the movie. When the footage is played back, the action will appear to be speeded up according to the speed factor you selected. No sound will be recorded.

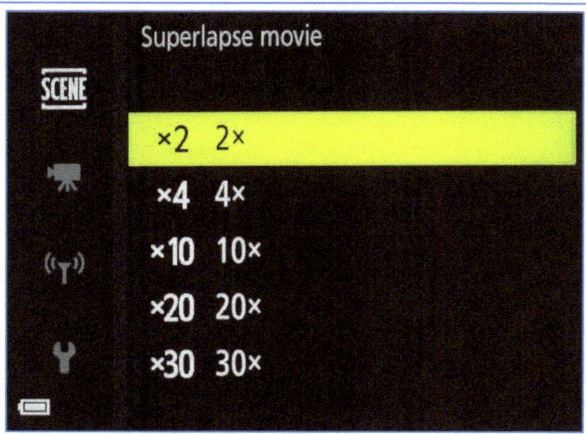

Figure 3-69. Superlapse Movie Menu Options Screen

Note that, as I will discuss in Chapter 8, the camera cannot record a movie for more than 29 minutes in any single sequence. So, for example, if you set the speed factor to 30x and record for the full 29 minutes, the final sequence will last less than one minute.

Creative Mode

Several notches around the mode dial from the SCENE mode is the slot indicated by a white camera icon with a stylized letter C, indicating the Creative shooting mode. This shooting mode provides you with interesting ways to alter the appearance of your images. These settings operate in a manner similar to the scene settings, but there are some differences. The Creative settings are not designed for particular types of subjects, such as portraits, landscapes, or fireworks, as the scene settings are. Instead, the Creative mode settings offer manipulations to change how an image looks, regardless of the subject.

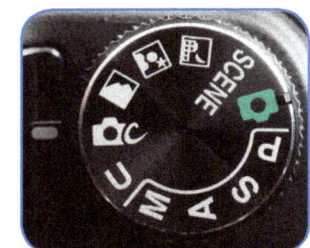

Figure 3-70. Mode Dial at Creative Mode

To select one of these settings, turn the mode dial to the Creative position, as shown in Figure 3-70, then press the OK button to go to the screen shown in Figure 3-71.

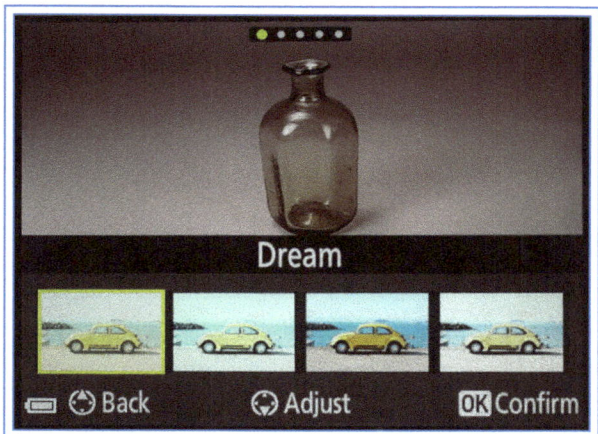

Figure 3-71. Effect Group Selection Screen for Creative Mode

This screen has five dots at the top, which represent the five overall effect groups for this mode, from left to right: Light, Depth, Memories, Classic, and Noir. To move among those groups, turn the command dial. Each time you turn that dial to select a different group, the thumbnail images at the bottom of the screen change, to show the four individual effects that are available within the currently selected group. To move among the individual effects within a group, you turn the multi selector dial or press the Left and Right buttons.

For example, if you start from the screen shown in Figure 3-71, above, and turn the command dial three times to the right, you will move from the Light group to the Classic group. If you then turn the multi selector dial to the right (or press the Left and Right buttons), you will move through the four individual effects within the Classic group: Sepia, Blue, Red, and Pink. If you keep turning the multi selector dial beyond the last (or first) setting within the current group, you will move to the first (or last) setting of the next group, either after or before the current group.

Once you have highlighted an individual setting you want to use, such as Sepia, you can press the OK button to select it without further adjustment. If you want to make some adjustments, press the Down button, which takes you to the screen shown in Figure 3-72. On this screen, you can turn the multi selector dial or press the Left and Right buttons to highlight Amount, Exposure Compensation, Contrast, Hue, Saturation, Filter, or Peripheral Illumination.

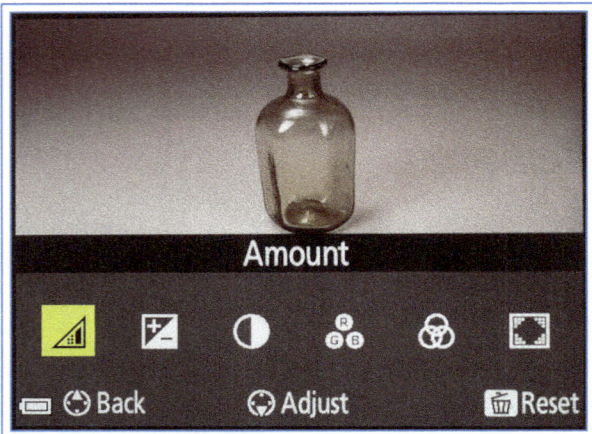

Figure 3-72. Parameter Selection Screen for Creative Mode

The available parameters vary according to the setting that is in use. The Hue parameter is not available with the Classic group settings, for example. The Filter parameter, which is available only for settings in the Classic and Noir groups, lets you simulate the addition of a yellow, orange, red, or green filter on the lens. The Peripheral Illumination parameter controls the amount of vignetting, or darkening of the corners, for an effect.

When you have highlighted the parameter you want to adjust, such as Contrast or Peripheral Illumination, press the Down button to move to the adjustment screen, shown in Figure 3-73.

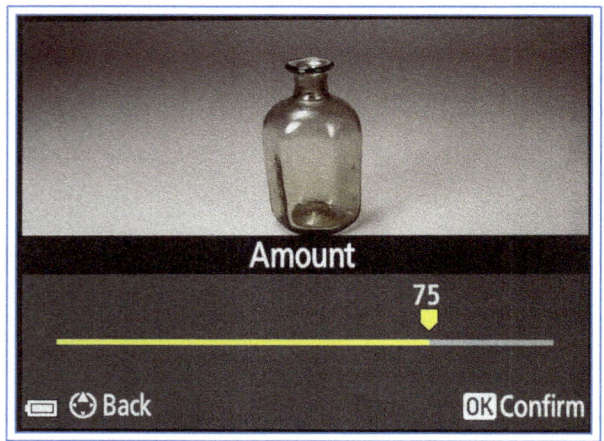

Figure 3-73. Parameter Adjustment Screen for Creative Mode

On that screen, use the Left and Right buttons or the multi selector dial to set the amount of the adjustment for the selected parameter and press the OK button to confirm. For example, with Peripheral Illumination, to increase the darkening, adjust the scale downward, to the left. To brighten the corners and reduce the vignetting, adjust the scale upward, to the right. If you want to reset the adjustments for an effect, press the Trash/Delete button, and the camera will ask you to confirm that operation. Select Yes, and the options will be reset to the default settings.

When you have finished adjusting all of the parameters you want to alter, press the Up button to select the effect, as adjusted. Or, just press the shutter button halfway to return to the shooting screen. You can then proceed to capture still images or videos with the chosen effect.

The camera's display will change according to the setting you have chosen. For example, if you select the Binary effect within the Noir group, the display will be in black and white and will appear grainy, with harsh contrast.

There are no other options available on the Shooting menu when you are using Creative mode, apart from Image Quality and Image Size. You cannot select any Raw options for Image Quality. However, you can use the buttons on the multi selector to choose exposure compensation, the self-timer, or a focus mode (autofocus or macro only). You also can choose a flash mode, if the flash unit is popped up.

You also can use the menu system to select an effect group and an effect within the group. However, in order to make adjustments to a setting, you still need to press the OK button to call up the options screen, and then press the Down button to make adjustments.

I will list the groups and their individual settings below, followed by a chart with sample images that use all of the various settings for the same subject, to give you a general idea of the effect that each setting produces. The settings are used here with no adjustments.

Light: Dream, Morning, Pop, Sunday

Depth: Somber, Dramatic, Silence, Bleached

Memories: Melancholic, Pure, Denim, Toy

Classic: Sepia, Blue, Red, Pink

Noir: Charcoal, Graphite, Binary, Carbon

Figure 3-74 is a chart that includes images of the same floral arrangement taken using each of the Creative settings, with no adjustments, to illustrate the general differences among these settings.

Creative Mode Comparison Chart

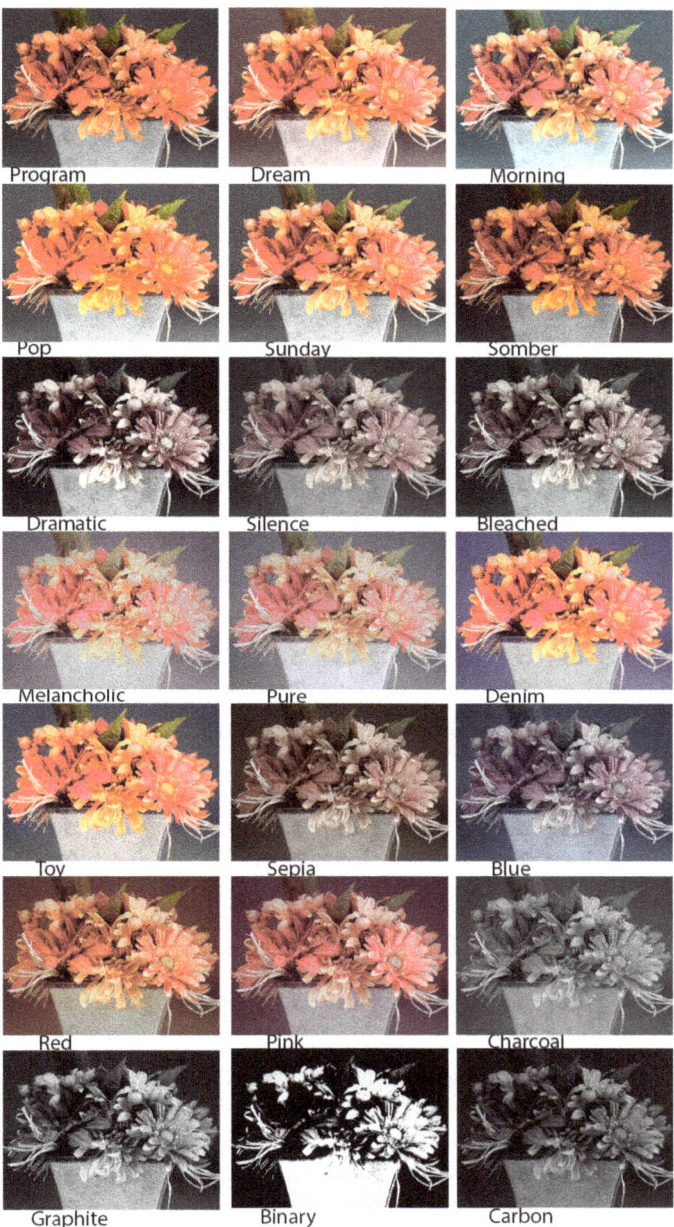

Figure 3-74. Creative Mode Comparison Chart

User Settings Mode

The last slot on the mode dial to be discussed is the U setting, as shown in Figure 3-75, which allows you, the user, to store a full set of your favorite or most often-needed settings for immediate recall.

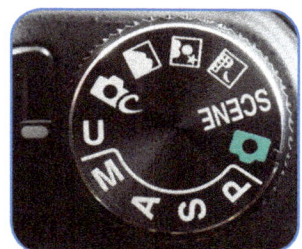

Figure 3-75. Mode Dial at User Settings Mode

When you turn the mode dial to the U position, you take advantage of a powerful feature of the Coolpix B700. You can set up the camera exactly as you want it, with shooting mode, zoom amount, flash mode, focus mode, self-timer, exposure compensation, function button assignments, and many Shooting menu settings, including options such as white balance, ISO, Image Quality, Image Size, and Picture Control, and then recall all of those settings instantly just by turning the mode dial to the letter U. The only shooting modes that you can save settings for are Program, Aperture Priority, Shutter Priority, and Manual; you cannot save them for the Auto, Scene, or Creative modes.

Here is how this works. First, set up the camera with the settings you will want to recall. For example, suppose you are doing street photography. You may want to shoot with a fast shutter speed, say 1/250 second, at ISO 1600 in black-and-white, using continuous shooting with autofocus, at the 16:9 aspect ratio with a large image size and Fine quality.

First, make all of these settings. Turn the mode dial to S for Shutter Priority, and use the command dial to set a shutter speed of 1/250 second. Then press the Menu button and choose Fine for Image Quality on the Shooting menu. For Image Size, select 5184 x 2920 pixels, which, as indicated to the left of those numbers, translates to a 16:9 aspect ratio with an image size of 15 megapixels. Then navigate in the menu to Picture Control and select Monochrome. Set the ISO menu option to 1600. Next, select the Continuous item on the Shooting menu and navigate to the next screen; on that screen, go down to the second option, Continuous H, marked with an H on a stack of frames, for high-speed shots. You also may want to push the zoom lever all the way to the left, for wide-angle shooting.

Once all of these settings are made, press the Menu button to call up the Shooting menu, and scroll down (or scroll up and wrap around to the bottom) to select the Save User Settings item, shown in Figure 3-76, and then press the OK button or the Right button; you will see a confirming message saying Done.

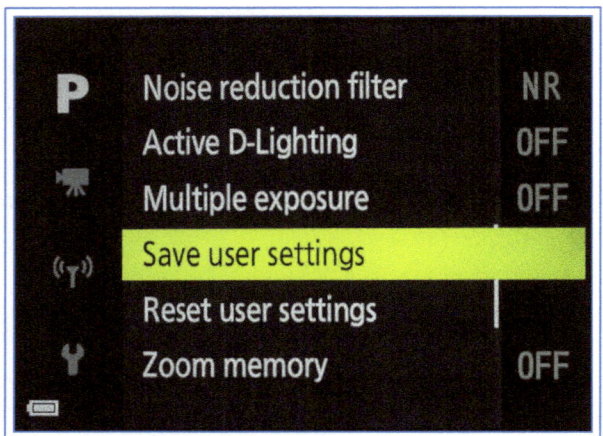

Figure 3-76. Save User Settings Item on Shooting Menu

Be sure you have all the settings the way you want them before you press OK or the Right button, because the camera does not ask you to confirm your choices; it just says "Done." I was a bit taken aback the first couple of times I used this feature, because in most other cases there's a chance to back out before you make your choices final; not here.

Now, to check how this option worked, try making some very different settings, such as Manual exposure with continuous shooting turned off, a shutter speed of one second, Picture Control set to Standard, the zoom lever moved all the way to the T for telephoto, ISO set to Auto, and Image Size set to the maximum, 5184 x 3888 pixels. Then turn the mode dial to the U setting, and you will see that all of the custom settings you made earlier have come back, including the zoom position, shutter speed, black-and-white shooting at ISO 1600, and everything else. This is really a wonderful feature, and more powerful than similar features on some other cameras, which can save menu settings but not settings such as shutter speed and zoom position.

The lone flaw I find with this mode is that there is only one slot for it on the mode dial, and therefore only one group of settings that can be saved at a time. But it's much better than nothing. I suggest you experiment to find one group of custom settings that is the most useful to you, and save it to the U mode for instant recall. Of course, you can change the settings that are stored as often as you like. You may want to jot down in a notebook some of your favorite groups of settings for various situations, so you can program the most appropriate set into the U slot when you're setting out for a particular type of shooting session.

Chapter 4: The Shooting Menu

Much of the power of the Nikon Coolpix B700 comes from the many options included in the Shooting menu, which helps you control the appearance of images and how they are captured. Depending on your preferences, you may not have to use this menu often. You may prefer to use the various scene settings or Creative mode selections, which choose many options for you, or you may prefer, at least on occasion, to use Auto mode, in which the camera makes its own choices. However, it's good to have this degree of control available if you want it, and it is useful to understand what items you can control.

As I discussed earlier, the menu options change depending on the setting of the mode dial on top of the camera. For example, if the mode dial is set to the green camera icon, for Auto mode, the Shooting menu options are limited, because Auto mode is for a user who wants the camera to make many decisions without input from him or her. If you have chosen one of the dedicated scene types with its own slot on the mode dial (Landscape, Night Portrait, or Night Landscape), the Shooting menu is re-named after the currently active mode.

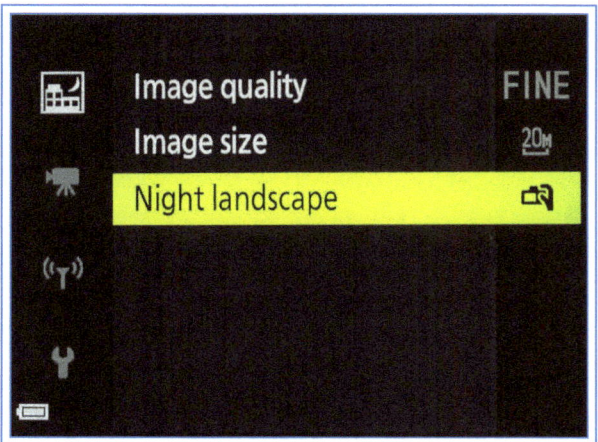

Figure 4-1. Night Landscape Menu

For example, if you select the Night Landscape mode from the mode dial and then press the Menu button, the menu that appears on the display is labeled Night Landscape, rather than Shooting, as shown in Figure 4-1.

These menus, as in Auto mode, are abbreviated versions of the Shooting menu; they include only a few items from the normal Shooting menu, usually Image Quality and Image Size. In addition, they may include a specific menu item for the mode that is in effect. In this case, there is a Night Landscape menu item, which lets you select either Hand-held or Tripod for shooting.

When the mode dial is turned to the SCENE setting, pressing the Menu button brings up another version of the Shooting menu, called the Scene menu. This menu provides a way to select either Scene Auto Selector or any one of the 20 specific scene types (Portrait, Moon, Sports, Sunset, etc.).

In addition, at the bottom of the Scene menu, just after the entries for Time-lapse Movie and Superlapse Movie, are the options for Image Quality and Image Size. (The Image Quality and Image Size menu options are unavailable for selection with certain Scene mode settings, such as Easy Panorama and Time-lapse Movie.)

When the Creative slot on the mode dial is selected, the menu becomes the Creative menu, which lets you choose an effect, along with Image Quality and Image Size.

Although the Shooting menu (or its equivalent, such as the Scene menu) presents you with some choices in all shooting modes, in the more automatic modes, including Auto and the various Scene mode types, there are only a few options, apart from options specific to a given mode, such as choosing a scene type. It is only when the mode dial is set to the P, S, A, or M setting for the Program, Shutter Priority, Aperture Priority, or Manual exposure mode, that the wide variety of Shooting menu options is available.

For the following discussion, I'm assuming you have the camera set to Program mode (mode dial turned to the P setting), because with that setting you potentially have access to all of the power of the Shooting menu. (Though some menu options will be unavailable in certain situations.)

After you turn the mode dial to P for Program mode, enter the menu system by pressing the Menu button. In the menu system, when the camera is in shooting mode, besides the Shooting menu (or Scene or Creative menu), there are the Movie menu, designated by a movie camera icon in the column at the left of the menu, the Network menu, marked by a wireless network icon, and the Setup menu, marked by a wrench icon. These icons are shown in Figure 4-2.

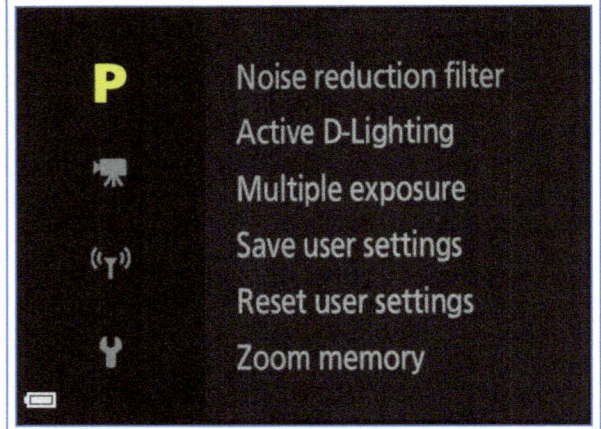

Figure 4-2. Menu Icons at Left of Menu Screen

When the camera is in playback mode, the three choices are the Playback, Network, and Setup menus. For now, I will discuss only the Shooting menu, which is designated by a capital letter or icon at the left standing for the current shooting mode: P, S, A, or M, or an icon for one of the more automatic modes.

On the Shooting menu (in Program mode), you'll see a fairly long list of options. Each option (such as Picture Control) occupies one line, with its name on the left and its current setting (such as the Standard icon) on the right.

You have to scroll through four screens to see all of the options. If you find it tedious to scroll using the Up and Down buttons, you can rotate the multi selector dial on the camera's back, which may help you speed through the menus a bit more quickly. Also, depending on which menu option you are trying to reach, you may be able to get there more quickly by reversing direction with the buttons or multi selector dial, and wrapping around

to reach the option you want. In other words, if you're on the top line of the first screen of the menu, you can scroll up to reach the bottom option of the last screen. Or, if the highlight is already near the bottom option of the last screen, you can scroll down to go back to the options at the top of the first screen of the menu. The first menu screen is shown in Figure 4-3.

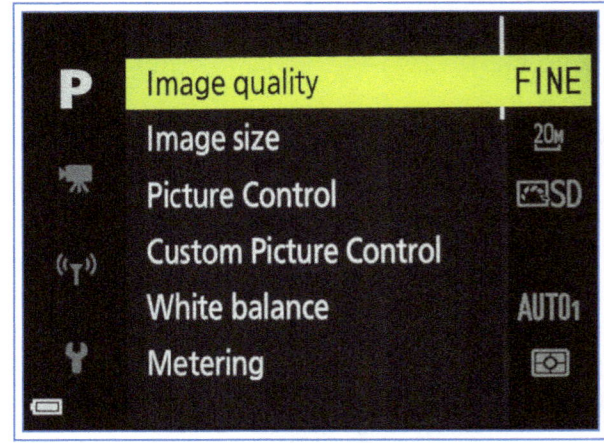

Figure 4-3. Screen 1 of Shooting Menu

Once you have highlighted the menu item you want, you can make any sub-selections by pressing either the OK button or the Right button, which will take you to the next screen (if one exists) for that menu item. To go back to a previous menu screen, press the Left button; to exit the menu system, press the Menu button. It is important to press the OK button to confirm your choice of a particular menu item selection; just highlighting it and then exiting from the menu screen will not activate that item.

On occasion you will find you are unable to select a certain menu option. That is, although an option will appear on the menu screen, you will not be able to navigate to it and select it. This situation occurs when there is an option in effect that is not compatible with the menu option you are trying to select. For example, if you have turned on Pre-shooting Cache by using the Continuous menu option, the Image Quality setting is fixed at Normal and the Image Size setting is fixed at one megapixel, so those settings cannot be changed (or even selected) in the menu system, as indicated in Figure 4-4, which shows those items dimmed on the menu screen.

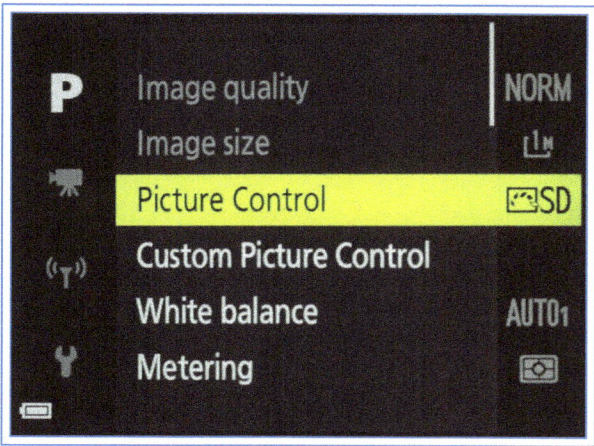

Figure 4-4. Two Items Dimmed on Shooting Menu

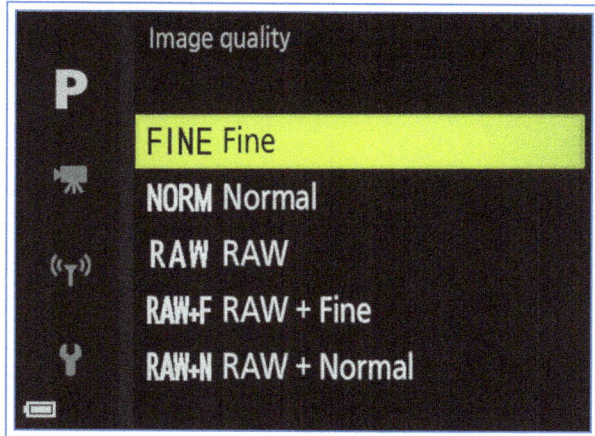

Figure 4-5. Image Quality Menu Options Screen

Or, if you have selected the option to shoot in monochrome from the Picture Control menu setting, you will not be able to get access to the White Balance menu option.

With the mode dial set to P you should have access to just about every option on the Shooting menu. If you find you can't select certain options, check to make sure you have set other options to compatible settings. For example, use the Continuous option on the Shooting menu to select single-shot exposures rather than a continuous setting, set ISO to Auto, and select Fine for Image Quality.

If you have trouble getting to some menu options and can't figure out what setting is causing the problem, you can go to the Setup menu (marked at the left of the screen by the wrench icon) and scroll down (or scroll up and wrap around) to the Reset All option, the next-to-last option on the menu. That action will reset all of the camera's basic shooting functions to their default values. In this way, you will undo whatever setting is causing a conflict with the setting you are trying to make. (You also will undo any custom settings you have made, so be sure you don't mind taking that step.)

Starting at the top line of the Shooting menu, I will discuss each option on the menu's four screens.

Image Quality

The Image Quality setting is one of the most important Shooting menu options for still images. The choices are Fine, Normal, Raw, Raw+Fine, and Raw+Normal, as shown in Figure 4-5.

The term "quality" in this context concerns the way in which digital images are processed. In particular, JPEG (Fine or Normal, and not Raw) images are digitally "compressed" to reduce their size without losing too much information or detail from the picture. However, the more an image is compressed, the greater the loss of detail and clarity in the image.

Raw files, which are in a class by themselves, are the least compressed of all and have the greatest level of quality, though they come with some complications, as discussed below. All other (non-Raw) formats for still images used by the Coolpix B700 (as with many similar cameras) are classified as JPEG, which is an acronym for Joint Photographic Experts Group, an industry group that created the JPEG standard. The JPEG files, in turn, come in two varieties on the B700: Fine and Normal. The Fine setting provides the least compression. Images captured with the Normal setting undergo more compression, resulting in smaller files with somewhat reduced quality.

Here are some guidelines for using these settings. First, you need to choose between Raw and JPEG images. Raw files are larger than other files, so they take up more space on your memory card, and on your computer, than JPEG files. But Raw files offer advantages over JPEG files. When you shoot in the Raw format, the camera records as much information as it can about the image and preserves that information in the file it saves to the memory card. When you open the Raw file later on your computer, your software can process that information in various ways. For example, you can change the exposure or white balance of the image when you edit it on the computer, just as if you had changed your settings while shooting. In effect, the Raw format gives you what almost amounts to a chance to travel back in time to improve some of the settings that you didn't get quite right when you pressed the shutter button.

For example, Figure 4-6 is an image I took with the B700 using the Raw format, with the exposure purposely set too dark and the white balance set to Incandescent, even though I took the picture outdoors on a sunny day.

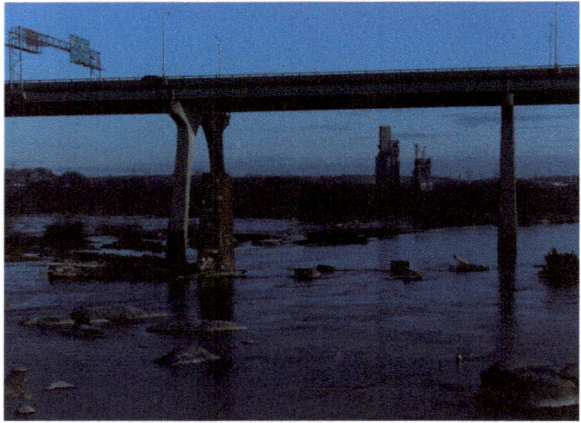

Figure 4-6. Raw Image Captured with Abnormal Settings

Figure 4-7 shows the same image after I opened it in Adobe Camera Raw software and adjusted the settings to correct the exposure and white balance. The result was an image that looked just as it would have if I had used the correct settings when I shot it.

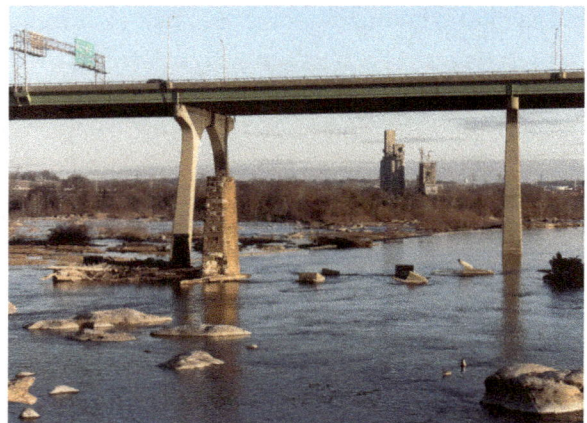

Figure 4-7. Raw Image After Corrections in Software

Raw is not a cure-all; you cannot fix bad focus or excessive exposure problems. But you can improve some exposure-related issues and white balance with Raw-processing software. You can use Nikon's Capture NX-D software to view or edit Raw files from the B700, and you also can use other programs, such as Adobe Camera Raw, that have been updated to handle Raw files from this camera.

Using Raw can have disadvantages, also. The files take up a lot of storage space; Raw images taken with the B700 are about 31.5 MB in size, while Large JPEG images I have taken are between about seven and nine MB, depending on the settings used. Also, Raw files have to be processed on a computer; you can't take a Raw image and immediately share it through social media or print it; you first have to use software to convert it to JPEG, TIFF, or some other standard format for manipulating digital photographs. If you are pressed for time, you may not want to take that extra step. Finally, some features of the B700 are not available when you are using the Raw format, such as the Scene mode and Creative mode settings, as well as the Continuous H settings of 120 and 60 fps, and the Multiple Exposure, Date Stamp, and Digital Zoom options.

If you're undecided as to whether to use Raw or JPEG, you have the option of selecting Raw+Fine or Raw+Normal, the fourth and fifth choices for the Image Quality menu item. With either of those settings, the camera records both a Raw and a JPEG image of the designated quality (Fine or Normal) when you press the shutter button.

The advantage with that approach is that you have a Raw image with maximum quality and the ability to do extensive post-processing, and you also have a JPEG image that you can use for viewing, sharing, printing, and the like. Of course, this setting consumes storage space more quickly than saving your images in just Raw or JPEG format, and it can take the camera longer to store the images, so there may be a slowdown in the rate of continuous shooting, if you are using that option. You also cannot use some menu options that conflict with the Raw setting.

When you choose Raw+Fine or Raw+Normal, you can select an Image Size setting that will apply only to the JPEG image; the Raw image, as noted earlier, is always at the maximum size. There are only a few Image Size settings available in this context, not the full range of those options.

The best way to preserve the quality of your images and your options for post-processing and fixing exposure mistakes later is to choose Raw files. However, if you want to use features such as Scene mode and Creative mode, which are not available with Raw files, then choose Fine or Normal, depending on your needs for using a small file size. If you do choose JPEG, I strongly recommend that you choose the Fine quality and the largest image size, unless you have an urgent need to conserve storage space on your memory card or on your computer. If you want Raw quality and are not concerned about storage space or speed of shooting, choose Raw+Fine. However, you will still not be able to use settings that conflict with the Raw format.

Image Size

The next option on the Shooting menu, Image Size, works together with Image Quality to determine the overall quality of your JPEG (Fine and Normal) images. (For Raw images, Image Size is always at the largest size and at the 4:3 aspect ratio.) With the Coolpix B700, Image Size actually has two components, which can be selected separately on some other cameras: resolution and aspect ratio. On the B700, these two components are not named, but their numerical values are listed on the Image Size menu. (The aspect ratio values are listed on the menu only for the settings that deviate from the normal aspect ratio of 4:3.)

The resolution of the image is the number of pixels it contains, given in a formula with the horizontal pixel count followed by the vertical pixel count. For example, the largest Image Size setting available on the B700 is 5184 x 3888, meaning the image has 5184 pixels horizontally and 3888 vertically.

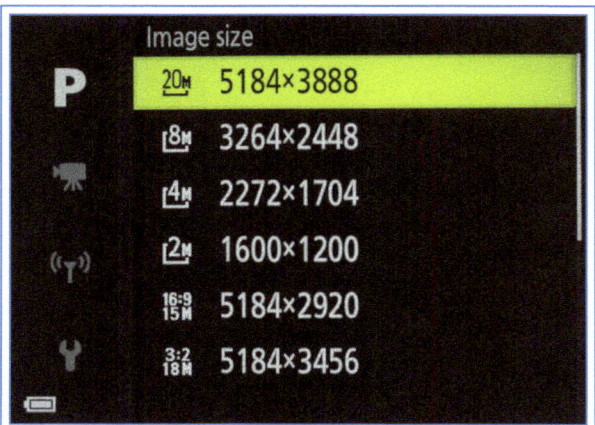

Figure 4-8. Largest Value Selected for Image Size

When you multiply these two numbers together, the result is about 20 million pixels, also written as 20 megapixels or 20M. So, as in Figure 4-8, you will see the figure 20M on the menu screen when you select this largest value for Image Size.

You can also determine the aspect ratio of the image from the Image Size setting. For example, the 5184 x 3888 setting yields an image four units wide for every three units tall, for a 4:3 aspect ratio. Most of the Image Size settings for the B700 are in that ratio, which is a standard one for digital images, being the same shape as the camera's LCD display. However, if you scroll down through the lines of the Image Size menu, you will see a few entries that note a different aspect ratio.

Specifically, just below the setting for 2M (1600 x 1200), there is the entry for 5184 x 2920 pixels. At the far left on the line for this entry, above the number of megapixels, 15M, the menu shows the notation 16:9, meaning this Image Size setting is in a 16:9 aspect ratio: 16 units wide for every nine units tall. This aspect ratio is another fairly common one, which corresponds to the shape of a widescreen HDTV set, and therefore often is called "widescreen."

Below that entry, another setting, 5184 x 3456, is labeled as 3:2, meaning its aspect ratio has three horizontal units for every two vertical ones. This is a common aspect ratio, which corresponds to the standard print size in the United States of six by four inches (15 by 10 cm).

Finally, the last entry on the Image size menu, 3888 x 3888 pixels, on screen 2 of the Image Size menu, is in an aspect ratio of 1:1, resulting in a square image. Some photographers like to use this aspect ratio because of its symmetry, or because it suits a particular composition.

With the Image Size menu setting, you have two choices to make. First, you can choose your images' resolution, or number of pixels (megapixels). The larger the number of pixels, the larger you can make high-quality prints on paper, and the more options you have for cropping the image to highlight particular details from the exposure. Second, although most of the choices on the menu are in the standard 4:3 aspect ratio, you have the option of selecting an aspect ratio of 3:2, 16:9, or 1:1 if you want.

Of course, you can always just shoot with the maximum image size of 5184 x 3888 and then crop the image down in software later; in that way, you can use any aspect ratio you want, including those listed here or any other. But, if you want to use a 1:1 aspect ratio for creative reasons, or you want your landscape photo to have the 16:9 widescreen look and you don't want to be bothered with changing the aspect ratio in software, you can select an Image Size setting that corresponds to your desired aspect ratio, so the final result will come straight out of the camera. In addition, you will have the advantage of seeing how the final image will be composed as you set it up on the camera's display screen or in the viewfinder.

If you choose Raw+Fine or Raw+Normal for Image Quality, you cannot choose an aspect ratio other than 4:3 for the Fine or Normal image that will be taken when you press the shutter button. The Image Size settings for those aspect ratios will be dimmed and unavailable for selection. However, you can still choose any of the first four options for Image Size.

Figures 4-9 through 4-12 were all taken at the same time and place; the only differences are that they were taken with different Image Size settings, resulting in different aspect ratios, as indicated in the captions.

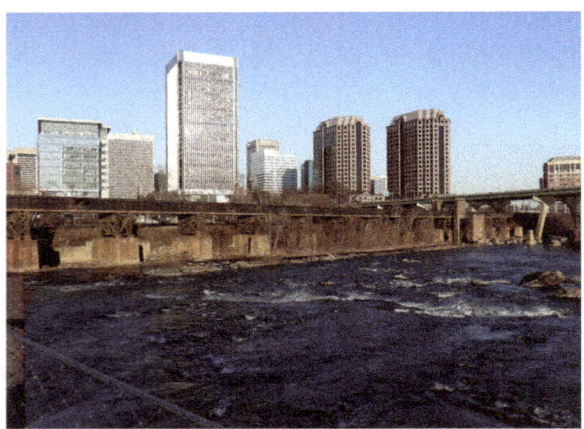

Figure 4-9. Aspect Ratio 4:3

Figure 4-9 was taken with the largest image size, which uses the 4:3 aspect ratio. With this setting, the camera captures the maximum number of pixels, and the resulting image includes all of the pixels included with the other aspect ratios, as well as some that are cut off with other settings.

Figure 4-10, taken with the 16:9, widescreen aspect ratio, includes all of the horizontal reach of the 4:3 image, but cuts off pixels at both the top and bottom of the image, as shown here.

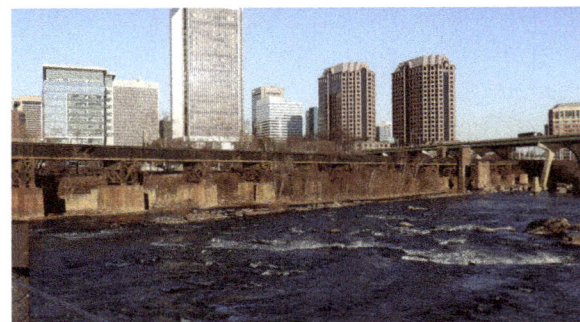

Figure 4-10. Aspect Ratio 16:9

Figure 4-11, taken with the 3:2 aspect ratio, also includes all of the horizontal reach of the 4:3 setting, but cuts off some pixels at the top and bottom of the image.

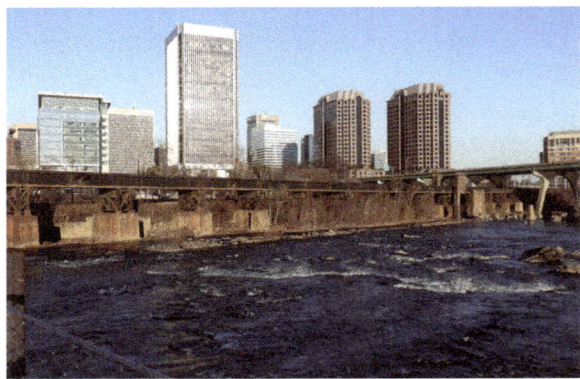

Figure 4-11. Aspect Ratio 3:2

Finally, Figure 4-12 illustrates the use of the 1:1 aspect ratio, with which the camera cuts off pixels at the left and right sides of the image to achieve a square shape.

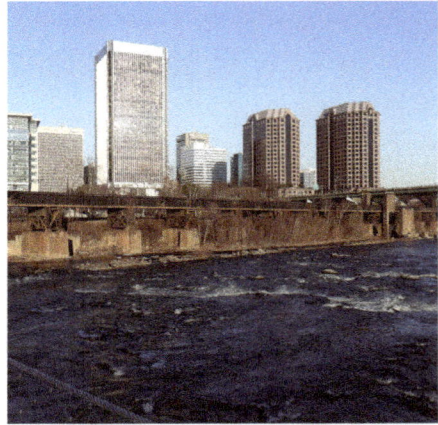

Figure 4-12. Aspect Ratio 1:1

Picture Control

This option provides you with four choices for the appearance of your images through in-camera processing: Standard, Neutral, Vivid, and Monochrome, as shown in Figure 4-13.

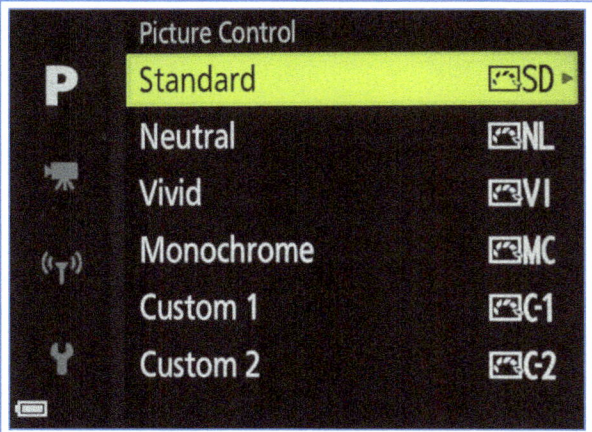

Figure 4-13. Picture Control Menu Options Screen

With these options, the camera provides varying degrees of adjustment to three basic parameters: sharpening, contrast, and saturation.

Following are descriptions of the settings, with a sample photo for each one showing the same portion of a colorful figurine, for comparison. Although the differences among the four settings are not all that dramatic (except for Monochrome), you should be able to see the general characteristics of each selection.

Standard

With this setting, as shown in Figure 4-14, you should see a normal rendering of the image, with no emphasis on any particular aspect.

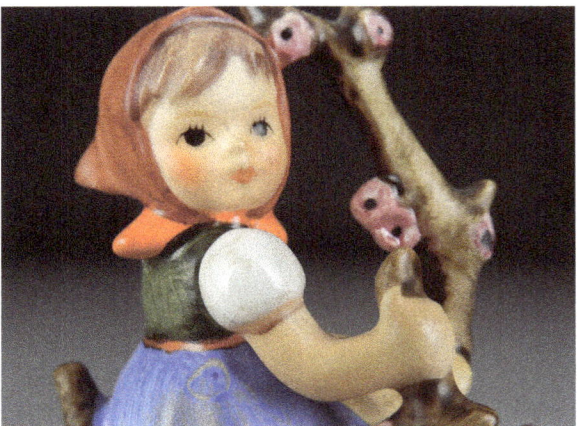

Figure 4-14. Standard Picture Control Example

The camera does some internal processing of the captured image to make it appear suitably sharp and contrasty for ordinary purposes. This is the setting you should use for everyday shooting when you have no interest in producing a specific effect.

Neutral

With the Neutral setting, illustrated in Figure 4-15, the camera does minimal internal processing of the image.

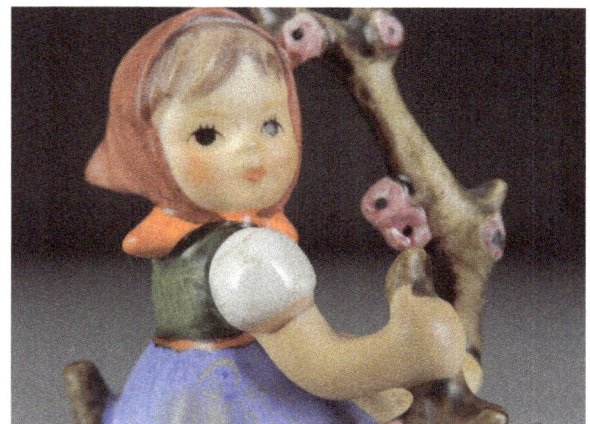

Figure 4-15. Neutral Picture Control Example

Therefore, the image may appear less sharp and contrasty, and have less color intensity, than you would like. The intent with this setting is for you to process the image after the fact in software such as Photoshop. With minimal processing, the camera is leaving the fine-tuning of the image up to you.

Vivid

Use this setting to increase the saturation, or intensity, of the colors in the image, as seen in Figure 4-16.

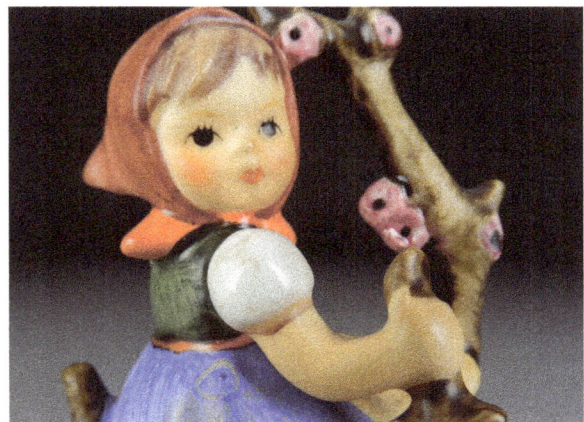

Figure 4-16. Vivid Picture Control Example

The Vivid setting also provides some increase in sharpening and contrast, with the result that the image may "jump" off the page or the screen at the viewer with increased impact.

Monochrome

The Monochrome setting gives you a quick way to set the camera to take black-and-white images. Of course, you can convert color images to monochrome using software such as Photoshop, but it is convenient to view your images in black-and-white on the camera's display before capturing them, and you may not want to devote time and effort to converting images on the computer. The Monochrome setting is illustrated in Figure 4-17. This setting is not available when AF Area Mode is set to Subject Tracking.

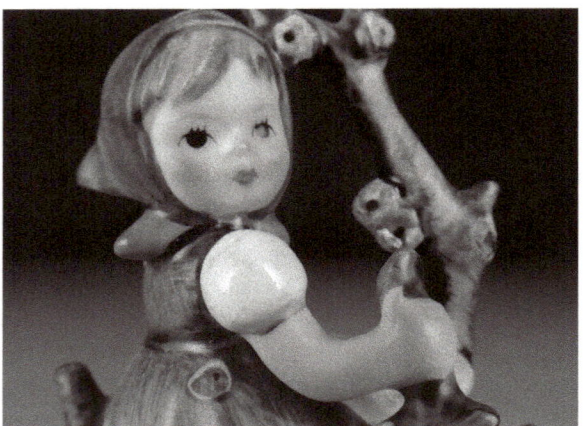

Figure 4-17. Monochrome Picture Control Example

Adjustments to Picture Control Settings

Once you have selected Standard or Vivid from the Picture Control menu, the camera will display a secondary menu screen with four lines: quick adjust, image sharpening, contrast, and saturation, as shown in Figure 4-18.

You can highlight any one of those four lines using the Up and Down buttons. When the line is highlighted, use the multi selector dial, the command dial, or the Left and Right buttons to change the settings. If you change the quick adjust value, you will see that the three values below it—image sharpening, contrast, and saturation—also move, but not all in the same amounts. Nikon has programmed the quick adjust feature to move the other three values in what Nikon considers to be "balanced" amounts, so that sharpening, contrast, and saturation may be adjusted upward or downward in amounts that work well with the adjustments to the other values.

If, instead of using the quick adjust option, you move the highlight to the specific line for image sharpening, contrast, or saturation, you can adjust any of those values individually. For example, suppose you like the punchy, aggressive look of images taken with the Vivid setting, but you don't want to have the colors quite so intense. You can set Picture Control to Vivid, and then, on the secondary screen, adjust the saturation value to a lower level, to reduce the intensity of the colors.

You also can set image sharpening, contrast, or saturation to the A position, for Automatic, in which case the camera adjusts that parameter according to its judgment based on current shooting conditions.

The contrast adjustment is unavailable if Active D-Lighting (discussed later in this chapter) is turned on in the Shooting menu.

If you select Neutral from the Picture Control menu, the secondary screen does not include the quick adjust option, but it does let you adjust image sharpening, contrast, and saturation individually.

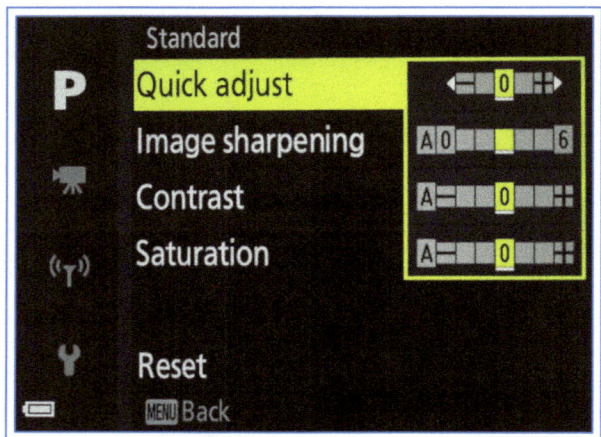

Figure 4-18. Picture Control Adjustments Screen

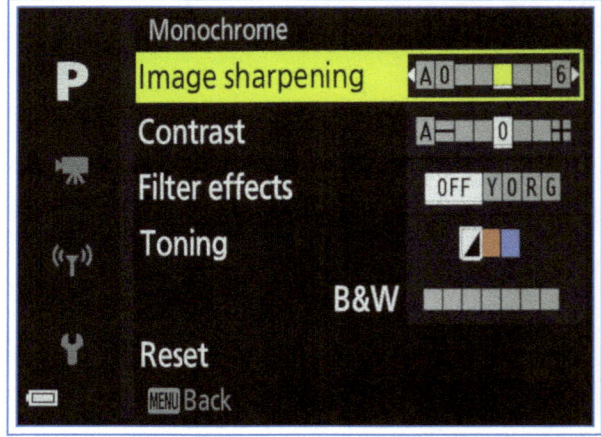

Figure 4-19. Adjustments Screen for Monochrome Setting

If you select Monochrome from the Picture Control menu, the secondary adjustment screen, seen in Figure 4-19, is even more different from the screens for the previous settings, all of which include saturation processing that affects colors.

With the Monochrome setting, adjustments are available for image sharpening and contrast, but not for saturation, and there is no quick adjust option. However, the Monochrome setting includes two other options: filter effects and toning. Contrast and sharpening work just the same as with the other Picture Control settings, as discussed above.

The third sub-option, filter effects, is a simulation of the use of a glass filter over the lens. If you select filter effects, you have four choices on the adjustment screen: Off, Y, O, R, and G, which stand for yellow, orange, red, and green. These settings are intended to mimic the effects of colored filters, which are often used with film cameras when taking photographs with monochrome films. The yellow, orange, and red filters can provide increasing levels of contrast that may, for example, darken the sky and enhance the appearance of a landscape scene. The green filter setting is intended to soften skin tones for use with portraits.

The toning options give you the ability to add a color cast to your monochrome shots. The three toning choices on the Monochrome menu are B&W (none), Sepia (brown), and Cyanotype (blue). Once you have selected either Sepia or Cyanotype, if you press the Down button, the cursor will move to a scale below the three options that includes gradations of intensity for the selected color tone. Using the direction buttons, the multi selector dial, or the command dial, move the cursor right for more intensity, or left for less. (The normal setting is level 4.) Then press the OK button to lock in the setting.

To illustrate the use of adjustments to the Picture Control settings, I took two pictures with adjustments to the Standard setting. In Figure 4-20, I adjusted contrast, sharpening, and saturation to their maximum values. In Figure 4-21, I adjusted those parameters to their minimum values. As you can see, the image in Figure 4-20 has a considerably sharper and punchier appearance than the one in Figure 4-21.

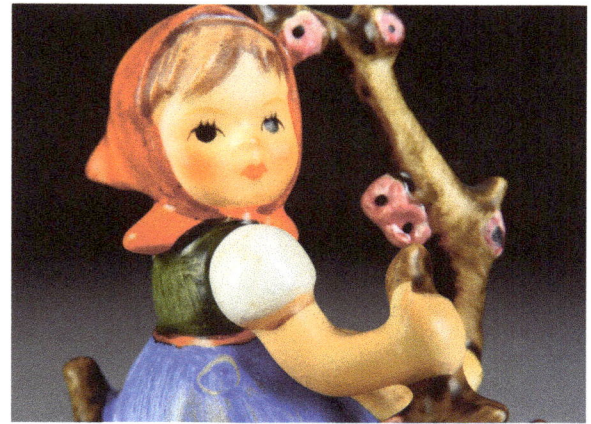

Figure 4-20. Picture Control Example: Maximum Adjustments

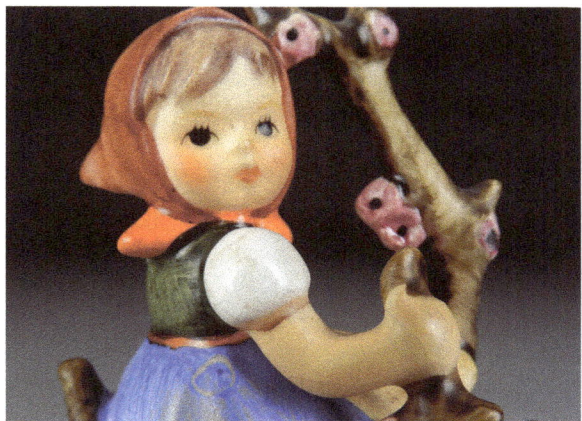

Figure 4-21. Picture Control Example: Minimum Adjustments

CUSTOM PICTURE CONTROL

The Custom Picture Control option lets you take one of the four available Picture Control settings (Standard, Neutral, Vivid, or Monochrome) and tweak the available parameters (sharpening, contrast, saturation, filter effects, and toning) to create a new setting that is crafted to your individual taste and that can be saved for later recall as an added selection for the Picture Control menu option.

To use this feature, select Custom Picture Control from the Shooting menu, then press the OK button or the Right button. You will first see a screen with the choices of Edit and Save or Delete. Select Edit and Save and press the OK or Right button again to bring up the menu with the four choices, as shown in Figure 4-22, and select one of them by pressing the OK button or the Right button.

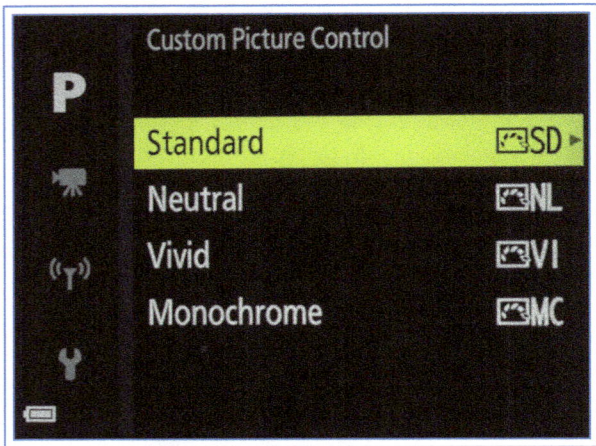

Figure 4-22. Main Options Screen for Custom Picture Control

When the adjustment screen appears, adjust the parameters, which are the same as for the Picture Control item, discussed above. As noted before, the options are different for some of the settings: Monochrome, for example, has no saturation adjustment, but does have filter effects and toning adjustments.

For example, suppose you have found a group of adjustments to the Neutral setting that produces an appearance you want to use whenever you take photographs of a certain waterfall. Go to the Custom Picture Control menu item, press OK or the Right button, and, on the next screen, shown in Figure 4-23, select Edit and Save. You then are taken to a screen with the four basic Picture Control settings. Select Neutral, and, on the next screen, make your adjustments to sharpening, contrast, and saturation. When you are done, press OK, and you are taken to a screen that lets you save this setting to the Custom 1 or Custom 2 slot.

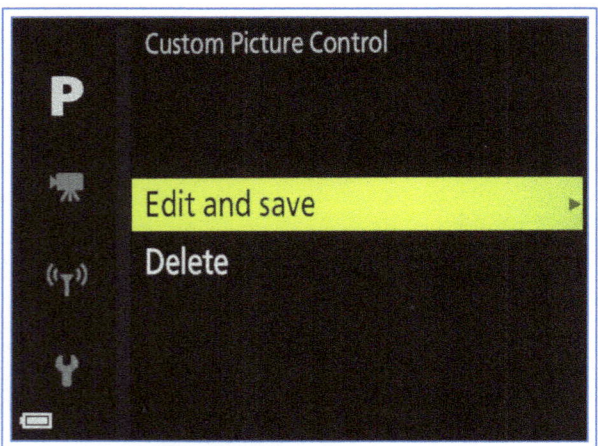

Figure 4-23. Custom Picture Control Menu Options Screen

Highlight one of those options and press the OK button to confirm the selection. The camera will display a Done message. Then, whenever you want to recall that setting, go to the Picture Control menu item and select Custom 1 (or Custom 2), which will appear as an option below Monochrome, as shown earlier in Figure 4-13.

If you later want to delete the Custom 1 or 2 option, use the Custom Picture Control item and select Delete instead of Edit and Save, as shown in Figure 4-23.

Here is one final note about the Picture Control settings. They are available for use when Image Quality is set to Raw. However, Raw images may not be displayed with a Picture Control effect, depending on what software is used to process them. For example, when I took a picture with Image Quality set to Raw and Picture Control set to Monochrome, the image appeared in color in Adobe Camera Raw software. However, when I opened it with Nikon's Capture NX-D software, the monochrome appearance was preserved. To be sure the Picture Control effect is not ignored, you can use a setting such as Raw+Fine to capture images using the Picture Control option.

White Balance

A digital camera's sensor reacts differently to colors than the human eye does. When you or I see a scene in daylight or indoors under various types of artificial lighting, we generally do not notice a difference in the hues of the things we see depending on the light source. However, the camera does not have this auto-correcting ability. The camera "sees" colors differently depending on the "color temperature" of the light that illuminates the object or scene in question.

The color temperature of light is a numerical value expressed in a unit known as kelvins (K). A light source with a lower kelvin rating produces a "warmer" or more reddish light. A light source with a higher kelvin rating produces a "cooler" or more bluish light. For example, candlelight is rated at about 1,800 K; indoor tungsten light (ordinary light bulb) is rated at about 3,000 K; outdoor sunlight and electronic flash are rated at about 5,500 K; and outdoor shade is rated at about 7,000 K.

If you are using a film camera, you may need a colored filter in front of the lens or light source to "correct" for the color temperature of the light source. Any given color film is rated to reproduce colors accurately at a particular color temperature (or, to put it another way, with a particular light source). So if you are using color film rated for daylight use, you can use it outdoors without a filter. But if you happen to be using that film indoors, you will need a color filter to correct the color temperature; otherwise, the resulting picture will look excessively reddish because of the imbalance between the film and the color temperature of the light source.

With a modern digital camera you do not need to worry about filters, because the camera can adjust its electronic circuitry to correct the "white balance," which is the term used in the context of digital photography for balancing color temperature. The Coolpix B700, like most digital cameras, has a setting for White Balance, which lets you choose the proper color correction to account for any given light source. Here is how to make this setting through the Shooting menu.

After you highlight the White Balance setting, which is the fifth item down on the first screen of the Shooting menu, press the OK button or the Right button to bring up the list of the following choices for the White Balance setting, each of them represented by an icon or a word or abbreviation: Auto (normal) [AUTO1]; Auto (warm lighting) [AUTO2]; Preset Manual [PRE]; Daylight [sun]; Incandescent [round light bulb]; Fluorescent [rectangular light bulb]; Cloudy [cloud]; Flash [lightning bolt]; and Choose Color Temperature [K]. The first six of these are shown in Figure 4-24.

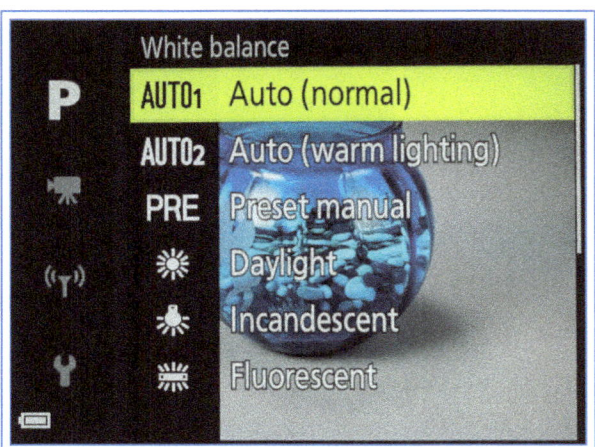

Figure 4-24. First Screen of White Balance Menu

The labels for these settings are self-explanatory. To select a setting, highlight it and press the OK button to confirm. If you select either Auto setting, you are done; there are no further adjustments available. With each of the other selections, though, you can fine-tune the setting, as described below.

If you highlight Daylight, Incandescent, Cloudy, or Flash, you cannot select it just by pressing the OK button. Instead, when you highlight one of these settings and press the OK button or the Right button, the camera displays a scale at the left going from -3 at the bottom to +3 at the top, as shown in Figure 4-25. To select the chosen setting (such as Daylight) with no adjustments, leave the zero point highlighted and press the OK button to confirm. Or, you can move the highlight to a positive or negative value using the Up and Down buttons or the multi selector dial.

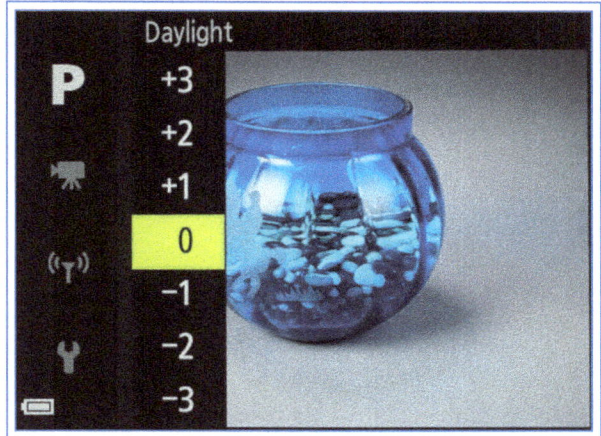

Figure 4-25. Adjustment Screen for Daylight White Balance

If the value is positive, the white balance is biased toward a bluish tint, and if it is negative, it is biased toward a reddish tint. You have to press the OK button while a number is highlighted in order to confirm the new white balance setting.

If you highlight the Fluorescent option, pressing the Right button or OK button brings up the further choices of 1, 2, or 3. These three sub-varieties of Fluorescent range from white to neutral to daylight. There are no other adjustments available with this setting.

Finally, if you select Preset Manual, you can set the white balance manually. Use this option when you are faced with mixed light sources, or a reddish or otherwise unusual light source. To make this setting, highlight Preset Manual, then press the OK button or the Right button. The next screen will present the options to Cancel or Measure, as shown in Figure 4-26.

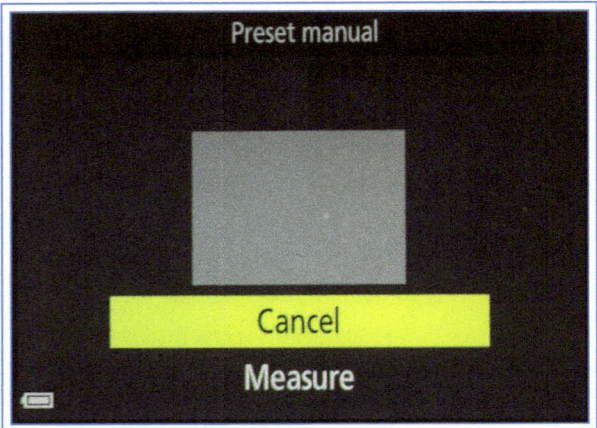

Figure 4-26. Screen to Measure Preset Manual White Balance

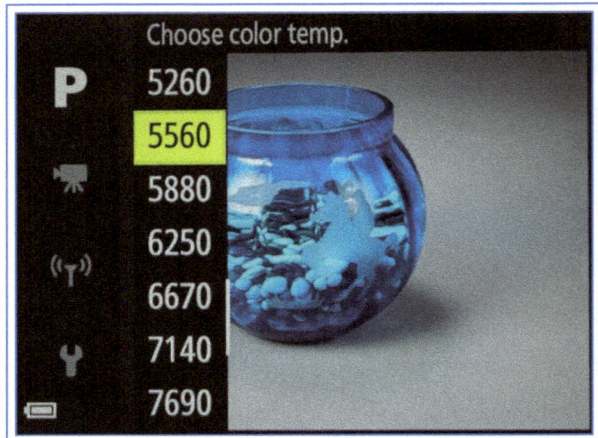

Figure 4-27. Color Temperature Selection Screen

Highlight Measure, and aim the square in the middle of the screen so it will be filled by a white or gray surface that is illuminated by the light source you will be using. Then press the OK button, and the camera will measure the white balance and store the setting. To use this setting now or in the future, turn back to the Preset Manual option at any time, even after the camera has been turned off and back on.

You can take advantage of the Preset Manual setting as a way to add a color tint to a scene for creative effect if you want. For example, you can set the white balance manually using a red or orange surface for the measurement, which will result in a pronounced blue tint for any pictures taken under the same light source that you used when setting that white balance value. Just be careful to turn the white balance setting back to Auto or another more normal setting when you don't want that special effect for your images.

The other choice for the White Balance menu option is Choose Color Temperature, the last item on the second screen of this menu item. Highlight this option, then press the OK button or the Right button to bring up the screen shown in Figure 4-27, with a scale of values at the left ranging from 2500 K to 10,000 K.

You can use this scale to set the numerical kelvin reading of your light source if you know it. You can determine this number using a color temperature meter, such as the one shown in Figure 4-28.

That meter works well when you need extra accuracy. If you don't use that option, you can still use the Choose Color Temperature method, but you will have to use guesswork or your sense of color.

For example, if you have lighting from incandescent bulbs, you can use 3,000 K as a starting point, then change the value and watch the camera's display to see how natural the colors look. As you lower the color temperature setting on the menu, the image will become more "cool," or bluish; as you raise it, the image will appear more "warm," or reddish. Once you find the best setting, leave it in place and take your shots.

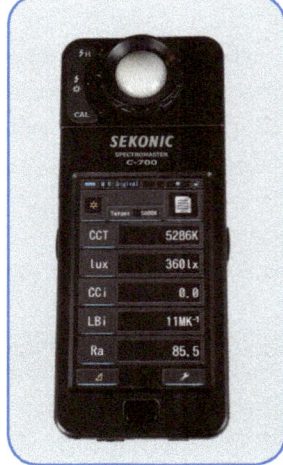

Figure 4-28. Sekonic C-700 Color Meter

Before I leave this topic, I'm going to include a chart of images showing how the various White Balance settings on the B700 affect the colors of your shots. All of these images, shown in Figure 4-29, were taken under the same indoor lighting, which was balanced for daylight; the only thing that changed from shot to shot was the camera's White Balance setting, as indicated.

In my opinion, Auto 1, Preset Manual, Daylight, Flash, and Choose Color Temperature (using a value of 5560K) all resulted in good color balance. Auto 2 and Cloudy would be acceptable. However, even the results of those

Chapter 4: The Shooting Menu | 63

settings could be improved if you were to make further adjustments to tweak them for more or less bluish and reddish tints. The only settings that probably would not be usable in this situation were Incandescent and Fluorescent. In practice, I usually leave White Balance set to Auto 1, but it is good to know that you have the option to make more individually crafted settings when the occasion calls for it.

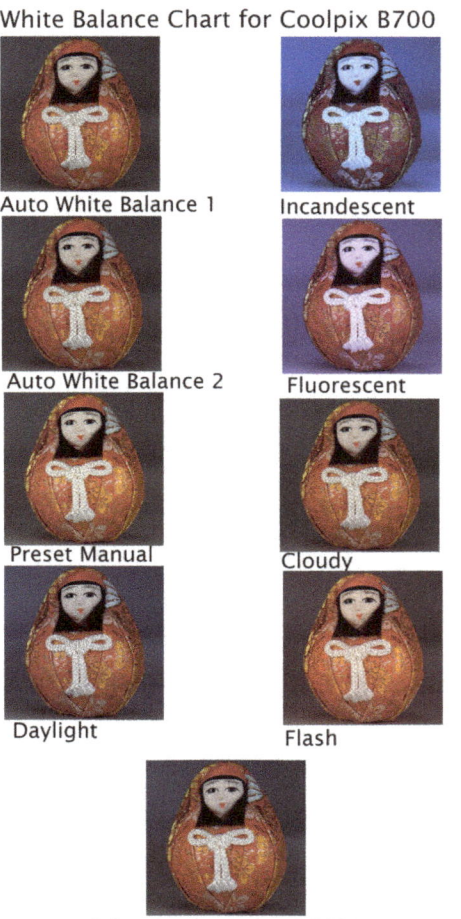

Figure 4-29. White Balance Comparison Chart

Metering

This next option on the Shooting menu lets you choose one of the three patterns of exposure metering offered by the Coolpix B700: Matrix, Center-weighted, and Spot. The menu selection screen is shown in Figure 4-30. This setting tells the camera's automatic exposure system what part of the scene it should evaluate when deciding how to set the exposure. If you choose Matrix, the default option, the camera uses the entire scene that is visible on the LCD and bases its exposure setting on the overall average brightness of the scene.

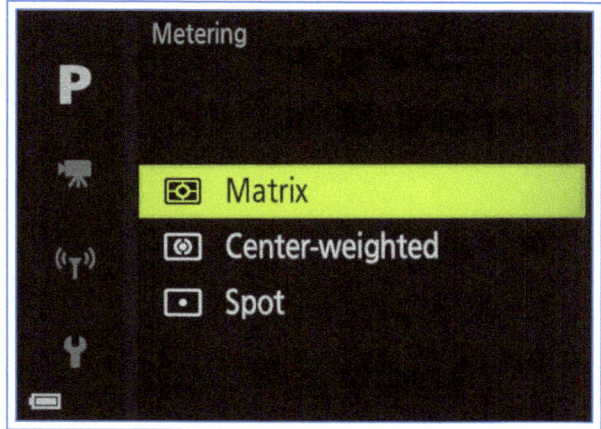

Figure 4-30. Metering Menu Options Screen

For example, if the camera is aimed at a landscape scene with trees, grass, buildings, sky, and people, the camera will measure the light being reflected from all of those parts of the scene and set the exposure accordingly. The resulting image is likely to look properly exposed.

If, instead of a standard landscape scene, the camera is aimed at a small, dark object in front of a large, white wall, the camera will take into account the large expanse of white and likely will set the exposure to be too dark to show the dark object properly. In that case, the Matrix metering mode probably will not work well. You can use exposure compensation or Spot metering to achieve a better exposure measurement.

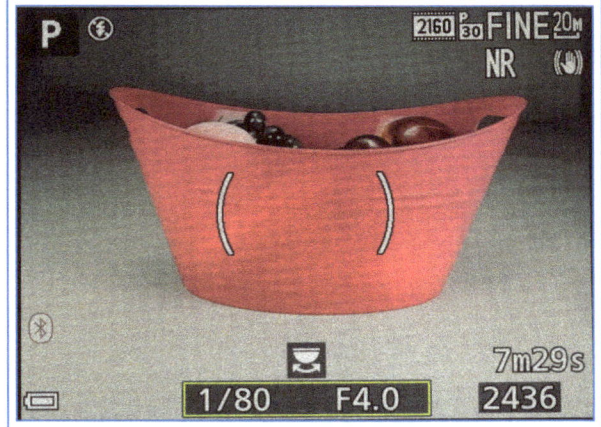

Figure 4-31. Center-weighted Metering in Use

If you choose Center-weighted for metering, as shown in Figure 4-31, the camera still considers all of the light from the scene, but it gives additional weight to the center portion of the image, on the theory that your main subject is in or near the center. The camera displays two large arcs to mark the area that is being emphasized. This is a good metering method to use when you have a subject in the center of the scene that is the most important item in your composition—for example, the subject of a portrait, or an antique that you are photographing for an auction.

With Spot, as illustrated in Figure 4-32, the camera considers only the light in the part of the scene covered by the small circle that appears in the center of the screen.

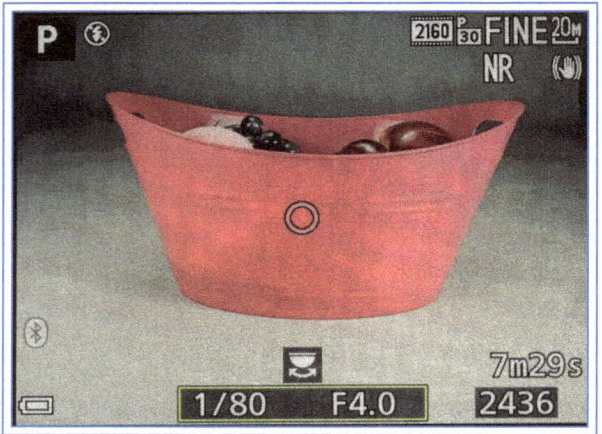

Figure 4-32. Spot Metering in Use

When you set the metering method to Spot, you can see the effects of the exposure system quite dramatically by setting the camera to the Program exposure mode and aiming the small circle at various points, some bright and some dark, and seeing how dramatically the brightness of the scene in the LCD changes. If you try a similar experiment by moving the camera around to aim at differently lit areas in Matrix mode, you will still see changes, but more subtle and gradual ones. The Spot setting is useful when you have a small subject for which proper exposure is particularly important, such as a collectible item that you are photographing for a catalog or online auction.

My preference is to use Matrix metering for outdoor shots with even lighting, such as landscapes, groups of people, and views of buildings, monuments, and the like. I use Spot metering on occasion, primarily when I am photographing an object whose brightness level contrasts sharply with the background. For example, if I am photographing a black camera against a white background, I may use Spot metering in order to avoid confusing the metering system because of

the expanse of bright white in the scene. The Center-weighted method is useful when taking portraits and other images with one central subject that is of primary importance to the scene.

With all of the metering settings, including Center-weighted and Spot, the autofocus mode has no effect on metering. So, even if you set AF Area Mode to Manual and move the focus frame away from the center of the screen, the camera will meter the area outlined by the Center-weighted arcs, the Spot circle, or the overall scene, depending on the metering setting.

The Metering menu option is dimmed and unavailable if Active D-Lighting is turned on at any level; in that case, the camera sets the metering mode to Matrix.

Next, I will discuss the items on screen 2 of the Shooting menu, shown in Figure 4-33, starting with Continuous.

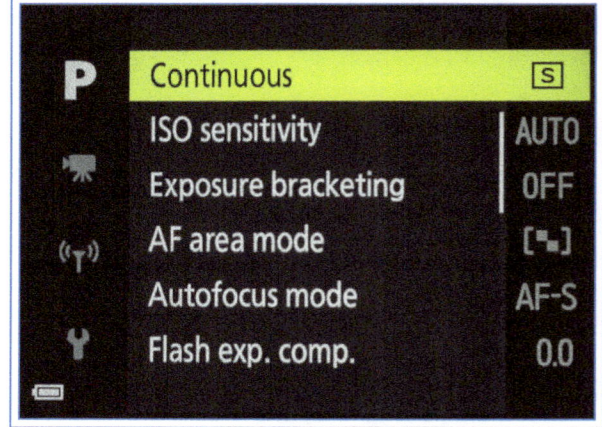

Figure 4-33. Screen 2 of Shooting Menu

Continuous shooting

With some cameras, there is a button you can use to turn on continuous (or "burst") shooting. With the B700, the numerous options for continuous shooting are located on the Shooting menu, under the item called "Continuous." However, as discussed in Chapter 5, the Fn1 button on top of the camera or the Fn2 button on the camera's back can be assigned to call up the Continuous option, so you can use a physical control for this purpose if you want to.

Regardless of whether you reach the continuous-shooting options through the menu system or by pressing a function button, when the camera is set to

Chapter 4: The Shooting Menu

a shooting mode in which burst shooting is available, the continuous options give you an impressive array of features. Before describing them, I will provide a brief introduction to continuous shooting.

The usefulness of rapid bursts of exposures is clearer in some contexts than in others. For example, when you're shooting sports, you can fire off a swift sequence of shots to catch the instant when a baseball player tags a runner, or to catch a soccer ball as it bounces off a player's head. But continuous shooting also can be helpful in more ordinary shooting, such as pictures of children at play. You have a better chance of capturing a fleeting smile or gesture if you keep the exposures rolling. Even if your subject is not moving, it can be useful to take multiple shots. When you're taking a portrait there may be subtle changes in the subject's expression, or in the way sunlight falls on a cheek. Taking a series of shots gives you some assurance that you won't come away from the photo session with no winning images.

In Figure 4-34, I used the Continuous high-speed option to capture a series of images of my family's dog as she ran back toward the house. As you can see, the speed of the continuous shooting froze the action in mid-air as she was leaping over some steps.

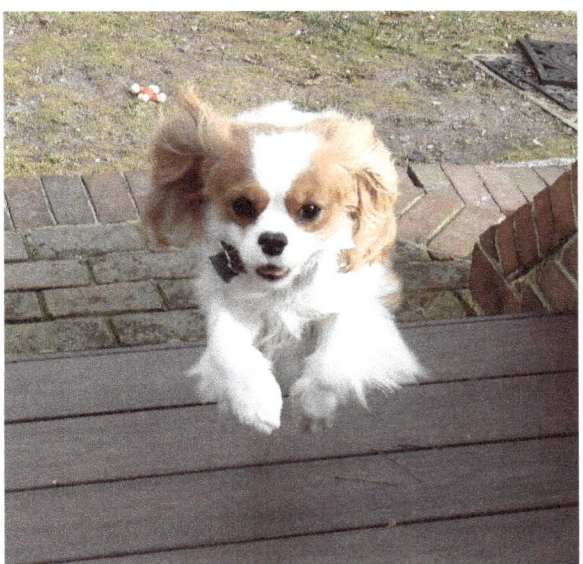

Figure 4-34. Continuous Shooting Example

The B700 provides an excellent set of continuous-shooting options. To get access to these options, the camera has to be in the Program, Aperture Priority, Shutter Priority, or Manual exposure mode. Select the Continuous menu item, press the Right button or the OK button, and the next screen will display the first six of the seven available settings, as shown in Figure 4-35. (Or, as noted above, press a function button if it is assigned to call up the Continuous menu option.)

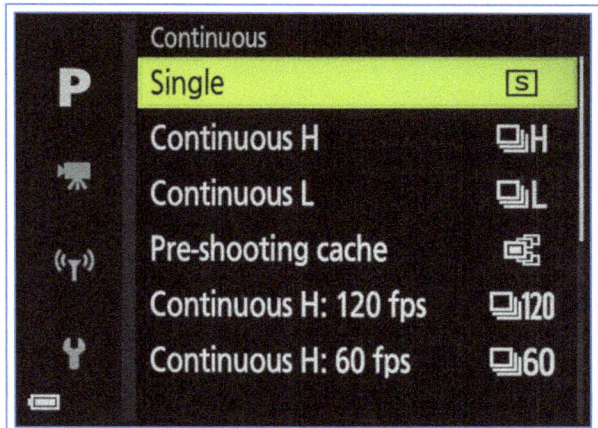

Figure 4-35. First Screen of Continuous Menu Options

These settings (except the first, for single shots, and the last, for intervals) offer various ways to take multiple shots while you hold down the shutter button. In each case, the exposure, focus, and white balance settings are fixed when the first image is taken, and they will not vary for later shots, even if conditions would require different settings. You cannot use flash for any of the multiple-shot settings except for the last one, the interval timer option.

You can activate the self-timer and it will work, but, with several continuous settings (Continuous H, Continuous L, and Pre-shooting Cache), when the timer triggers the shutter only one image will be taken, even if you continue to hold down the shutter button. So there is no point in turning on continuous shooting and the self-timer at the same time, with those settings. However, with the Continuous H: 120 fps and Continuous H: 60 fps settings, the camera will take the full 60 shots when the self-timer triggers the shutter. Also, the self-timer can be used to start a sequence using the interval timer setting.

There are limitations on the shutter speeds the camera can use or that you can set when any of the continuous options is in effect, other than the interval timer. The slowest speed you can set is 1/30 second, 1/60 second, or 1/125 second, depending on what setting is selected.

As I discuss in Chapter 6, playing back continuous shots in this camera can be confusing. In playback mode, you will see the first image of a continuous series displayed on what looks like a stack of frames, indicating that this is the first of a continuous set, or, using Nikon's terminology, the "key" image of a "sequence."

There also will be a message at the bottom of the screen indicating that you have to press the OK button to display the full set of images, as shown in Figure 4-36.

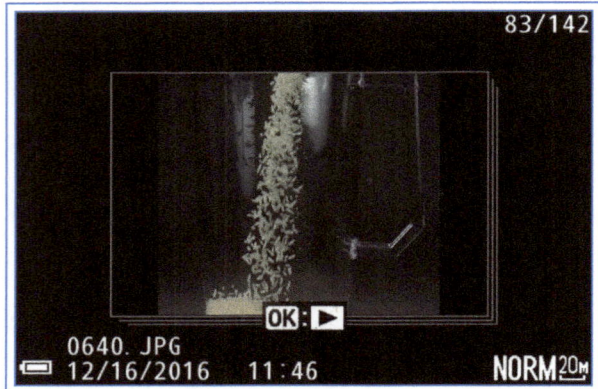

Figure 4-36. Screen for Playback of Burst of Images

After you press the OK button, you can move through the images in this group of continuous shots using the normal navigation tools—the Left and Right buttons and the multi selector dial. To return to the main playback screen, press the Up button. You can then keep navigating through the other individual shots and sequences on the memory card.

Following are details about the choices for continuous shooting with the Coolpix B700, as shown in Figure 4-35. The first option at the top of the Continuous menu screen is designated by an icon with an S, for single shots. This is the default option. In effect, choosing this first option turns off continuous shooting.

The second option on the menu is the first selection for multiple shots. It is marked by an icon that looks like a stack of rectangular frames with the letter H inside, representing high-speed continuous shooting. With this option, the camera shoots up to five shots at a speed of up to five frames per second, depending on factors such as image size, image quality, lighting conditions, and the like. You can use any settings for the image quality and size, including the maximum Raw+Fine with Image Size at its highest setting.

The next icon, marked by an L for low-speed shooting, provides a capability similar to that for high-speed shooting, except that there is a trade-off of increased capacity versus slower speed. That is, you can take up to 200 images, but at a speed of no more than about two frames per second.

The next icon on the list looks like a stack of frames branching out in two directions. Selecting this icon activates an interesting feature called Pre-shooting Cache. With this option, the camera actually captures several images before you press the shutter button to take pictures.

In practice, this option has its limitations, though it is still a welcome innovation. When you press the shutter button halfway to evaluate exposure and focus, the camera will capture up to five images before you press the button the rest of the way down, and up to 15 more as you hold the button down to take the images. The Pre-shooting Cache icon on the display turns green while images are being recorded to the cache; once you press the shutter button all the way down, the last five of those cached images are saved to the memory card, along with up to 15 shots taken while the shutter is pressed all the way down. The maximum rate is a speedy 15 frames per second, but the catch is that the images are fixed at a small size of one megapixel, or 1280 x 960 pixels, and at Normal quality.

Pre-shooting Cache is a tool to use when you are monitoring a scene and waiting for just the right moment to catch a particular action or expression that may come up very quickly, and possibly will fade away quickly as well. When it looks as if the action is about to happen, you can press the shutter button halfway down to get ready, and, if the action comes up faster than expected, you won't miss it because of slow reactions. You can then press the shutter button all the way down to capture the rest of the sequence.

If you don't mind a reduction in the resolution of your images, this is an interesting option to have available. I have found it useful when photographing birds. When I see an interesting bird that has landed on a fountain, I activate Pre-shooting Cache and wait for the bird to take off. With this feature, I have a good chance of catching the bird in flight. Without this feature, I would not be able to react quickly enough to activate the camera when the bird starts to fly.

Be sure to note one possible pitfall here: If you press the shutter button down halfway but never press it all the way to take any pictures, the contents of the pre-shooting cache will be discarded and no pictures at all will be recorded. You have to press the shutter button down all the way at some point in order to "lock in" the pre-shooting images. Note also that, when you have finished shooting, you may see the circulating-block icons on the screen, indicating that the camera needs time to process the contents of the cache as well as the contents of the other images you have taken.

The next choice on the menu, Continuous H: 120 fps, is indicated by the number 120, representing the extremely rapid rate of 120 frames per second. With this setting, the camera emphasizes both speed and quantity, giving you 60 images at this super pace, but at a drastic reduction in quality down to 640 x 480 pixels (VGA), which is the resolution of an old-fashioned computer monitor.

Images shot using this option may look fine on your computer, but they will be quite grainy and will not be suitable for any degree of enlargement. Still, if you need to analyze a golf swing or otherwise shoot a sequence of many pictures over a period of about one-half second, this is the choice for you. Here again, you will almost certainly see the time-delay icons after shooting, as the camera processes the large quantity of image information that it sucked in like a vacuum cleaner.

The next option, Continuous H: 60 fps, is similar to the previous one, except that the camera takes 60 shots at a somewhat larger resolution of two megapixels, or 1920 x 1080 pixels, and at the slower speed of 60 frames per second. Use this option if you need a very speedy sequence of shots, but need a bit better quality or prefer the 16:9 widescreen aspect ratio of this setting.

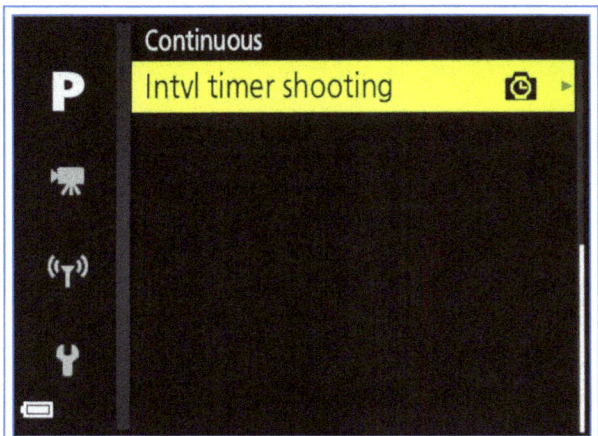

Figure 4-37. Second Screen of Continuous Shooting Menu Options

The single option on the second screen of the Continuous menu, shown in Figure 4-37, is interval timer shooting, which gives the Coolpix B700 a second capability for time-lapse shooting. (The other time-lapse option is found under Scene mode, as discussed in Chapter 3.) When you select this option and move to the next screen, the camera displays two blocks, as shown in Figure 4-38—one for minutes and one for seconds.

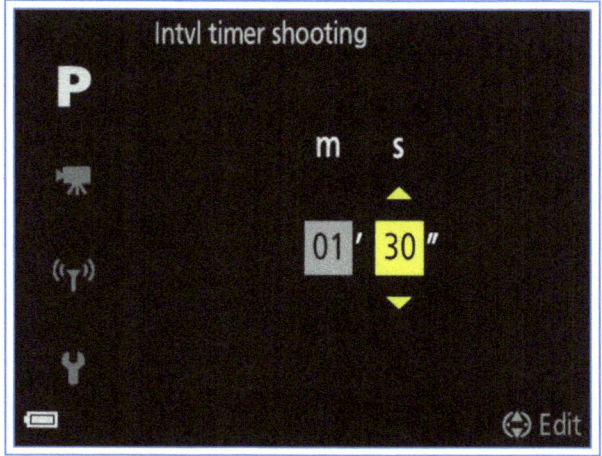

Figure 4-38. Interval Timer Shooting Settings Screen

You can select a value from 0 to 60 minutes for the interval, and either 0 or 30 for seconds, up to a maximum time of 60 minutes overall. So, the shortest interval available is 30 seconds; the next is one minute; then one minute 30 seconds, and so on.

There is no setting for the number of shots; the camera will keep taking shots at the specified interval until the memory card is filled up, or you stop the process.

When you press the shutter button all the way, the camera takes the first image, then blanks out the display. The green light around the power button will blink slowly while interval shooting is active. Shortly before each interval elapses, the display will turn back on, and at the specified time the camera will take the next shot. To interrupt the sequence of shots before it is complete, you can press the shutter button and the sequence will end.

You can use any image size and quality settings with this option, including Raw or Raw+Fine. The resulting images are stored on your memory card in specially designated folders with the letters INTVL in their names. For example, a folder with shots from one shooting session might be labeled as 102INTVL, and the images inside it might be labeled DSCN001.jpg, DSCN002.jpg, and so on.

The interval timer option gives you excellent opportunities for creative photography. For example, you can aim the camera at a construction site and record all work that is done over a period of time, then play back the images as a time-lapse movie using software such as Adobe Premiere Elements or iMovie. If you set the camera to shoot one image every minute for 24 hours, you would end up with 1440 images. If you play them back at the standard video rate (in the United States) of 30 frames per second, the video showing 24 hours of action would play back in just 48 seconds. You may have seen sequences of this sort on television showing weather patterns unfolding at rapid speeds or speeded-up views of crowds gathering for events.

Also, interval shooting can be used to operate the camera remotely, as when you place it on a pole or other location that is out of your reach, to record images from a high or otherwise inaccessible vantage point. If you are able to attach the B700 to a remote-controlled aerial drone, you could turn on interval shooting to capture images from the air.

When you use interval shooting on the ground, you need to set the camera on a sturdy, steady tripod; the slightest motion of the camera will be obvious when the sequence is played back. Also, you need to be able to keep the camera powered on continuously. The camera does turn off its display between shots, so its battery power is conserved to some extent. However, if you are shooting a sequence that lasts several hours or more, you should use the Nikon AC adapter designated for this camera, model number EH-67A, which is discussed in Appendix A. Finally, it's generally a good idea, if it is practical under the circumstances, to use Manual exposure mode and to set the white balance and ISO to definite settings rather than to Auto settings, so there is no distracting flickering among the images when the camera adjusts these settings automatically.

As noted above, the Coolpix B700 offers a similar feature called Time-lapse Movie as one of the scene types for Scene mode, as discussed in Chapter 3. With that option, you select a preset type of time-lapse subject, such as Sunset or Cityscape, and the camera makes all of the necessary settings for you. Three of those settings, Cityscape, Landscape, and Sunset, use intervals shorter than 30 seconds, so you would not be able to duplicate those settings using the interval timer feature. On the other hand, the interval timer option has no upper limit for the number of shots to be taken, whereas the Time-lapse Movie option will only take a preset number of shots. And, of course, with interval timer, you can make many settings on the Shooting menu, but you cannot do so with the Time-lapse Movie option. So, with interval timer and Time-lapse Movie, you have two different features available with different capabilities.

ISO Sensitivity

ISO is a measure of the light sensitivity of photographic film or digital sensors. The higher the ISO rating, the more sensitive the film or sensor is to light. If you shoot an image or video using a high ISO value, you will not need as much light to achieve a normal exposure as you would with a lower value. One result is that you can use a faster shutter speed, narrower aperture, or possibly both, than with a lower ISO.

The trade-off is that, with higher ISO values, the sensor is likely to produce visual "noise" that affects the image with an appearance of graininess. Camera makers have made considerable strides in creating sensors that can use high ISO values without too much noise, but there still is some drop-off in quality, especially at the highest ISO values.

Generally speaking, you should shoot your images with the camera set to the lowest ISO possible that will allow the image to be exposed properly. (One exception to this rule is if you want, for creative purposes, the grainy look that comes from shooting at a high ISO value.) For example, if you are shooting indoors in low light, you may need to set the ISO to a high value (say, ISO 800) so you can expose the image with a reasonably fast shutter speed. Otherwise, if the camera uses a slow shutter speed, the resulting image would likely be blurry and possibly unusable.

To summarize: Shoot with low ISO settings (usually 100 with the B700) when possible; shoot with high ISO settings (400 or higher, up to 1600 or even 3200) when necessary to allow a fast shutter speed to stop action and avoid blurriness, or when desired to achieve a creative effect with graininess.

Chapter 4: The Shooting Menu

To set ISO on this camera, press the Menu button and move to the ISO Sensitivity line on the second screen of the Shooting menu, then press the Right button to get to the screen that lets you select either ISO Sensitivity or Minimum Shutter Speed, as shown in Figure 4-39.

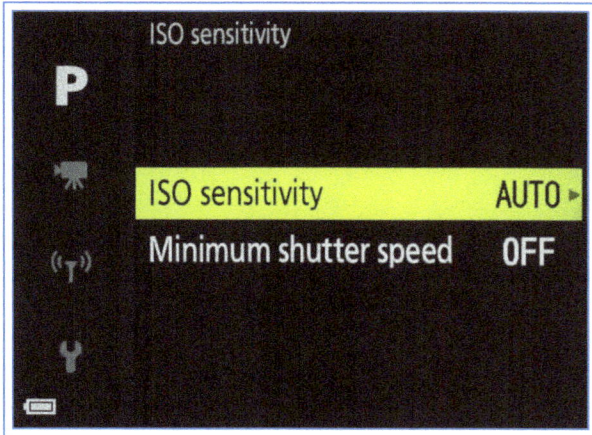

Figure 4-39. ISO Sensitivity Main Options Screen

For now, select ISO Sensitivity and press the Right button again to get to the ISO Sensitivity options. This first of the two ISO menu screens is shown in Figure 4-40. (As an alternative, you can assign the Fn1 or Fn2 button to call up a menu with the ISO settings, as discussed in Chapter 5.)

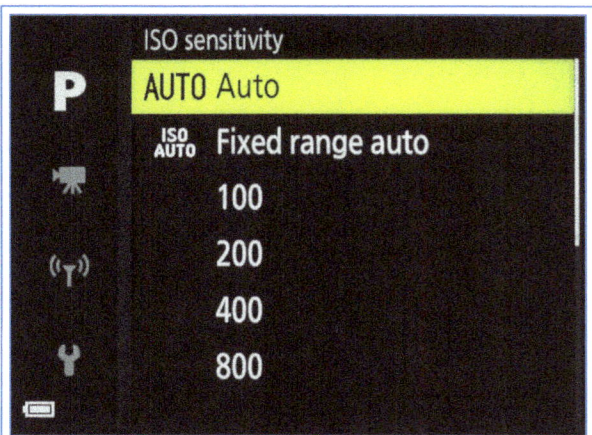

Figure 4-40. First Screen of ISO Sensitivity Values

With the first option at the top of this list, Auto, the camera sets the ISO to 100 in relatively bright light, and it will raise the level as high as 1600 as the light grows dimmer. The second option, Fixed Range Auto, has two choices, reached by pressing the OK button or the Right button. You can choose from two ranges, 100-400 or 100-800, if you want to limit the camera's ISO choices to a fairly narrow range of possibilities. You can use this approach if you want the camera to use some flexibility, but you want to make sure the ISO value does not go high enough to cause noticeable noise in your images.

Finally, instead of choosing Auto or Fixed Range Auto, you can choose any one of the individual ISO values, to specify exactly what ISO setting the camera uses. The choices are 100, 200, 400, 800, 1600, and 3200.

I usually leave the ISO setting at Auto for everyday shooting. However, in some cases, I will choose a specific numerical value. When I am using a tripod and want to have the best possible quality, I will set the ISO to 100. Because of the tripod, I am not concerned about image blur from camera motion if a slow shutter speed is needed. However, if I am shooting fast action such as sports or shooting in dim light, I often will set the ISO to a high value, such as 800 or 1600, so the camera can use a fast shutter speed to stop the action or to make a bright enough exposure.

I rarely set ISO as high as 3200, because of the negative effect of such a setting on image quality. Figure 4-41 is a composite image containing enlarged areas from two shots I took of a firefighter figurine, the top image shot at ISO 100, and the bottom one at ISO 3200, to illustrate the difference in quality that often results from using such a high ISO value.

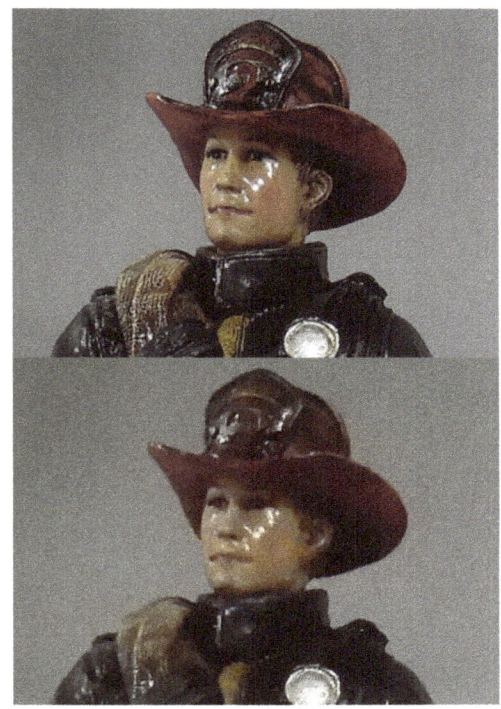

Figure 4-41. Top Image: ISO 100; Bottom Image: ISO 3200

As you can see, the bottom image, taken with the high ISO setting, shows considerable graininess and distortion of the picture of the figurine. That image would not be usable for many purposes. The top image, taken at ISO 100, is a higher-quality image that depicts the subject clearly.

Minimum Shutter Speed

Going back to the first branch on the ISO menu screens, you can use the Minimum Shutter Speed setting to specify the slowest shutter speed the camera will use when it is set to the Program or Aperture Priority mode and either of the Auto ISO settings (Auto or Fixed Range Auto) is in effect, before it starts to increase the ISO sensitivity. The first screen of values for this setting is shown in Figure 4-42. The other three values (1/30, 1/60, and 1/125 second) are on the second screen of this menu option.

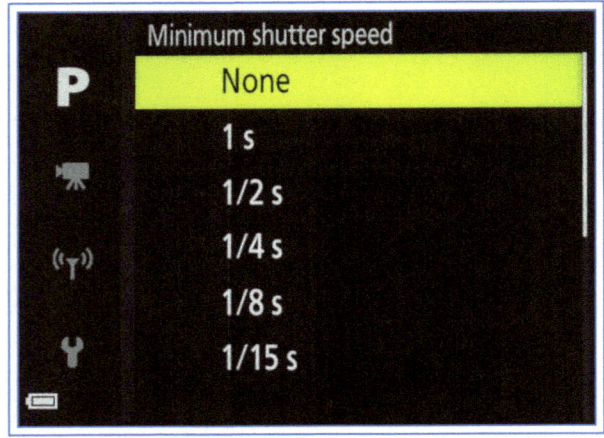

Figure 4-42. First Screen of Values for Minimum Shutter Speed

To understand this setting, it's helpful to consider an example. Set the camera to Program mode and the ISO Sensitivity setting to Auto. Press the Menu button, use the multi selector dial or the Up and Down buttons to highlight ISO sensitivity on the display, and press the Right button to get to the next screen. Then highlight Minimum Shutter Speed, press the Right button, and select 1/30 second from the list on the second screen of the menu option. Press the OK button to confirm.

With those settings, the camera will attempt to expose the image properly using a shutter speed no slower than 1/30 second, your Minimum Shutter Speed setting. If the Auto ISO setting increases to its maximum limit and the image is still too dark, then the camera will drop to a slower shutter speed in order to achieve a good exposure. So, in effect, this setting forces the camera to try to keep the shutter speed at 1/30 second or faster, but if that's not possible, the camera will then change to a slower shutter speed. You may want to use this setting to avoid using slow shutter speeds that are likely to result in blurred photos because of camera motion, or to capture images of moving subjects, such as children playing. If you use a setting such as 1/125 second (the fastest setting possible) for Minimum Shutter Speed, along with an ISO setting such as Auto, which allows the ISO to go as high as 1600, you are likely to be able to take all of your exposures using the 1/125 second shutter speed (if the light is bright enough), preserving your ability to avoid camera shake and to capture ordinary action.

Here are some more notes on the Auto ISO options. If you select either of the two Auto ISO settings in Manual exposure mode, the camera will set the ISO to 100. You can select any other numerical ISO value if you want, and the camera will use that setting. However, and I find this confusing, the camera will let you set Auto or Fixed Ranged Auto for ISO, but it will actually set the ISO to 100, despite the Auto setting.

In Auto mode, Creative mode, and with all of the scene settings, Auto ISO is automatically set and you cannot adjust the ISO setting. With several continuous-shooting options (Pre-Shooting Cache, Continuous H:120 fps, and Continuous H:60 fps), ISO is automatically set to Auto.

Exposure Bracketing

Exposure bracketing is a feature that lets you take three pictures with one press of the shutter button, with different exposure settings, giving you an added chance of getting one good, usable image. In addition, exposure bracketing is a good way to take three pictures that can be merged in software to produce an HDR (high dynamic range) composite, which shows clear details in both the highlights and the shadows throughout the image by combining the best-exposed parts of each shot.

To use exposure bracketing, the camera must be set to the Program, Aperture Priority, or Shutter Priority mode. Navigate to the third line on screen 2 of the Shooting menu and press the OK or Right button to go to the next screen, which lists four choices: ±0.3, ±0.7, ±1.0, and Off, as shown in Figure 4-43.

Chapter 4: The Shooting Menu

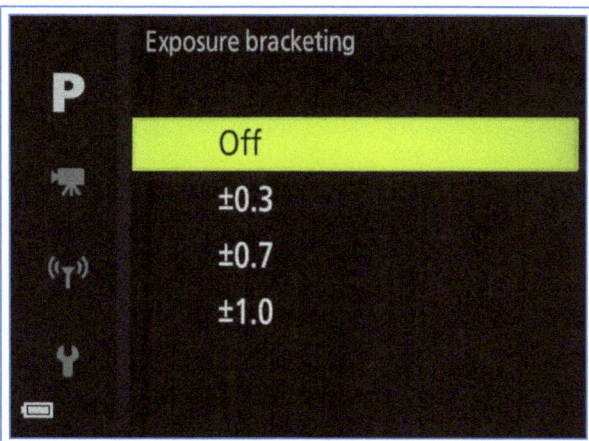

Figure 4-43. Exposure Bracketing Menu Options Screen

Use the multi selector dial or direction buttons to highlight your choice, then make the selection by pressing the OK button. When you exit to shooting mode, the camera's display will include a notation such as BKT±0.7, unless you left bracketing turned off. (Press the Display button if necessary to show this detail on the screen.)

This notation means the camera will take three exposures separated by the indicated amount of exposure value (EV, a standard measure of brightness). Adding a single unit of EV, +1.0, has the same effect as opening the aperture by one full f-stop.

When you are ready to shoot, press and release the shutter button and hold the camera steady (or use a tripod) while it takes the three exposures. The first picture taken is always at the metered level, or 0 change in exposure value (EV); the second is at the lower EV (darker), and the third is at the higher EV (brighter). If you have added exposure compensation, the bracketed exposures are taken at three levels relative to the adjusted exposure.

The flash cannot be used when bracketing is in effect. If you press the Flash button (Up button) when bracketing is turned on, nothing will happen. If the flash was previously set to forced on (Fill Flash) or any other mode in which the flash might fire, the camera will turn the flash off when bracketing is selected. Also, the self-timer cannot be used with bracketing.

Be sure to cancel exposure bracketing when you are done using this feature; otherwise, it will stay in effect even after you turn the camera off and back on again.

AF Area Mode

This next option on the second screen of the Shooting menu gives you several options for controlling how the autofocus frame is set up when the camera is in autofocus mode.

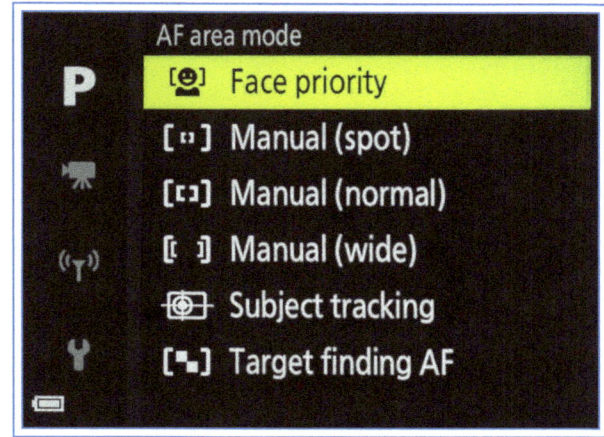

Figure 4-44. AF Area Mode Menu Options Screen

Once this menu option is highlighted, press the OK button or the Right button to display the next menu screen, and then use the multi selector dial or the Up and Down buttons to select one of the six options shown in Figure 4-44, as follows:

Face Priority

With this option, the camera looks for human faces. If it detects one or more faces, it puts a yellow, double-bordered frame on the closest face, and single-bordered frames on other faces, as shown in Figure 4-45.

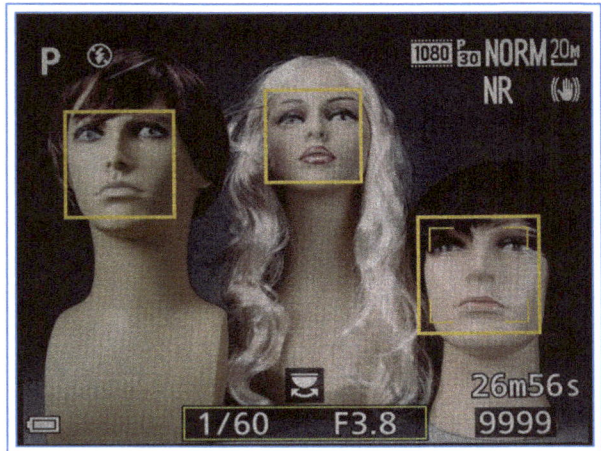

Figure 4-45. Face Priority Setting in Use

When you press the shutter button halfway, the camera will focus on the main face, place a double-bordered green frame on it, and set the exposure and white balance for that face. If no faces are detected, the camera focuses on an object closest to the camera within nine focus areas.

This is a good option to choose when you're at a picnic or other group function and you need to take a quick snapshot with focus fixed on people's faces rather than on trees, buildings, or other objects. In other situations, you may want to take more time and select the focus point and other options yourself.

Manual (Spot, Normal or Wide)

The next three options for AF Area Mode are variations of the Manual setting. The only differences are the sizes of the focus frames used—Spot, Normal, or Wide. The illustrations here all use the Normal size.

If you select one of the three Manual settings for AF Area Mode, the camera displays a focus frame of the chosen size in the center of the screen, with arrows pointing in each direction outside the frame. You can now use the four direction buttons or the multi selector dial to move the focus frame to any of 99 possible locations around the screen, as shown in Figure 4-46. (There are only 81 locations available if you are using the 1:1 aspect ratio, with Image Size set to 3888 x 3888.)

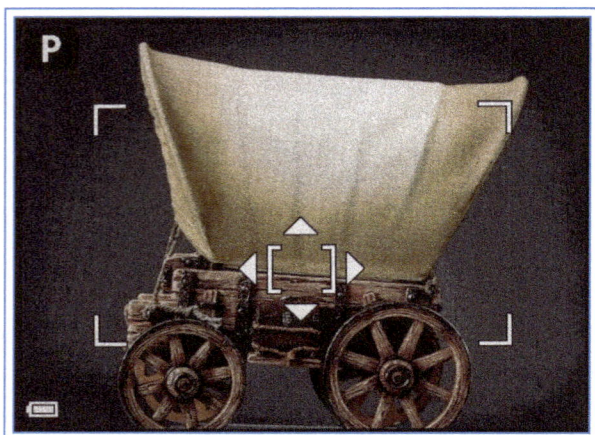

Figure 4-46. Movable Focus Frame on Shooting Screen

This is a good option if you are shooting a scene with items at varying distances from the camera and you want to focus on an item that is not in the center of the scene. Of course, this option is useful only if you have time to select it and move the focus frame to the location where you want it. If you don't have time, it may be easier just to place the center of the frame over the item you want to focus on, press the shutter button halfway to lock focus, and then move the camera back to compose the image as you want it. That method is also the best way to proceed when you are using Center-weighted or Spot for the Metering option and you want to lock both focus and exposure for a subject that is not in the center of the scene. If you use the Manual setting for AF Area Mode in that situation, the focus will be set on the off-center subject, but the exposure will be based on the subject in the center of the screen.

If you do have the time to use the Manual option for AF Area Mode, here is how to use it. Once you have located the focus frame where you want it using the buttons or dial, press the shutter button to lock focus and then take the picture. The focus frame will stay in this location even after the camera is powered off and back on, so be sure to reset it to the center when you no longer need it in an off-center position. When the frame is in the center of the screen, a dot will appear in the center of the frame while it is movable.

If you need to use one of the four direction buttons for another purpose while using Manual AF Area Mode, press the OK button to return the buttons to their other functions (flash mode, focus mode, self-timer, and exposure compensation); then, after using a button for another function, press OK again to return the buttons to controlling the location of the focus frame.

Subject Tracking

This next AF Area Mode option is designed for situations in which you need to track a moving subject, such as a sports competitor, a pet, or a child at play.

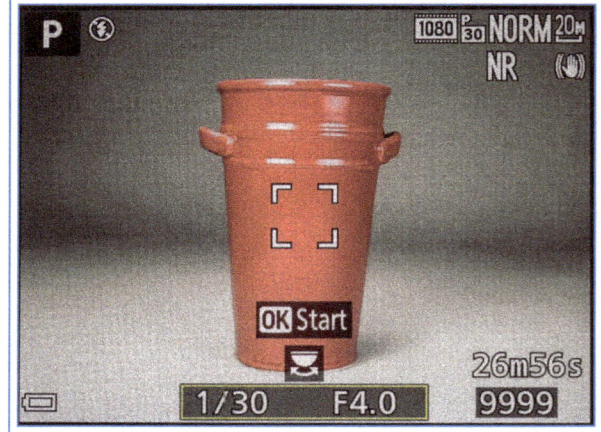

Figure 4-47. Subject Tracking Setting Ready for Use

Once you have selected this mode, you will see a small, white, square-shaped bracket in the center of the screen, with the words OK Start below it, as shown in Figure 4-47.

Aim this square bracket at the subject you want to track and press the OK button. The frame will change to a yellow square with corner brackets, as shown in Figure 4-48, which the camera will try to keep centered over the subject, even as the subject (or the camera) moves.

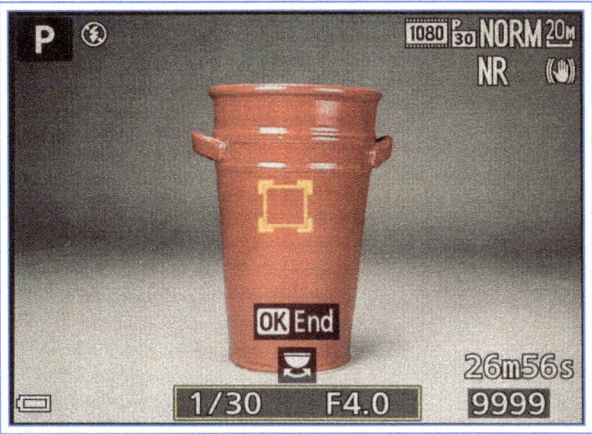

Figure 4-48. Subject Tracking Setting Activated

When you press the shutter button halfway to check exposure, the frame turns green to confirm exposure, and tracking stops. To start tracking again, release the shutter button. To end tracking without taking a picture, press the OK button. Press the shutter button all the way down when you are ready to take the picture.

When Subject Tracking is selected, the Monochrome setting for Picture Control is not available.

Target Finding AF

The final option for AF Area Mode, Target Finding AF, is the default setting and the one the camera uses when it is set to the Auto shooting mode. With this option, the B700 uses its programming to try to select the main focus point(s). The camera does not have any focus frame on its display screen at first. As you aim the camera at a scene, though, the camera may display yellow frames of varying sizes and shapes on the screen, as shown in Figure 4-49, as it tries to detect the subject to focus on.

When you press the shutter button halfway to lock focus, the camera will try to select the "main" subject. It will first look for a human face, then for a subject that matches programmed factors, such as size, position, and color. If it has not found a main subject, it will focus on the items closest to the camera within nine focus blocks in the central part of the display. It will display one or more green rectangles on the screen to show the point(s) it chose for focusing, as shown in Figure 4-50.

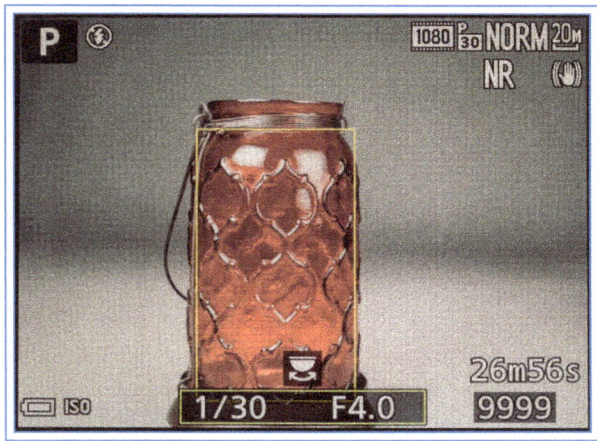

Figure 4-49. Target Finding Subject Detection Frame

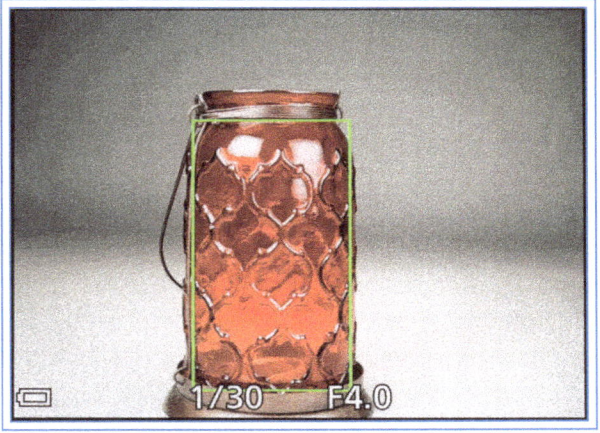

Figure 4-50. Target Finding Frame After Focus Achieved

This focusing mode is good for shots of general scenes when you don't have time to choose a focus point yourself or when there is not much doubt about where the camera will set its focus. For example, if you are taking a snapshot of a person in front of a scenic view, you can safely assume that the camera will focus on the person. If the scene includes multiple objects fairly close to the camera, you might be better off using one of the Manual settings to make sure the subject you want to focus on is inside the focus frame.

Autofocus Mode

This feature, whose menu screen is shown in Figure 4-51, lets you decide whether the camera will focus just once, when you press the shutter button halfway, or will focus continuously before you press the button halfway.

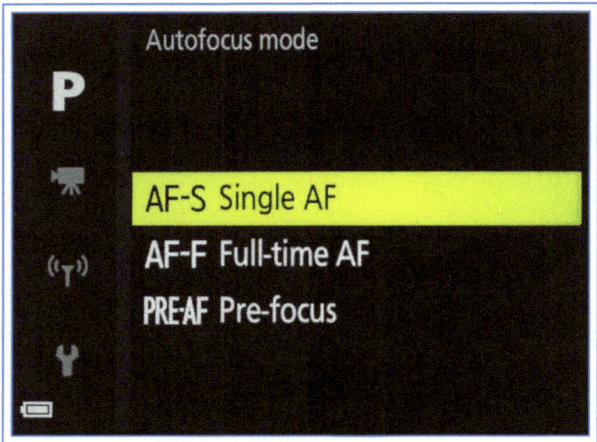

Figure 4-51. Autofocus Mode Menu Options Screen

Choose Single AF if you want to conserve the battery and wait until you are ready to take the picture before the camera uses its autofocus mechanism; choose Full-time AF if you want the camera to focus continuously. Choose Pre-focus (the default setting) if you want the camera to focus continuously when it detects motion, or when the camera is moved to a new position.

Although the Full-time option and the Pre-focus option will use up your battery more quickly than Single AF, they have the advantage of keeping the image in focus as you move the camera or your subject moves around; in that way, when you are ready to capture the image, the camera can make the final focusing adjustments quickly when you press the shutter button.

This Autofocus Mode option does not apply for shooting movies; you need to select an autofocus mode from the Movie menu for that situation. When manual focus is in effect, you can select this menu option, but it will not have any effect until you switch the camera to autofocus or macro focus. With the infinity focus mode, the camera uses Single AF for Autofocus Mode, regardless of the setting for this option. When the Smile Timer is active, the camera uses Single AF in all cases.

For ordinary photography, I prefer Single AF. But, if I am shooting subjects that are in constant motion, Pre-focus or Full-time AF can be useful.

Flash Exposure Compensation

This option works in similar fashion to standard exposure compensation, discussed in Chapter 2. It is available only in Program, Aperture Priority, Shutter Priority, and Manual exposure modes. You can dial in an amount of positive or negative flash exposure compensation up to two EV units in either direction, in increments of 1/3 EV. When you do that, the camera will increase or decrease the output of the flash, unless it was already using its maximum or minimum power.

I find this setting of most use when I'm taking a portrait with flash. I like to use some negative flash exposure compensation to make sure the flash does not wash out the image with excessive brightness.

To use this setting, go to its entry on the second screen of the Shooting menu and press the OK button or the Right button to get to the adjustment screen. At that screen, turn the multi selector dial or use the Up and Down buttons to dial in up to +2.0 EV or -2.0 EV, as shown in Figure 4-52.

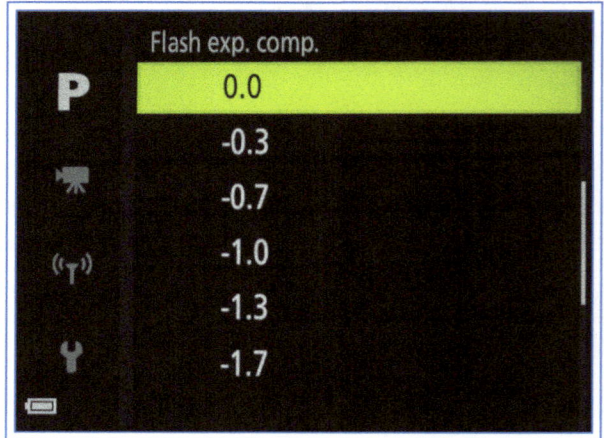

Figure 4-52. Flash Exposure Compensation Adjustment Screen

Press the OK button to confirm your selection when the value you want to choose is highlighted in the yellow bar. When you have activated a positive or negative amount of flash exposure compensation, that value will appear on the camera's display in the lower right-hand corner, but only when the flash is popped up. That value will remain in effect even after the camera has been powered off and back on, so be sure to cancel it when you no longer need the compensation.

The third screen of the Shooting menu is shown in Figure 4-53.

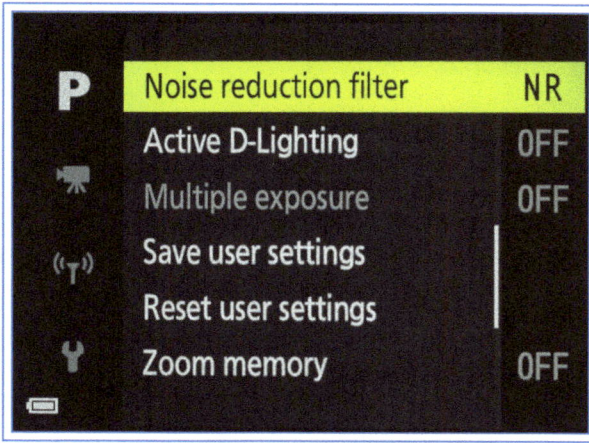

Figure 4-53. Screen 3 of Shooting Menu

Noise Reduction Filter

Noise reduction is an electronic feature built into the B700's programming to compensate for visual noise in your images, which can be caused by long exposures or by the use of high ISO settings. By default, this option is set to Normal, which causes the camera to use a moderate amount of noise reduction. If you want to have larger or smaller amounts of noise reduction applied in every case, you can switch the setting to High or Low, as shown in Figure 4-54. You cannot turn noise reduction completely off.

Excessive noise reduction can reduce the detail and other positive features of an image. However, if you will be using the shots straight from the camera and you are shooting with long shutter speeds or high ISO settings, you may well want to turn this setting up to High to avoid the graininess that comes from visual noise.

Active D-Lighting

This second entry on screen 3 of the Shooting menu can help avoid problems with excessive contrast in your images. Such problems arise because digital cameras cannot easily process a wide range of dark and light areas in the same image—that is, their "dynamic range" is limited. So, if you are taking a picture in an area partly lit by bright sunlight and partly in deep shade, the resulting image is likely to have dark areas in which details are lost in the shadows, or areas in which highlights, or bright areas, are excessively light, or "blown out," so, again, the details of the image are lost. One approach to this problem is to use HDR techniques, with which multiple photographs of the same scene with different exposures are combined into a composite image that is properly exposed throughout the entire scene. I discussed that technique in Chapter 3, in connection with the Backlighting/HDR setting of Scene mode.

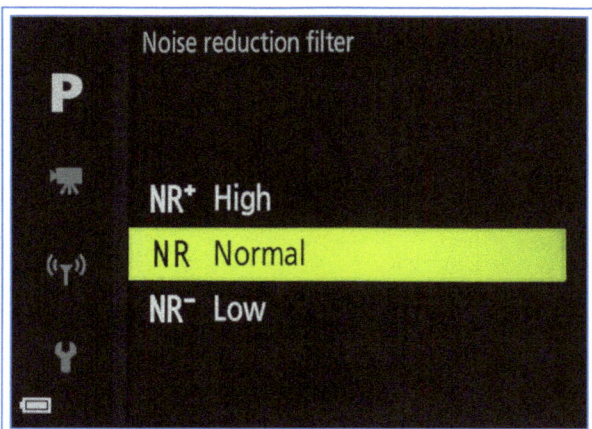

Figure 4-54. Noise Reduction Filter Menu Options Screen

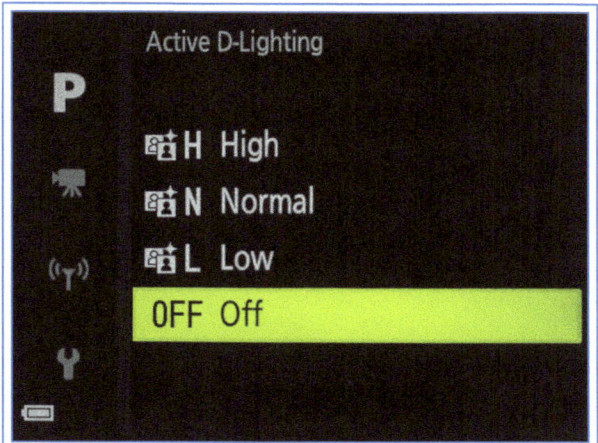

Figure 4-55. Active D-Lighting Menu Options Screen

What you do with this setting is a matter of personal preference. If you are going to be using Photoshop, Photoshop Elements, or similar software to process the images on your computer, you may want to leave Noise Reduction Filter set to Low, so you can minimize the amount of processing done in the camera.

The Active D-Lighting menu option gives you another way to approach the problem of uneven lighting. This feature uses processing in the camera to boost details in dark areas and reduce overexposure in bright areas, resulting in a single image with better exposure than would be possible otherwise. If you turn this option on, the camera reduces the overall exposure and performs digital processing as it records the image, resulting in some restoration of details in the shadows and in the highlights, to even out the lighting. This menu option, as shown in Figure 4-55, provides three levels of this processing: High, Normal, and Low, as well as Off, the default setting.

To illustrate the effects of this setting, I took a pair of photographs of a group of stones on a wooden platform, partly in sun and partly in shade. For the first sample image, Figure 4-56, Active D-Lighting was turned off; for Figure 4-57, it was turned on to the High setting.

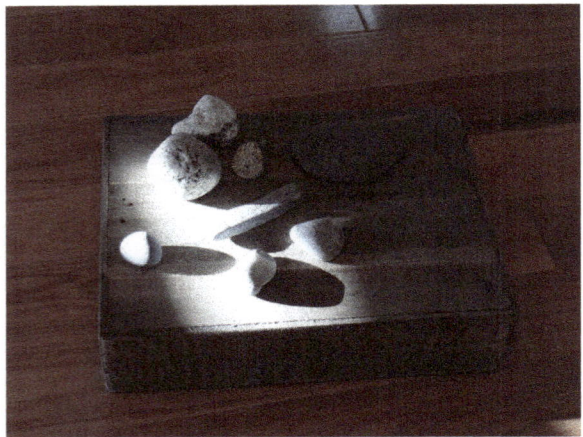

Figure 4-56. Active D-Lighting Turned Off

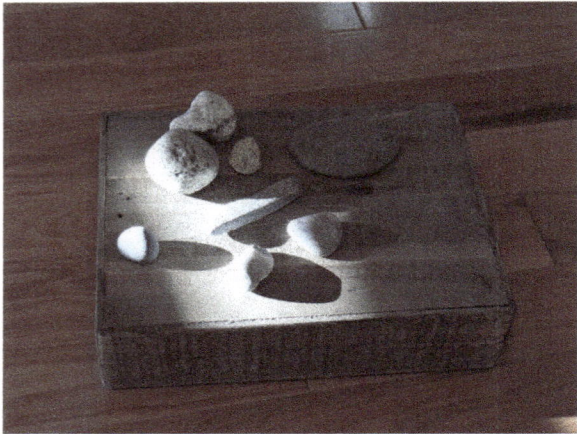

Figure 4-57. Active D-Lighting Set to High

With Active D-Lighting set to High, the camera noticeably reduced the overexposure in the brighter part of the image. It also brought some details out of the shadowed areas. My recommendation is to turn on Active D-Lighting when you are shooting a subject that is partly in the sun and partly in the shade. In those cases, if you have time, I would try setting this menu option to its various levels to see how the results compare. You also might want to use exposure bracketing, discussed earlier in this chapter, and merge those three exposures using HDR software, as discussed in Chapter 3. Or, you can use the Backlighting/HDR setting of Scene mode, also discussed in Chapter 3.

The Metering option is not available on the Shooting menu when Active D-Lighting is turned on to any level; in that case, the camera uses the Matrix setting for metering. Also, when Active D-Lighting is in effect, you cannot adjust the contrast parameter for the Picture Control settings.

The Coolpix B700 has a related feature called simply D-Lighting, which is used in playback mode for images that have already been taken. I'll discuss that feature in Chapter 6.

Multiple Exposure

This menu option lets you shoot multiple exposures in the camera. You can shoot either two or three images on the same digital frame. When you highlight this option and press the OK button or the Right button, the camera displays the options screen, shown in Figure 4-58.

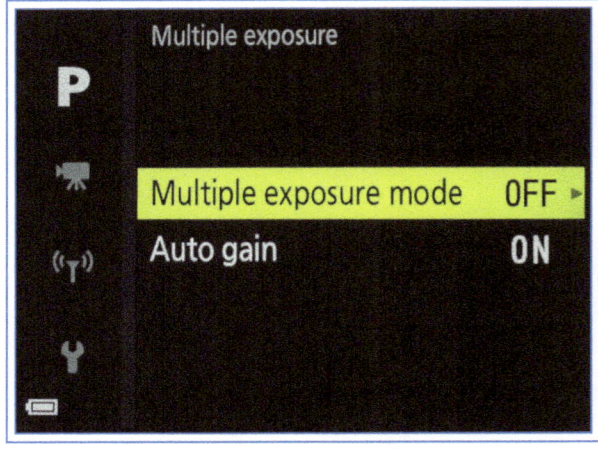

Figure 4-58. Multiple Exposure Menu Options Screen

To shoot multiple exposures, set the Multiple Exposure Mode option to On. You can set the Auto Gain option to On, which is the default setting, or turn it off. If it is turned on, then the camera will use its programming to adjust the relative brightness of the multiple images as it sees fit. If you turn this option off, then the images will be recorded without adjustment. In my experience, this feature produces better results with Auto Gain turned on, though certain situations may yield better results with it turned off.

After you turn Multiple Exposure Mode on, aim at your first subject. If you will be shooting the second subject to appear beside the first one, then be sure to plan ahead, leaving space where the second subject will appear. Plan for a third subject also, if you will be shooting three items. If the positioning of the subjects will be critical, you should use a tripod to keep the composition precise. It usually works best to shoot any given object only once in the same position to avoid making it too prominent in its exposure. So, if an object has been photographed in one position in the first exposure, I usually remove it before taking the second exposure.

Press the shutter button to take the first picture. After some processing time, the camera will display the first image on the screen in a translucent mode, so you can continue to view the first image while you compose the next one. Now, line up the second image while viewing the first one, as illustrated in Figure 4-59.

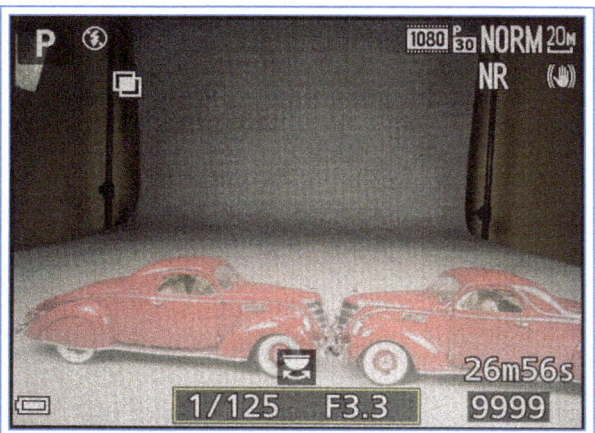

Figure 4-59. Multiple Exposure Shooting Screen After First Shot

When this composition looks right, press the shutter button to take the second image. The camera will take even longer to process this exposure. When it has finished processing, the double exposure will be displayed on the screen. It will still appear translucent, because you now have the option of taking one more picture to add to the composition. If you want to do that, go ahead and line up the third shot and press the shutter button. After the camera finishes processing the third image, it will return to the shooting screen. To view the final composition with three exposures, press the Playback button and you will see the finished product.

If you want to use only two exposures, then, after you take the second one, go to the Menu system and turn off the Multiple Exposure Mode option, or just turn the mode dial to a shooting mode that does not provide access to this menu item, such as Auto, Scene, or Creative.

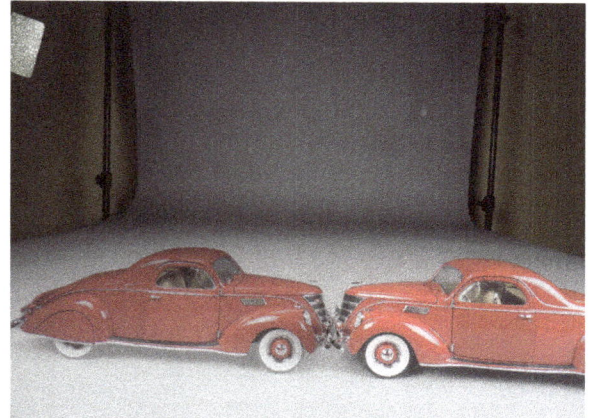

Figure 4-60. Multiple Exposure Final Image

Figure 4-60 shows the finished image from two shots of a model car. I had Auto Gain turned on and I added some contrast and sharpening to the final image in Photoshop, because it looked a bit faded as it came out of the camera. The camera also saves each individual image you take for the multiple exposure composite.

Save User Settings

I discussed this feature in Chapter 3 in connection with the User Settings shooting mode, marked by the letter U on the mode dial. To save your current shooting settings for instant recall with the U slot on the dial, navigate to the Save User Settings option on the Shooting menu, as shown in Figure 4-61, and press the Right button or the OK button to save the settings.

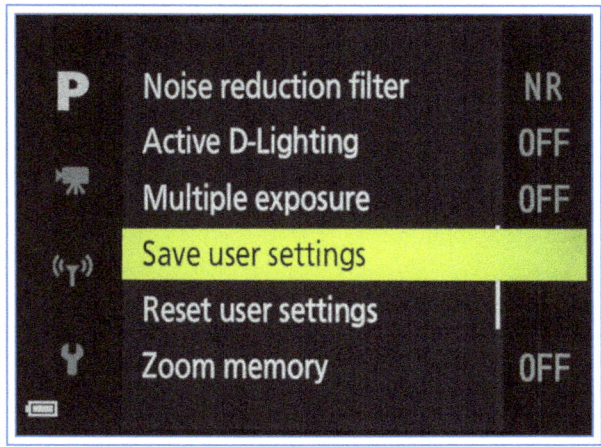

Figure 4-61. Save User Settings Item on Shooting Menu

Be sure the settings are as you want before you press the button, because the camera does not ask you to confirm your choice; it just displays a "Done" message once you press the button.

Reset User Settings

This option on the Shooting menu, shown in Figure 4-62, lets you reset the settings that have been saved to the User Settings mode (U slot on the mode dial).

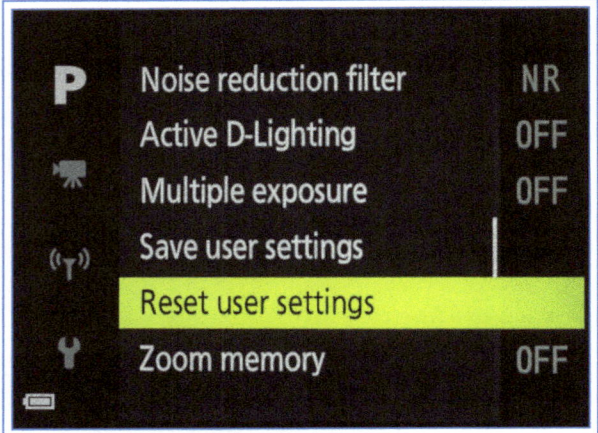

Figure 4-62. Reset User Settings Item on Shooting Menu

When you select this option, the camera resets all of those items to their default values without your having to adjust each one in the menu screens. For example, choosing this option sets the shooting mode to Program, the flash mode to Auto, exposure compensation to 0.0, the zoom lens to its wide-angle position, and all items on the Shooting menu to their default settings.

Note that this option affects only the settings saved for the User Settings shooting mode; if you want to reset all settings for the camera for all modes, you have to use the Reset All menu option, which is found on the Setup menu, as discussed in Chapter 7.

Zoom Memory and Startup Zoom Position

The next two items on the Shooting menu are closely related, so I will discuss them together. The first of these, the last option on screen 3 of the menu, is Zoom Memory. This feature lets you control whether the zoom lever zooms the lens in continuous increments or in distinct, separate steps. By default this option is turned off, so you can zoom continuously in any amounts, zooming either in or out.

If you turn the Zoom Memory setting on using this menu option, the camera displays a list of focal lengths, from full wide-angle to full telephoto: 24mm, 28mm, 35mm, 50mm, 85mm, 105mm, 135mm, 200mm, 300mm, 400mm, 500mm, 600mm, 800mm, 1000mm, 1200mm, and 1440mm, the maximum range of the optical zoom. It takes three menu screens to include all of these values; the first screen is shown in Figure 4-63.

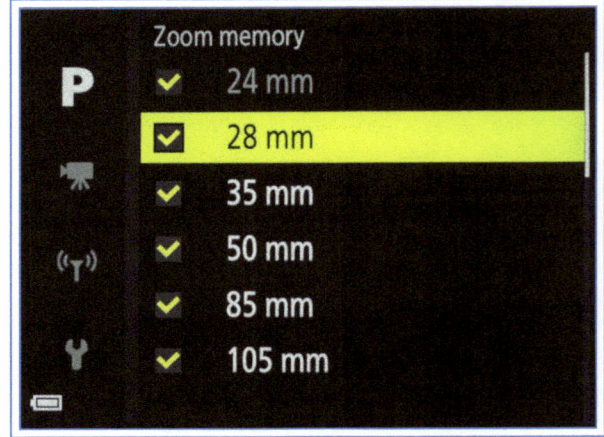

Figure 4-63. First Screen of Zoom Memory Menu Options

Each value initially has a check box to its left. Move the selection bar through the list using the Up and Down buttons, the multi selector dial, or the command dial. As each value is highlighted, you can press the OK button to check or un-check its box. If the box is checked, the camera will include that focal length in the zoom memory.

When you have finished checking the boxes for the focal lengths you want to include in the zoom memory, exit from the menu screen by pressing the Menu button or by half-pressing the shutter button. Now, when you press the zoom lever, each press will take the lens to the next focal length that was checked for Zoom Memory.

For example, suppose you turned on the Zoom Memory option and checked the boxes for 50mm, 200mm, and 1000mm. Then, if you are starting from the full wide-angle position of 24mm, when you press the zoom lever to the right, the lens will zoom in to 50mm. If you press it again, the lens will zoom to 200mm, and one more press will take it all the way to 1000mm. You will not be able to zoom the lens to any other positions. The same will be true when you zoom back out by pressing the zoom lever in the other direction. It does not matter if you use a very quick press of the lever or hold the lever in position; it will zoom only to the next level that has its box checked on the Zoom Memory menu screen.

As I noted above, the Zoom Memory option is closely related to the Startup Zoom Position option, which is the first item on screen 4 of the Shooting menu, shown in Figure 4-64.

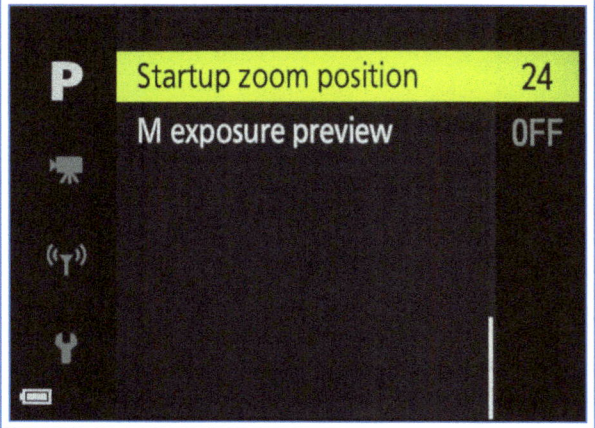

Figure 4-64. Screen 4 of Shooting Menu

With that menu item, whose first screen of options is shown in Figure 4-65, you select the single focal length to use when you first turn the camera on.

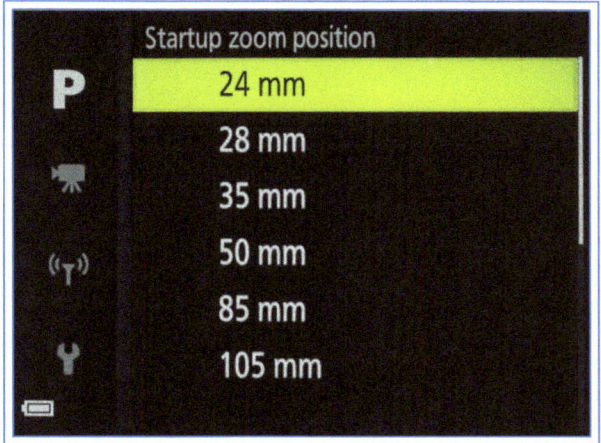

Figure 4-65. First Screen of Startup Zoom Position Menu Options

In this case, the choices are more limited: 24mm, 28mm, 35mm, 50mm, 85mm, 105mm, or 135mm. You select this menu option, then move to the next screen and position the yellow selection bar on the focal length you want to choose. Press the OK button to confirm. Then, when you turn the camera on the next time, the lens will automatically zoom to that focal length.

Here is how these two settings are related. When you select a focal length for Startup Zoom Position, you will see that that value is automatically checked on the screen for Zoom Memory, and its menu item is dimmed, meaning you cannot alter it. In other words, you cannot un-check the box for that focal length, because the camera is going to start up at that focal length. For example, suppose you select 50mm for Startup Zoom Position, and 35mm, 200mm, and 1000mm for Zoom Memory. The next time you turn on the camera, the lens will zoom to the 50mm position. If you press the zoom lever to zoom out, the lens will move to 35 mm. If you then press the lever to zoom in, the lens will zoom back to 50mm, the startup position. From there, it will zoom to 200mm, then 1000mm.

These two menu options, working together, give a good deal of control over how the lens zooms. Of course, you don't need to have that degree of control; you may be content to use the default settings, using the startup position of 24mm and allowing the lens to zoom to any setting, without using the Zoom Memory option. However, it can be convenient to know what focal length you are using for a given shot. If you turn Zoom Memory off, the camera's display will not show you what focal length it is using; if you turn it on, you will see the focal length briefly displayed at the top of the display, when you press the zoom lever.

The Zoom Memory option does not control the operation of the side zoom control, the switch on the left side of the camera that can be used for zoom (or for manual focus if that function is assigned through the Setup menu). So, even if you have turned on the Zoom Memory menu option, thereby restricting the zoom lever to certain focal lengths, you can still use the side zoom control to zoom the lens continuously, as long as that control is assigned to the zoom function.

Manual Exposure Preview

This last item on the Shooting menu has a narrow, specific purpose—to control whether the camera's display reflects the brightness of the image that will result from current settings when the camera is in Manual exposure mode. When you select this item and press the OK button or the Right button to move to the next screen, you will see a screen with options to turn this feature on or off. If you leave it at its default setting of Off, then, when the camera is in Manual exposure mode, the camera's display will show a normally exposed view of the scene, even if the current settings of aperture, shutter speed, and ISO would result in a heavily underexposed or overexposed image.

If you turn this feature on, then, when you adjust the shooting settings in a way that would result in an unusually dark or bright image, the camera's display will become darker or brighter also, so you will have notice that the image may be improperly exposed.

There are limits to the operation of this feature. First, it works only in Manual exposure mode, although the menu option is also available for selection in Program, Aperture Priority, and Shutter Priority mode. Second, even when the option is turned on, it does not show the full effect of severe underexposure or overexposure. For example, I turned this option on and set the camera to 1/500 second at f/7.1, with ISO set to 100. With these settings in a normally lighted room, the screen grew somewhat dark, as shown in Figure 4-66.

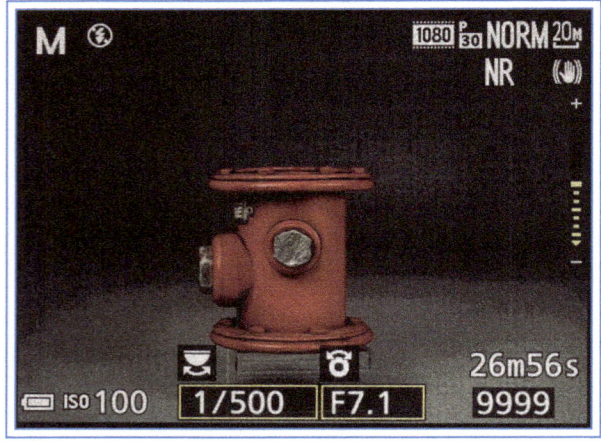

Figure 4-66. Manual Exposure Preview in Use

When I took a picture with these settings, though, the resulting image was completely black. And, when I used settings causing strong overexposure, the display showed only mild overexposure.

There is one more aspect of this option to point out. As discussed in Chapter 7, you can turn on a shooting-mode histogram using the View/Hide Histograms option of the Monitor Settings item on the Setup menu. If you do that, the camera will display a histogram to help you gauge the exposure level of your shot. However, in Manual exposure mode, this histogram will not be accurate unless you turn on the Manual Exposure Preview menu option.

If you leave this option turned off, the histogram will reflect the exposure level seen on the camera's display, which, in most cases, will look normal, even if the exposure settings would result in a heavily underexposed or overexposed image. If you turn the Manual Exposure Preview option on, the histogram will reflect the actual shooting conditions more accurately, though it does not fully reflect the effects of extreme exposure settings, in my experience. (The histogram is not displayed at all in Auto shooting mode, when AF Area Mode is set to Target Finding, or when a movie is being recorded.)

Because this feature does not give you a fully accurate representation of underexposure or overexposure, even with the histogram, I prefer to leave it off and rely on my own judgment, or to take a test shot to see how the final image will look.

Chapter 5: Physical Controls

The Coolpix B700, like many compact cameras, does not have very many physical controls, relying heavily on its menus for changing settings. But the B700 is an advanced camera, and it has more controls than many cameras. In this chapter, I'll discuss each of these controls and how they can be used to best advantage.

The controls on the top of the camera are shown in Figure 5-1.

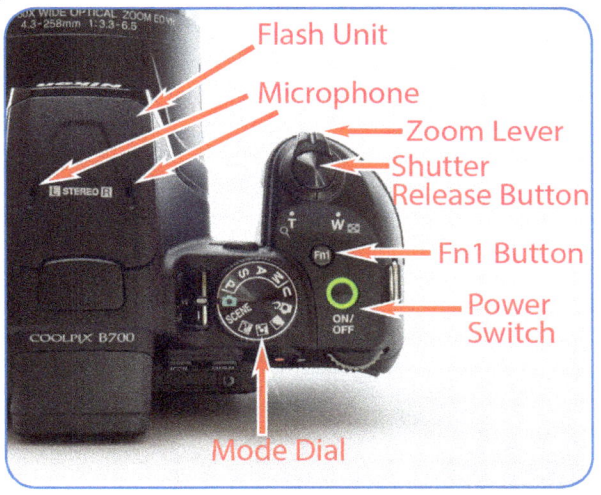

Figure 5-1. Controls on Top of Camera

Power Switch

The power switch, at the right side of the camera's top, has only one function—to turn the camera on and off. When the camera is turned on, the green light around this button illuminates. When the camera enters standby mode to save power, the green light starts to blink, and does so for about three minutes. During that time, you can press the power switch, the Playback button, the shutter release button, or the Movie button, to cancel standby mode and restore the camera to full power. After the three minutes, the camera turns itself off and you have to use the power switch to turn it back on again. When the battery is charging, the light around the switch flashes slowly; it turns off when the battery is fully charged.

Shutter Release Button

The shutter release button (often called simply the shutter button) is the single most important control on the camera. When you press it halfway down in most shooting modes, the camera evaluates exposure and focus (unless you're using manual focus). Once you are satisfied with the settings, you press the button all the way down to record the image.

When you press the button halfway, it locks exposure and focus. If you need to set the camera's automatic exposure and focus for a subject that is not in the center of the scene, you can aim the center of the display at that subject, half-press the shutter button to lock exposure and focus, and then move the camera back to the position for taking the picture, with that subject at one side. This procedure is useful when you are using Center-weighted or Spot for the metering mode and Manual for the Autofocus Area mode, with the focus frame set in the center of the screen. In those situations, the camera will lock both focus and exposure on the object away from the center of the scene when you press the shutter button halfway.

When the camera is set for continuous shooting, you hold this button down to make the camera take a burst of images.

You can press this button halfway to switch from playback mode to shooting mode so you can resume taking pictures. You also can press it to exit from a menu screen. Note, though, that you have to press the OK button to confirm your selection on the menu screen before pressing the shutter button; otherwise, your setting on the menu screen will not be saved. You can press the shutter button to re-awaken the camera from standby mode, which it enters after a period of inactivity. You have to press this button (or the Playback button, power switch, or Movie button) while the green light around the power switch is flashing.

The shutter release button also can be used to take still images while recording a movie, with some restrictions. There are more details about this option and its limitations in Chapter 8.

Mode Dial

The mode dial, located at the right side of the camera's top, is central to the operation of the camera. Its function is to change from one shooting mode to another.

Zoom Lever

The zoom lever is a small ring with a handle, surrounding the shutter release button. The lever's main function is to change the focal length of the lens to various settings between its wide-angle setting of 24mm and its telephoto setting of 1440mm. If you have the camera set for digital zoom, the lever will boost the focal length to a maximum of 5760mm. (That impressive-sounding zoom amount is illusory, though, because the quality will be degraded by the electronic enlargement of the image.) If you move the lever sharply to either side, the zoom range will adjust quickly; if you move it more gradually, the range will change more slowly.

When you turn the camera on, the lens moves to whatever position has been selected with the Startup Zoom Position option on the Shooting menu, as discussed in Chapter 4. And, the behavior of the zoom lever will change depending on the settings for the Zoom Memory menu option, also discussed in that chapter. As a brief reminder, if Zoom Memory is turned on, pressing the zoom lever will move the lens to the next zoom level that was selected through the Zoom Memory menu option. If that feature is turned off, pressing the lever will zoom the lens continuously through its full range of focal lengths.

In playback mode, moving the zoom lever to the left repeatedly (pointing to the W setting) produces index screens with increasing numbers of images, and moving the lever to the right enlarges the current image. These functions are discussed in Chapter 6.

Function Button 1 (Fn1)

One very useful control on the camera's top is the function button—the small, recessed button marked with the Fn1 designation, behind the shutter release button. (There is a second function button on the camera's back, discussed later in this chapter.)

As discussed in Chapter 4, many of the most important settings on the Coolpix B700 are located in the Shooting menu, including Image Size, ISO, White Balance, Continuous, and others. It can be inconvenient to change these settings when you have to press the Menu button, navigate to the Shooting menu, and move to the proper line on the menu before you can make a change.

With this small button, you can program any one of nine of the most useful settings of the camera into a physical control. The options available for choice are Image Quality, Image Size, Picture Control, White Balance, Metering, Continuous, ISO, AF Area Mode, and Vibration Reduction. When you press the function button, a menu for the item that is assigned to the button pops up on the screen, as shown in Figure 5-2, letting you quickly change the setting.

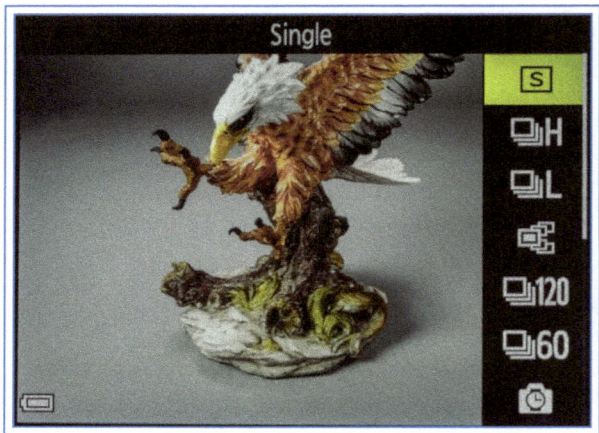

Figure 5-2. Function Menu for Continuous Shooting

You can move through that menu using the Up and Down buttons or by turning either the command dial or the multi selector dial; you have to press the OK button to confirm your selection once it is highlighted.

The default choice for this button, as shown here, is Continuous, for continuous shooting, which I tend to prefer as the item assigned to this button. If I want to fire off a burst of shots, it is convenient to press the

Fn1 button and switch to one of the burst modes for a short time. When that situation has passed, I can just as quickly press the button again and reset the camera to single-shot mode. However, it also can be very useful to be able to adjust ISO quickly or to choose Raw or Fine for Image Quality. Of course, the option you choose to assign to this button will depend on your own particular circumstances. If you like to experiment with various image-processing settings, you might want to program Picture Control as the function. In any event, it is good to have this option available.

To change the feature assigned to the Fn1 button, use the same menu that pops up when you press the button. The last item, on the second screen of the menu, is the Fn item, shown in Figure 5-3.

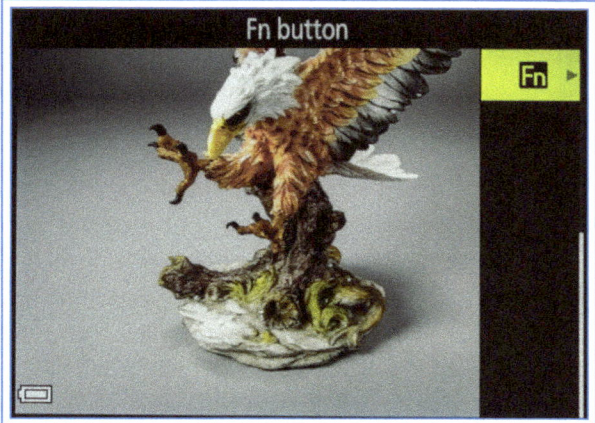

Figure 5-3. Fn Item for Changing Fn1 Button Assignment

After you press the Fn1 button, the easiest way to reach this menu item is to press the Up button and wrap around to the bottom of the menu, to the Fn item.

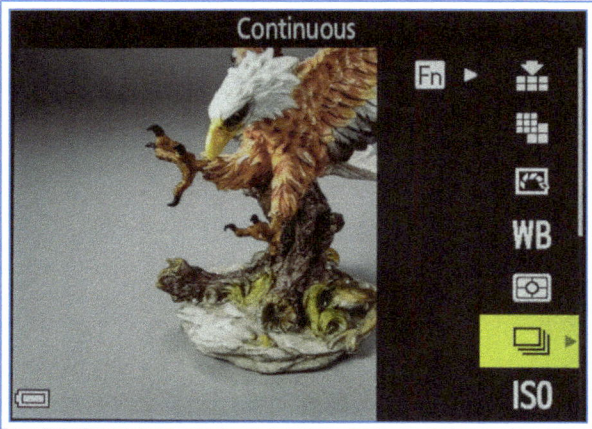

Figure 5-4. Menu of Items Assignable to Fn1 Button

Then, press the Right button or the OK button, and the camera will display a menu of the options available for assignment to the function button, with a yellow highlight block marking the one that is currently assigned, as shown in Figure 5-4.

Using the Up and Down buttons, the multi selector dial, or the command dial, scroll through the choices until the one you want to select is highlighted, and then press the OK button. The next time you press the function button, the menu for the newly selected item will appear on the display.

The Fn1 button operates only when the camera is in shooting mode and the mode dial is set to the Program, Aperture Priority, Shutter Priority, Manual exposure, or User Settings mode.

The controls on the left side of the camera are shown in Figure 5-5.

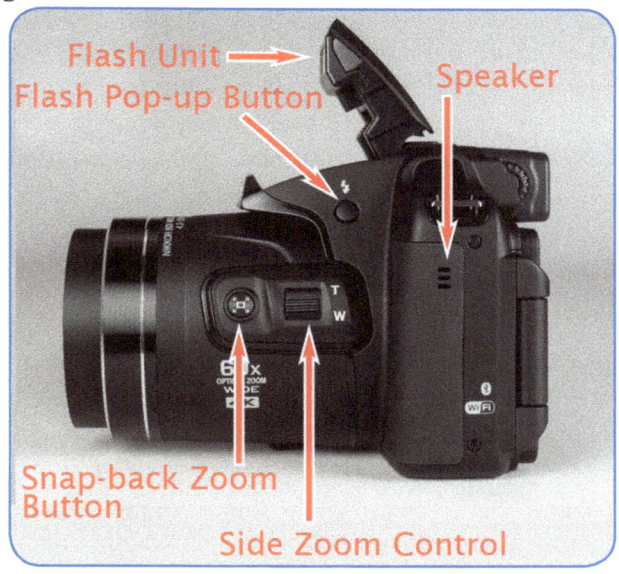

Figure 5-5. Controls on Left Side of Camera

Flash Pop-up Button

This small round button on the left side of the flash housing has one simple purpose—to release the built-in flash unit so it will pop up and be available for use. If you expect you will be using the flash, you need to press this button to make the unit available; if you don't press the button, the flash will not pop up and cannot fire. If you select a shooting mode that requires use of the flash, such as Night Portrait, the camera will display a message prompting you to raise the flash. When you have finished with the flash unit, press it gently back down until it clicks into place.

The requirement that you press this button to pop up the flash has one clear advantage: When you are in a museum or other location where photography is permitted but the use of flash is prohibited, you can just leave the flash unit stowed away and you can be sure it will never pop up by itself and send out a flash that proves to be embarrassing. (With some compact cameras, the flash is always available to fire, and you have to remember to set the flash mode properly to avoid having it go off unexpectedly.)

Side Zoom Control

A useful feature of the B700 is this second zoom switch, located on the left side of the lens barrel as you hold the camera in shooting position. One reason for having this alternative control available is to free up your right hand to hold the camera firmly, rather than having to reach up to the standard zoom lever on top of the camera. With the super-powerful zoom range of the B700, you need to hold the camera as steady as possible when zooming in to the longer ranges.

The functioning of the side zoom control is not governed by the Zoom Memory setting on the Shooting menu. That is, when Zoom Memory is turned on, restricting the operation of the zoom lever to certain focal lengths, the side zoom control can still be used to zoom the lens continuously to any focal length.

Another helpful aspect of the side zoom control is that its function is assignable. That is, you can use this switch to control manual focus instead of zoom, depending on your preference. To do this, you use the Assign Side Zoom Control item on screen 2 of the Setup menu, as discussed in Chapter 7. If you do so, you still can use the multi selector dial or the command dial to control manual focus, but you then have the option of using the side switch as an alternative. You may like the feel of pressing this control for adjusting the focus rather than turning either of the other dials.

Snap-back Zoom Button

The button marked with a square with arrows coming out of its corners, next to the side zoom control, operates a function that Nikon calls "snap-back zoom."

When you have zoomed the lens in to a powerful telephoto setting, you can press and hold the snap-back zoom button to "snap" the focal length back to a wider view in a preset amount. Continue to hold down the button while you frame the subject, using the inset frame that is displayed on the screen. While holding down the button, you can widen the view further by pressing the zoom lever toward the W position. When you have finished locating and framing the subject, release the snap-back button and the lens will zoom back in to its original position.

In essence, with the snap-back button, when you have the camera zoomed in for a magnified, telephoto view, you can experiment with different telephoto settings. You can "snap" the lens back out, which can help you get a sense of your ultimate subject by seeing a wider view. Afterwards, you can snap the camera back to its original telephoto setting without having to use trial and error; that setting has been preserved precisely for you in the camera's "snap-back" memory.

I did not see much need for the snap-back function at first. However, when I was trying to photograph a bird at a long distance using the superzoom lens, I found this feature very useful. When the lens was zoomed all the way in, I found it hard to locate the bird. I eventually realized that I could quickly snap the lens back to a wider view until I found the bird in my field of view. Once I had the bird centered again, I could release the snap-back zoom button to snap the zoom back to the full-power telephoto view. To help you stay oriented when using the very long focal lengths available with the B700, the snap-back control can be quite useful. This function is available for shooting still images only, not for movies.

The next controls to be discussed are those located on the camera's back, on the right side of the LCD screen, as shown in Figure 5-6.

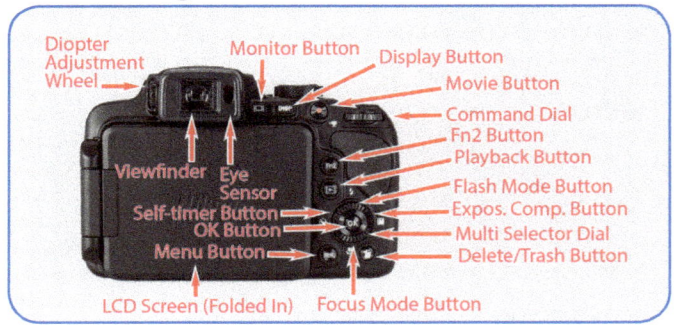

Figure 5-6. Controls on Back of Camera

Fn2 Button

This second function button works the same as the Fn1 button, discussed above. You can assign the Fn2 button to carry out any one of the functions listed earlier for the Fn1 button.

Playback Button

This button, marked with a triangle, is used to put the camera into playback mode, which allows you to view your images on the LCD (or in the viewfinder) and lets you get access to the Playback menu by pressing the Menu button. You also can use the Playback button instead of the power switch to turn the camera on, placing it immediately into playback mode. You might want to do this if you're only going to view your recorded images and won't be using shooting mode.

If you turn the camera on using the Playback button, pressing it again will not turn the camera off; it will just switch the camera into shooting mode. When the camera is in playback mode, you can always press the shutter button down halfway to change into shooting mode.

When the camera enters power saving mode and the light around the power button starts to blink, you can press the Playback button, among others, to stop the camera from powering off. If you do that, the camera will then be in playback mode.

Diopter Adjustment Dial

The small wheel on the left side of the viewfinder is used to dial in optical correction to the viewfinder, so you can see a sharply focused image in the viewfinder window. As discussed in the next section below, when you fold the LCD screen in the closed position, press the Monitor button, or (depending on a menu setting) move your head near the eye sensor, the viewfinder is activated. You can then turn this little wheel in either direction until the image in the viewfinder is at its clearest for your eyesight. If you wear glasses, you may be able to dial in enough of an adjustment that you can take your glasses off and still see the image clearly through the viewfinder. (I am a glasses wearer, and this works for me, though I usually just keep my glasses on.)

Monitor Button and Eye Sensor

The button with a monitor icon, directly to the right of the viewfinder, is used to switch the view of shooting screens and playback screens between the LCD screen and the viewfinder. Press the button to toggle those views. This button can switch the views only when the LCD screen is unfolded and visible; when it is folded in against the camera in its protective position, the viewfinder is activated and the Monitor button will not change the view.

If you want the camera to switch automatically between using the viewfinder and using the LCD screen, use the EVF Auto Toggle option on the first screen of the Setup menu. When that option is turned on, the view will switch from the LCD screen to the viewfinder when your head (or another object) approaches the eye sensor, the small slot built into the right side of the viewfinder. When that menu option is active, you can still switch the view with the Monitor button or by folding in the LCD screen.

Display Button

The button marked DISP, to the right of the Monitor button, switches among the various displays of information on the camera's LCD or viewfinder, in both shooting and playback modes. In shooting mode, there are three displays available that are called up by successive presses of the Display button.

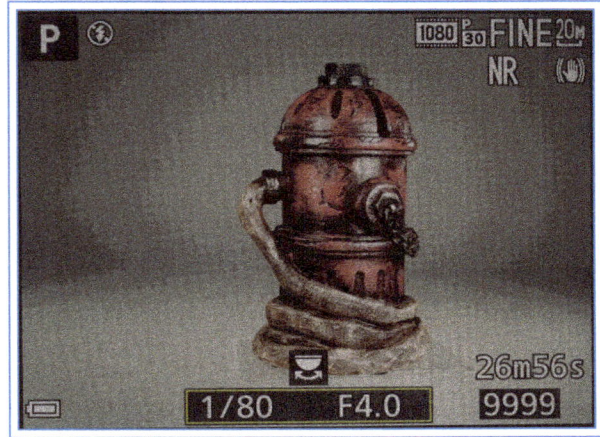

Figure 5-7. Shooting Screen with Basic Information

The display screen that I use the most, seen in Figure 5-7, shows the live view overlaid with icons for shooting mode, flash mode, shutter speed, aperture, image size and quality, images remaining, and a few other items.

Another press of the Display button produces a similar screen, illustrated in Figure 5-8, which includes the same shooting information with a frame overlaid that shows the area of the image that would be used for shooting a movie, based on the current setting for movie format.

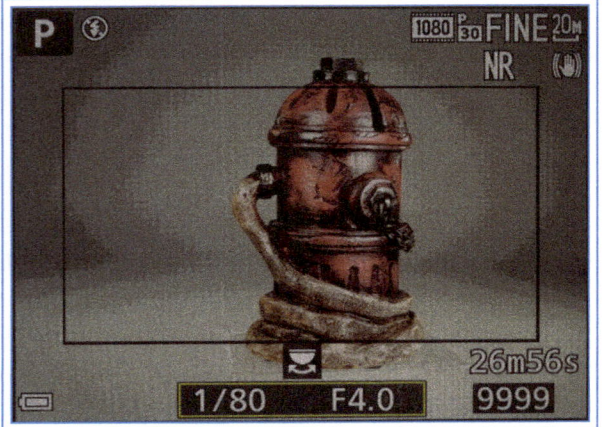

Figure 5-8. Shooting Display Screen with Movie Frame

One more press of the button produces a view of the image alone, with no information.

Through the Monitor Settings item on the Setup menu, as discussed in Chapter 7, you can add a histogram and a framing grid to the shooting information display if you want to. The grid, if activated, will appear on all three shooting screens described above, including the screen with no shooting information. In addition, the camera will add a fourth screen with no information and no grid.

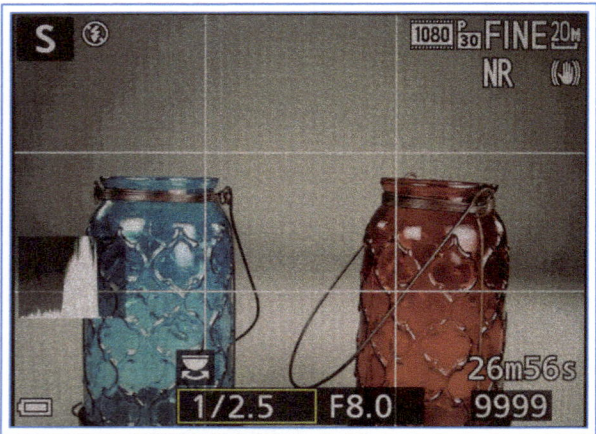

Figure 5-9. Shooting Display with Grid and Histogram

The grid can be useful in framing your composition with a level horizon and for arranging a composition according to the Rule of Thirds, which states a preference for having the subject located at a point one-third of the way from an edge of the frame.

Similarly, the histogram, if activated, will appear on the three screens described above, and the camera will add a fourth screen with no information and no histogram. The histogram is discussed in Chapter 6, in connection with the histogram that displays in playback mode. Figure 5-9 shows the shooting screen with both the framing grid and the histogram in use.

In playback mode, there are four screens available through presses of the Display button. The first option provides photo information, which shows the recorded image overlaid with icons and figures showing the date and time the picture was taken, its identification number, image quality and size, and which image is being shown out of how many total images. The battery status icon also is shown.

The second view is a detailed display of information, including a thumbnail image and a histogram that shows the brightness values in the image along with details including shooting mode, shutter speed, aperture, exposure compensation, ISO value, white balance, and image number. The third view, available for still images only, shows Comment, Artist, and Copyright information fields with any data that has been added to those fields. The fourth and final view is of the image only, with no information added. I will provide information about the playback displays in Chapter 6.

Movie Button

The red Movie button, at the top of the camera's back just below the mode dial, has one major purpose: to start and stop recording of your videos. Press it once and release it to start recording; press it again to stop recording. I will discuss movie recording options in Chapter 8. You also can press this button to cancel standby mode when the monitor turns off in connection with the camera's power-saving function.

Command Dial

This wheel, sticking out near where your right thumb is likely to grip the camera at the top of the right side of the camera, has several functions. It is used to adjust shutter speed in the Shutter Priority and Manual exposure modes. In the Program exposure mode, this dial is used to activate the Flexible Program function, which causes the camera to choose an alternative pair of shutter speed and aperture values. If you use the Toggle Av/Tv Selection option on the Setup menu, these functions of the command dial are switched with those of the multi selector dial, so the command dial controls aperture rather than shutter speed, and the command dial no longer controls the Flexible Program feature. In all shooting modes, this dial can control manual focus.

In playback mode, you can turn the command dial to magnify and shrink an image once you have pressed the zoom lever to begin enlarging it. The dial also can be used to move forward or backward through a movie.

Also, the command dial can be used to adjust values in the on-screen menus after you have pressed one of the direction buttons to bring up a menu on the display. That is, the command dial can be used to adjust exposure compensation after the Right button has been pressed to put that scale on the display; it can be used to select a focus mode after the Down button has been pressed to put the focus mode menu on the display; it can be used to set the mode for the self-timer after the Left button is pressed; and it can be used to select a flash mode after the Up button is pressed. This dial also can be used to select an item from the menu that appears on the screen when the Fn1 or Fn2 button is pressed, or to select a group of effects in the Creative shooting mode, once you have pressed the OK button to activate the selection process.

mode, the icon, which looks like a white half-disk with a curved arrow below it, is positioned above the value for shutter speed. This means you can turn the command dial to adjust that setting.

Menu Button

The Menu button, to the lower left of the multi selector, is straightforward in its function. Press it to enter the menu system, and press it once more to return to whatever mode the camera was in previously (shooting mode or playback mode). There are several different menus available, depending on what mode the camera is in. The main menu systems are for Shooting, Playback, Network, and Setup, but there also are more specific menus for various shooting modes, including Scene, Creative, Night Landscape, Night Portrait, and others, as well as a separate menu for Movie mode. I discuss the menu systems in Chapters 4, 6, 7, 8, and 9.

Delete/Trash Button

To the right of the Menu button is a round button marked with a trash can icon. This is the Delete button, which can also be called the Trash button. This control's primary function is to delete images and videos. When the camera is in playback mode, press the Trash button and the camera will display a short menu of choices: Current Image, Erase Selected Images, or All Images, as shown in Figure 5-11.

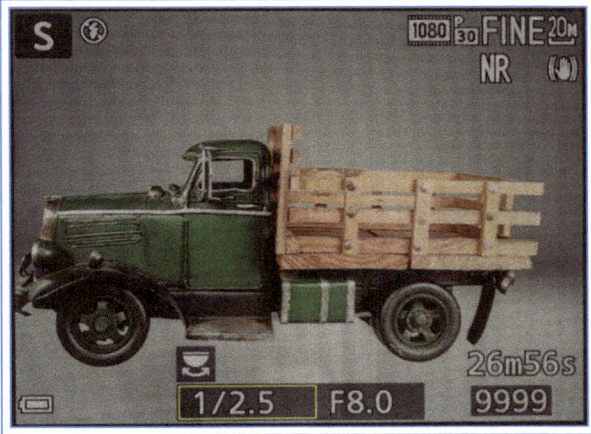

Figure 5-10. Icon Showing Command Dial Controls Shutter Speed

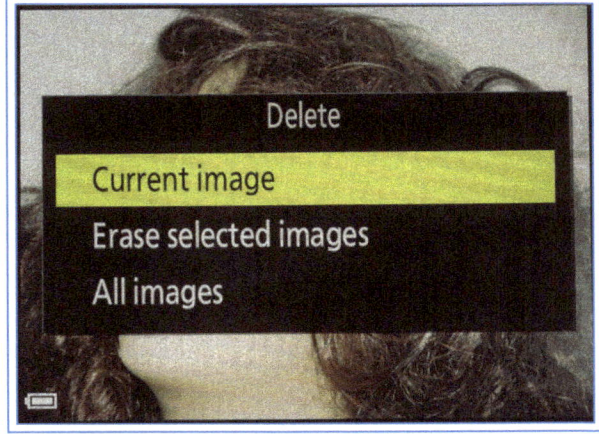

Figure 5-11. Delete/Trash Button Options Screen

In some situations, the Coolpix B700 puts an icon on the screen representing the command dial when there is a value that can be adjusted by turning the dial. For example, as shown in Figure 5-10, in Shutter Priority

Use the multi selector dial or the Up and Down buttons to highlight your choice. If you select Current Image and press the OK button, the camera will display a message asking you to confirm. You can cancel out of this (or any) deletion message by pressing the Menu button. When you are using the Trash button to delete images in this way, if the image displayed is the key image for a sequence of continuous shots (see discussion of continuous shooting in Chapter 4), choosing Current Image will delete all images in the sequence. If the images in the sequence are displayed individually, only one image at a time will be deleted.

If you choose Erase Selected Images and press OK, the camera displays an index screen showing thumbnail versions of recorded images and movies. Navigate through them with the Left and Right buttons or the multi selector dial, and press either the Up or Down button to mark (or unmark) any item to be deleted. Trash can icons will appear below items marked for deletion. You can enlarge any thumbnail to get a better view of an image or video by turning the zoom lever toward the telephoto position; turn it back the other way to reduce the image back to the thumbnail size. When all selected items have been marked, press the OK button and the camera will ask you for one final confirmation before deleting the selected images and videos.

If you choose All Images, the camera will delete all images and movies that are not protected using the Protect function, as discussed in Chapter 6.

When the camera is in shooting mode, if you press this button, the camera will ask if you want to delete one image. If you select Yes, the last image or video to be saved will be deleted and the camera will return to shooting mode. With this option, you will not get a chance to preview the image (or movie) before deletion, so be careful when using it. I recommend switching the camera to playback mode before deleting images.

In addition, the Delete button can be used to reset the settings for an adjustment that has been made to a setting in Creative mode, as discussed in Chapter 3.

Multi Selector and its Buttons and Dial

The most prominent set of controls on the back of the camera is contained within the perimeter of the multi selector, the circular dial with raised edges. Those edges also function as the four direction buttons. Each of these four buttons also has another purpose designated by an icon next to the button. In the center of the multi selector is the OK button, and the round, ridged wheel functions as a rotating dial. I will discuss each of these controls in turn.

Multi Selector Dial

The ridged wheel that surrounds the OK button, known as the multi selector dial, is easy to operate, because you can catch it on your thumbnail or just engage it with the flesh of your thumb or finger and spin it freely. One of its duties is to control the aperture setting when you are shooting in Aperture Priority mode or Manual exposure mode. If you turn on the Toggle Av/Tv Selection option on the Setup menu, then the functions of this dial and the command dial are reversed, so the multi selector dial controls shutter speed for Shutter Priority mode and Manual exposure mode. In addition, if that menu option is activated, the multi selector dial controls the Flexible Program feature, which ordinarily is controlled by the command dial.

When you are using the menu systems, this dial moves up and down the lists of menu options, and it can be used to highlight an item from the menu that appears when the Fn1 or Fn2 button or one of the direction buttons is pressed. In playback mode, the dial navigates through individual images, and it moves through the screens of images when index screens are displayed. You can use it to move forward or backward through a movie. As with the command dial, the camera displays an icon representing the multi selector dial when the dial controls a given function, such as aperture.

OK Button

This button in the center of the multi selector serves as a selection, confirmation, or "set" button when you choose certain options. For example, whenever you highlight a desired menu option, you press the OK button to confirm and set your selection. Similarly, when you press the focus mode button (Down button) and then highlight a focus mode on the pop-up menu (autofocus, macro focus, infinity focus, or manual focus), you press the OK button to confirm that choice. You also can use this button to get access to sub-menus. For example, after you highlight Image Quality on the Shooting menu, you can press the OK button to bring up the sub-menu with the list of choices: Fine, Normal, Raw, Raw+Fine, and Raw+Normal. Then, you can press the OK button again to confirm the actual selection after you highlight it.

In manual focus mode, pressing this button toggles the shooting screen between an enlarged view and the normal-sized display.

In playback mode, when the first frame of a movie is displayed on the screen, the OK button is used to start the movie playing. The button also is used to select any one of the playback controls that appear at the bottom of the screen during movie playback. (You use the direction buttons to highlight one of these controls, such as play, stop, or rewind, and then press OK to choose that function.)

The OK button is also used to "open up" a sequence of continuous shots so you can view them individually. The button can be used to pause and resume video recording, and to start a panorama scrolling across the display screen at a larger size. When you are using the Moon or Bird-watching scene setting, you press the OK button to zoom the lens in for a magnified view, after fixing the subject in the small frame that appears on the display. Pressing this button returns an enlarged image to normal size in playback mode.

Direction Buttons

Each edge—Up, Down, Left, and Right—of the multi selector dial is a "button" you can press to get access to a setting or operation. This may not be immediately obvious, and sometimes it can be tricky to press the dial in exactly the right spot, but these four direction buttons are important to your control of the camera. You use them to navigate through menus and screens for settings, whether moving left and right or up and down.

You also use them in playback mode to move through your images and, when you have enlarged an image using the zoom lever, to scroll around within the magnified image.

Besides these navigational duties, the direction buttons have functions in connection with various settings. When you are navigating in the menu system, you can use the Left button to move back one screen in the system. When you are on the main screen of a given menu system (Shooting, Playback, Scene, etc.), pressing the Left button moves the yellow selection highlight to the left column of the screen, which contains the icons that identify the currently available menus. You can navigate up and down through these icons to select the symbol for the menu you want to use. You can then press the OK button to select that menu.

The Right button also can be used to move to the sub-menu screens within the menu system. In most cases, you can press either the OK button or the Right button to move to the sub-menu screen that contains further options for a given menu item.

The Up button also has a non-obvious extra function. When you are viewing a "sequence" of continuous shots as individual images, as discussed in Chapters 4 and 6, pressing the Up button returns the camera to normal playback mode, in which you view only the "key" image from the continuous set (assuming the Sequence Display Options setting on the Playback menu is set to show key images rather than individual images from sequences).

Finally, each of the four direction buttons has its own separate identity, as indicated by the icon that appears next to each of the buttons, as discussed below.

Up Button: Flash Settings

When the camera is in shooting mode with the flash unit popped up, pressing the Up button displays a small menu showing the options for setting the behavior of the flash unit, as shown in Figure 5-12.

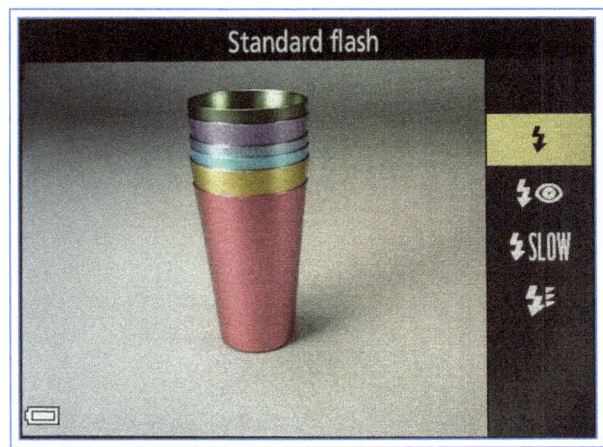

Figure 5-12. Flash Mode Menu

(If the flash unit is not popped up, the camera will display a message telling you to raise the flash.) Depending on the shooting mode, these options may include Auto, Auto with Red-eye Reduction, Fill Flash/ Standard Flash, Slow Sync, and Rear-curtain Sync, or in some cases only three or four of those options. In other cases, such as when you have selected Night Landscape, Landscape, or Night Portrait on the mode dial, pressing the Up button will not bring up any menu, even if the flash unit is popped up. The camera makes all flash decisions for you in those modes.

Once the menu with options has appeared, you need to press the Up and Down buttons or turn the command dial or multi selector dial to highlight your choice, and press the OK button to select it. With this menu, as with all four of the menus that are summoned by the direction buttons, you have to make your selection quickly, because the menu disappears in about five seconds if you don't take some action with a control button or dial. I will discuss the working of these flash options further in Chapter 9.

Right Button: Exposure Compensation

When not acting as the Right button, this control serves as the exposure compensation button. As I discussed in Chapter 2, you press this button to bring up an EV (exposure value) scale on the screen, as shown in Figure 5-13, and then press the Up and Down buttons on the multi selector or turn the command dial or multi selector dial to adjust the value.

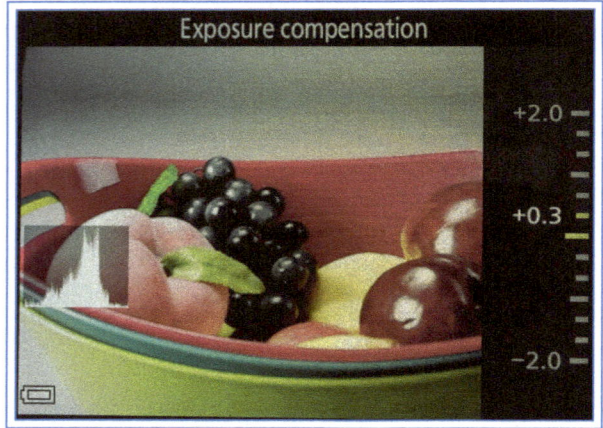

Figure 5-13. Exposure Compensation Scale on Shooting Screen

You do not need to press the OK button to select the value; just let the pop-up menu disappear and the setting will take effect. This adjustment is available in all shooting modes except for Manual exposure mode, the Fireworks and Multiple Exposure Lighten settings of Scene mode, and the Night Sky and Star Trails settings for Time-lapse Movie in Scene mode.

When you are using manual focus, you can press the Right button to force the camera to use its autofocus mechanism on the subject in the center of the screen. You can then adjust the focus further manually by turning the multi selector dial or the command dial.

When you are recording a video, the Right button can be used to lock exposure, as discussed in Chapter 8.

Down Button: Focus Mode

In shooting mode, press this button to bring up the menu of options for the camera's focus mode: autofocus, macro focus, infinity focus, and manual focus, as shown in Figure 5-14.

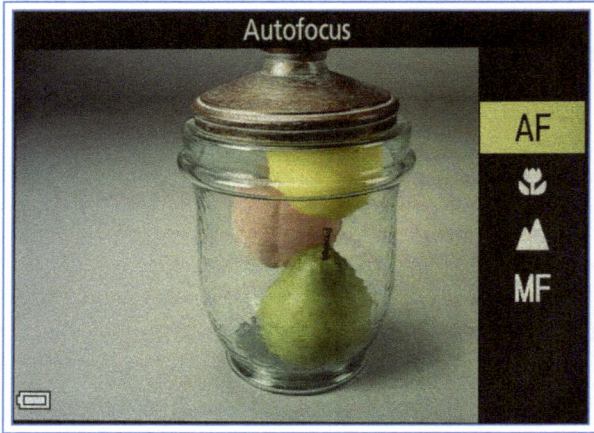

Figure 5-14. Focus Mode Menu

You have limited choices for focus mode in Auto mode, Creative mode, and some of the scene modes. You can make the full range of choices in the Program, Aperture Priority, Shutter Priority, and Manual modes. After you press the Down button, use the Up and Down buttons or the multi selector dial or command dial to navigate to the icon for the desired mode, then press the OK button to confirm. I discussed manual focus in Chapter 2; I'll discuss macro focus (for close-up shots) in Chapter 9.

For most purposes, the normal autofocus setting works well. Use the infinity setting when you want to force the camera to focus in the distance. Note that the flash is disabled when infinity is selected for the focus mode. Note, also, that there is another infinity setting, symbolized by a mathematical infinity sign, that is automatically selected in Scene mode with the following settings: Fireworks Show, the Night Sky and Star Trails settings of Time-lapse Movie, and the Star Trails setting of Multiple Exposure Lighten.

Left Button: Self-timer; Smile Timer; Pet Portrait Release

Press this button in shooting mode and the camera displays the menu of available choices for setting the self-timer, as shown in Figure 5-15.

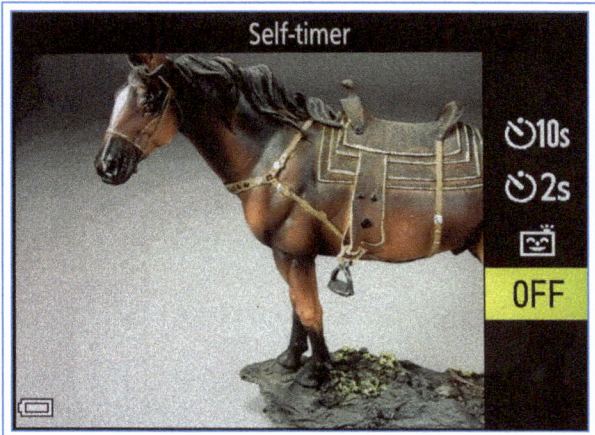

Figure 5-15. Self-timer Options Screen

Those choices may include 10 seconds, two seconds, Smile Timer, and Off, or, in some cases, such as with most scene modes and all Creative mode settings, all of these except the Smile Timer. (The Smile Timer is available with the Portrait and Night Portrait settings of Scene mode, but not with the other Scene mode settings.)

If you set a self-timer delay of either 10 seconds or two seconds, the camera will delay the specified amount of time before taking the picture, after you press the shutter button. Choose 10 seconds if you need a substantial delay so you can get into a group picture after pressing the shutter button; choose two seconds if you just need to avoid touching the camera during the exposure, to minimize the camera shake that can accompany a shutter press. You might need to use the two-second delay when you're taking extreme close-ups, because any camera motion could be magnified by the closeness to the subject. Also, the two-second delay can help when you're shooting in dim light and a slow shutter speed is needed, because any camera motion during the long exposure could blur the image.

When you turn on continuous shooting through the Shooting menu, the self-timer is available to a certain extent, though it is not useful with some of the settings. When you select Continuous H, Continuous L, or Pre-shooting Cache, you can turn on the self-timer and it will operate. However, when it triggers the shutter, only one shot will be taken, despite the continuous-shooting setting. So, in effect, the self-timer cannot be used for this sort of continuous shooting.

However, the self-timer will trigger a full series of continuous shots when used with the super-fast settings of Continuous H: 120 fps and Continuous H: 60 fps. With either of these settings, the camera will take the full set of 60 shots when the shutter is triggered by the self-timer. Of course, as was discussed in Chapter 4, the quality of these images is quite low, particularly for the 120 fps option, whose images are recorded at the very low quality image size of 640 x 480 pixels. The images taken at the 60 fps speed are larger in size, at about two megapixels. So, if you are trying to capture a series of images of your own golf swing, for example, you can set the self-timer and then stand in front of the camera as it captures a rapid series of shots at this reasonable level of quality. The self-timer functions properly with the interval timer setting as well.

You also can use the self-timer normally with movie recording, so you can turn on a delay with the self-timer and then press the red Movie button to start a movie recording after the specified delay.

When the camera is set to the Auto, Program, Aperture Priority, Shutter Priority, or Manual exposure mode, or to the Portrait or Night Portrait Scene mode setting, the Smile Timer is added to the options on the self-timer menu.

The Smile Timer is a special feature that fires the shutter automatically when the camera detects a smile. This function works together with the camera's face detection system, which is automatically turned on when the Smile Timer is selected. In addition, the focus mode is automatically set to autofocus when the Smile Timer is turned on.

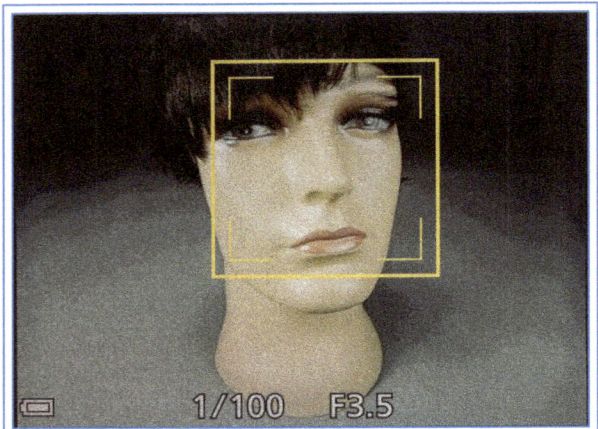

Figure 5-16. Shooting Screen with Smile Timer in Use

Once the Smile Timer is turned on, the camera places the Smile Timer icon in the upper left corner of the screen. Then, whenever the camera detects one or more faces, it focuses on the face closest to the center of the image and places a yellow double border around that face, as shown in Figure 5-16. The self-timer lamp starts blinking slowly to indicate that a face has been detected.

The Smile Timer icon is no longer displayed once a face has been detected. The shutter is triggered automatically if the face inside the double border smiles. Once that happens, the lamp will blink rapidly to indicate that a picture of a smile has been taken.

This feature is more of a novelty than a particularly useful function in my opinion, though it may be of use in some situations, such as when you need to encourage a child to smile by telling him or her that a smile will automatically trigger the camera. Also, the Smile Timer acts as a sort of remote control for the camera; each time the subject smiles, the camera is triggered again. So, if you are taking self-portraits, you can stand in front of your B700 on its tripod, and control the camera's operation with your smile as it takes repeated portraits.

In the Pet Portrait setting in Scene mode, the Left button has a special function of turning on or off the Pet Portrait Auto Release function. As was discussed in Chapter 3, when that feature is turned on, the shutter is automatically triggered when the camera detects the face of a cat or dog. In that situation, neither the self-timer nor the Smile Timer is available.

When the camera is using manual focus, pressing the Left button switches among views that are of normal size or enlarged two times or four times. Press the OK button to exit the manual focusing screen; press it again to recall the focusing screen to continue focusing manually.

When you are recording a video with the Autofocus Mode option on the Movie menu set to AF-S, and autofocus mode in effect, you can press the Left button to cause the camera to refocus on the subject.

Finally, there are a few other controls that are located in their own particular areas, described below.

AF Assist/Red-eye Reduction/Self-timer Lamp

The small lamp on the front of the camera, shown in Figure 5-17, has multiple functions. Its reddish light blinks to signal the operation of the self-timer and smile detection when the Smile Timer is used, and it also turns on in dark environments to assist with autofocusing. In addition, when the flash mode is set to Auto with Red-eye Reduction, this lamp lights up for a second or two before the flash fires, in an effort to cause a human subject's pupils to narrow. In that way, there should be less chance of the flash's light bouncing off of the person's retinas to cause the unsightly "red-eye" effect.

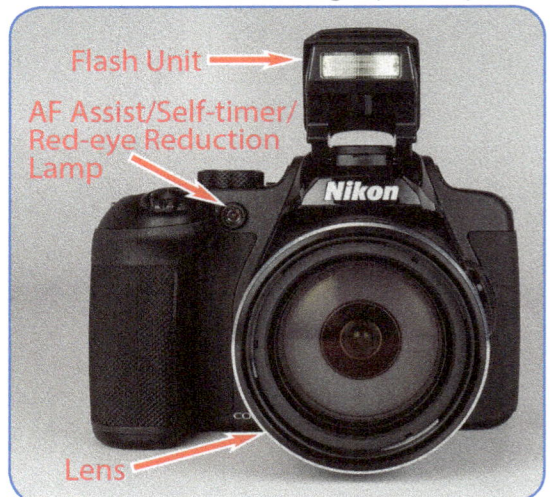

Figure 5-17. Items on Front of Camera

You can control the use of the lamp for autofocusing with the AF Assist item on screen 2 of the Setup menu, as discussed in Chapter 7. You might want to disable it when taking pictures during a religious ceremony or in another environment where this rather bright light could be distracting. Even if you disable it for purposes of autofocus, though, the lamp will still light up to indicate the functioning of the self-timer and Smile Timer, and it will still function as the Red-eye

Reduction lamp if the flash mode is set to use the lamp for that purpose.

Ports on Right Side of Camera

The USB port and HDMI port are small openings under a flap on the right side of the camera, shown in Figure 5-18. These ports have several functions. The larger one at the top, the micro-USB port, is where you plug in the charging cable when you charge the battery inside the camera. It also is where you connect the camera to a computer to transfer your photos and movies using the supplied USB cable, and you can use this port to connect the camera to a printer to print images directly from the B700, as discussed in Chapter 6.

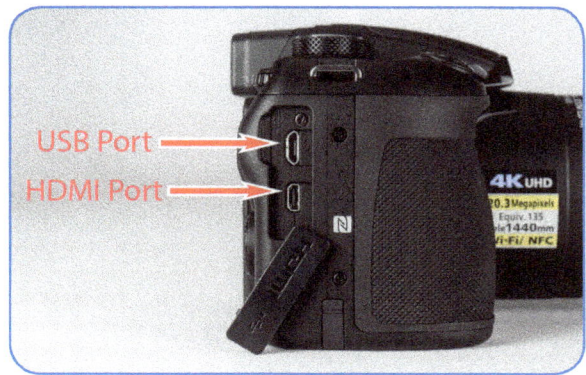

Figure 5-18. Ports on Right Side of Camera

The smaller, lower port is where you plug in an HDMI cable (which you need to purchase as a separate option) to view photos or videos on an HDTV set. The end going into this port is a micro-HDMI connector; the end going to the HDTV needs to be a standard HDMI connector. These cables are available through various sellers, including Amazon.com. I discuss connecting the B700 to TV sets in Chapter 9.

The NFC active area, below and to the right of the two ports and marked with a fancy letter N, is the spot where you can touch an Android smartphone or tablet that uses the near field communication protocol, to initiate a wireless connection between the camera and that device. When you touch the camera and device together, the connection should be made automatically, without having to use a menu option. I discuss this process in Chapter 9.

Tilting and Swiveling LCD Screen

The last item to be discussed in this chapter is not really a "control," but it does allow physical adjustment, so I will discuss it here. This is the articulated LCD display on the back of the camera. This screen, even without its tilting and swiveling ability, is a notable feature of the camera. It has a diagonal span of three inches (7.5 cm) and provides a resolution of 921,000 dots, giving a clear view of your images before and after you capture them.

With its ability to pivot both horizontally and vertically, the monitor gives you a considerable amount of flexibility for your shooting.

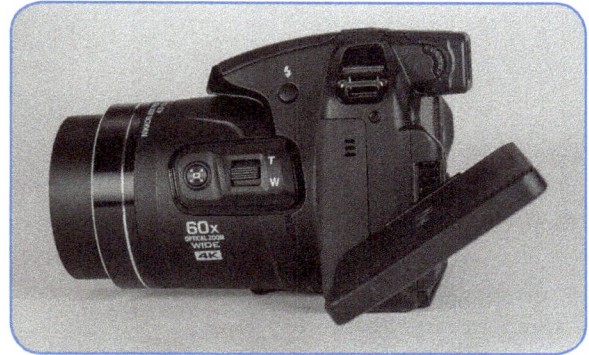

Figure 5-19. LCD Screen Tilted Downward for Overhead Shots

If, as shown in Figure 5-19, you pull it out to the left and then rotate the screen so it aims downward, you can hold the camera high above your head and view the scene as if you were an arm's-length taller, or were standing on a small ladder.

If you attach the camera to a monopod or other support and hold it up in the air, you can extend the camera's vertical height even farther and still view the LCD screen quite well. You can activate the self-timer before raising the camera up in the air to take the photo, or you can turn on interval timer shooting with the Continuous option on the shooting menu, and set the camera to take a series of shots at 30-second intervals while it is raised up in the air. You also can use the camera's Wi-Fi capability, discussed in Chapter 9, to trigger the camera by remote control from a smartphone or tablet while the camera is raised overhead.

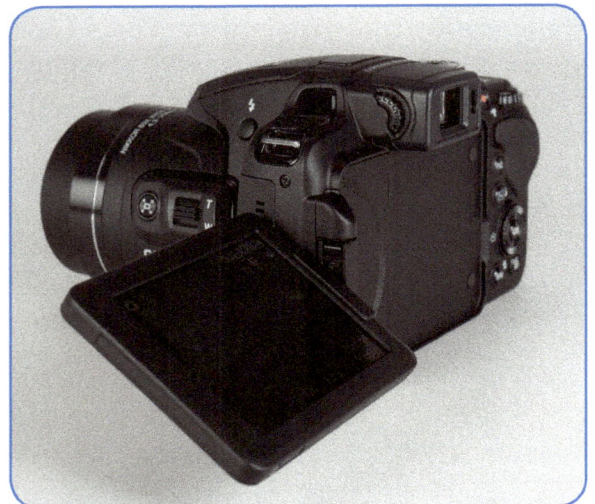

Figure 5-20. LCD Screen Tilted Upward for Low-angle Shots

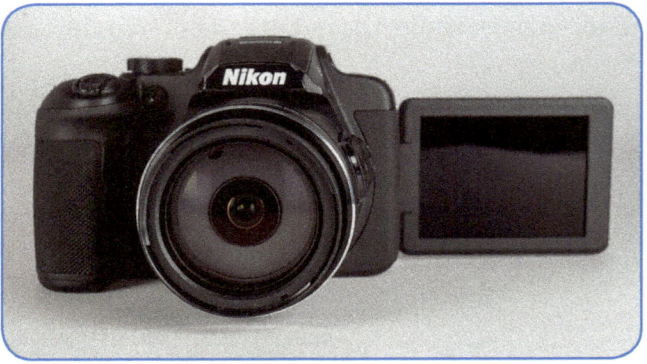

Figure 5-21. LCD Screen Facing Forward for Self-portraits

On the other hand, if you need to take images from a very low vantage point near ground level, you can rotate the screen so it tilts upward toward your eye, as shown in Figure 5-20, and hold the camera down as far as you need, to get a mole's-eye view of the world. Also, as I discuss further in Chapter 9, this angle is useful for street photography, because you can look down at the camera while taking long-zoom photos of people on the streets without drawing attention to yourself.

If you fold the screen out so that its viewing area faces in the same direction as the lens, as shown in Figure 5-21, you can take a self-portrait while observing your image on the display.

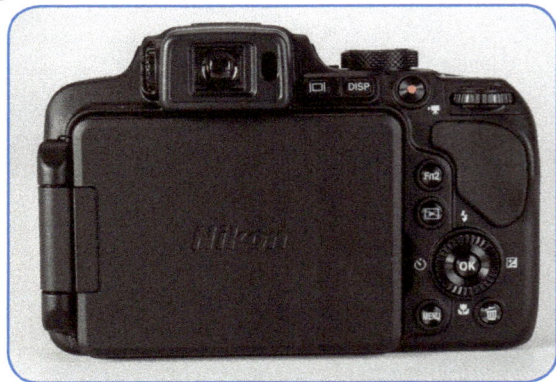

Figure 5-22. LCD Screen Folded in for Protection

Finally, the screen can be folded in so its viewing surface is hidden. In this configuration, shown in Figure 5-22, the LCD display is protected against damage, and the camera automatically switches to using the viewfinder.

Chapter 6: Playback

In this chapter, I will discuss the various playback functions of the Nikon Coolpix B700. I'll also discuss options for printing images.

Normal Playback

When you take a new photo, the recorded image stays on the screen for about one second for review. If your major concern with viewing images in the camera is to check them right after they are taken, this feature is helpful, but the review time is very brief and there is no way to adjust its duration. You can turn this feature off using the Monitor Settings/Image Review item on screen 1 of the Setup menu, as discussed in Chapter 7. If you want to view your images in more detail, you need to use the features that are available in playback mode.

For ordinary image review in playback mode, press the Playback button, marked by a right-facing triangle, to the right of the LCD screen on the camera's back. Once you press that button, the camera is in playback mode and you will see the most recent image or video saved to the memory card that is in the camera. To move back through older images, press either the Left button or the Up button or turn the multi selector dial (the dial that surrounds the OK button on the camera's back) to the left. To move through the increasingly more recent images, use the Right button or the Down button, or turn the multi selector dial to the right. To scroll through your images rapidly, hold down the Left or Right (or Up or Down) button. To delete images, press the Trash button and follow the prompts on the screen, as discussed in Chapter 5.

Index Views, Calendar View, and Enlarging Images

In playback mode, you can press the zoom lever on top of the camera to view an index screen of your images or to enlarge a single image. When you are viewing an image, press the zoom lever once to the left (toward the W setting), and you will see a screen showing four images, one of which is outlined by a yellow frame, as shown in Figure 6-1.

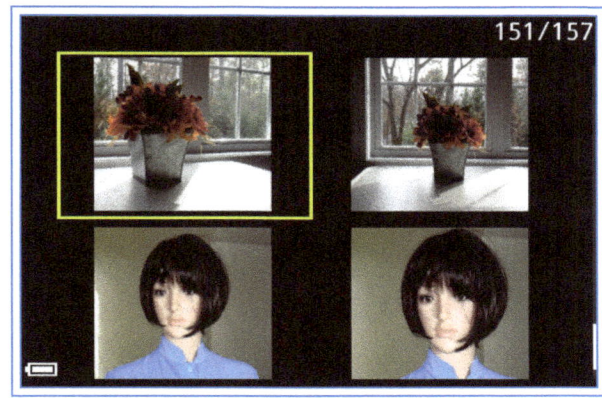

Figure 6-1. Index Screen with Four Images

You can then press the OK button to bring up the outlined image as the single image on the screen, or you can move through your images with the four-image index screen by pressing the four direction buttons or by turning the multi selector dial.

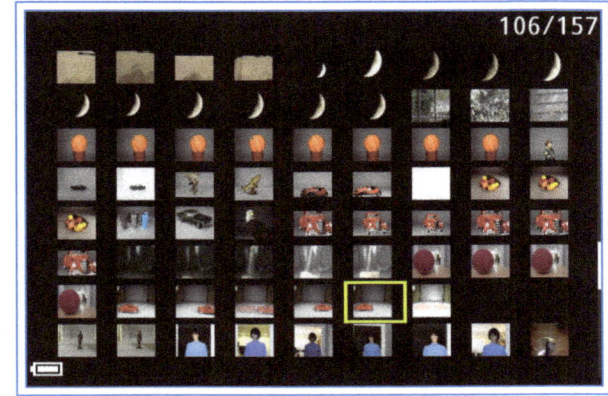

Figure 6-2. Index Screen with 72 Images

If you move the zoom lever to the W mark once more, the camera will show an index screen of nine images; another press brings 16 images; and another press brings a 72-image screen, as seen in Figure 6-2 (assuming in each case that you have that many images; if not, there will be blank spaces on the screen).

If you press the lever in the same direction one more time, the camera displays a calendar screen with a thumbnail image on each date for which images exist, as shown in Figure 6-3.

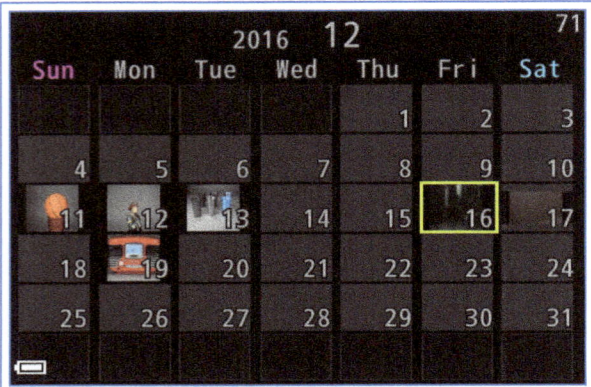

Figure 6-3. Calendar Screen

You can maneuver through any of the index screens to select a single image for viewing. If you want to reduce the number of images per screen, press the zoom lever to the right (toward the T position) repeatedly to reverse the progression of index screens. On the calendar screen, highlight a date and press the OK button to display images from that date.

When you are viewing a single image, a press of the zoom lever to the right enlarges that image, as seen in Figure 6-4.

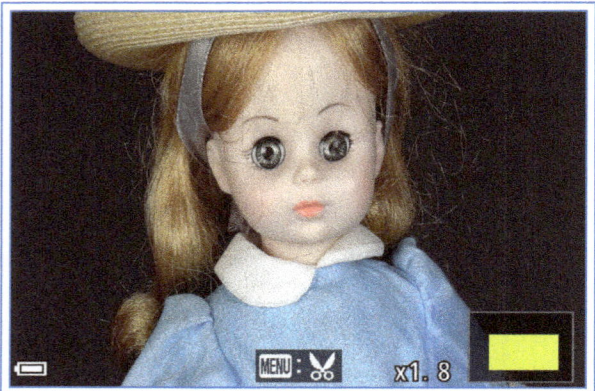

Figure 6-4. Enlarged View of Image

You will see a display in the lower right corner with an inset yellow block that represents the portion of the image that is now filling the screen in enlarged view. The display with the inset block appears only if you are viewing the screen with basic information or the screen with copyright information. With the screen that includes no information and the screen with detailed information with a histogram, the image will be enlarged, but without the inset block.

If you press the zoom lever to the right repeatedly, the image will be enlarged up to a maximum of about 10 times normal. For example, Figure 6-5 shows an image enlarged to a higher level than in Figure 6-4.

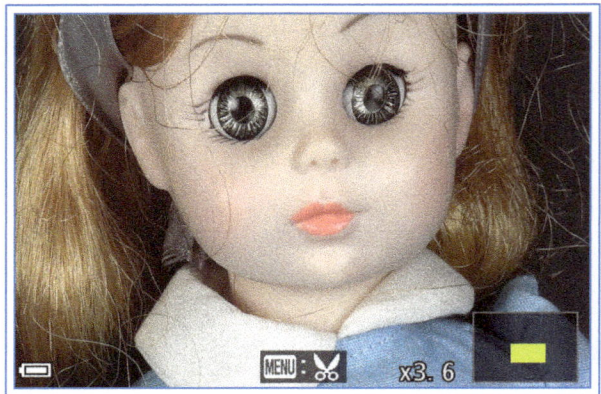

Figure 6-5. Image Enlarged to Higher Level

While the image is magnified, you can scroll around within it using the four direction buttons; you will see the inset yellow block move around within the white rectangle that represents the whole image. To reduce the image size again, press the zoom lever to the left as many times as necessary. You can also increase or decrease the zoom level by turning the command dial right or left. Press the OK button to restore the image to its original size immediately.

While the image is enlarged, the word MENU appears on the screen with a scissors icon, as shown in Figures 6-4 and 6-5. When you see that display, you can press the Menu button to save the enlarged area as a separate file. This feature gives you a rough-and-ready way to edit your images in the camera. So, if you want to crop a group photo to save just the face of one person, you can enlarge the image and scroll it around using the direction buttons until just that face is visible. Then press the Menu button to save a separate file with that face as the only subject. This process is no match for editing with a computer, but it could come in handy when no computer is available and you need a particular part of an image for a special purpose, such as a business presentation. This function is not available with panoramas.

Various Playback Screens

When you are viewing a still image in playback mode, pressing the Display button repeatedly cycles through four screens.

These are the full image with no added information except movie format for movies (not shown here); full image with basic information, including date and time it was taken, file name, image number, image size and quality (Figure 6-6); a reduced-size image with detailed recording information, including aperture, shutter speed, ISO, recording mode, exposure compensation, and other data, plus a histogram (Figure 6-7); and the image with an overlaid block with information slots for Comment, Artist, and Copyright, which you can enter using the Image Comment and Copyright Information options on the Setup menu (Figure 6-8).

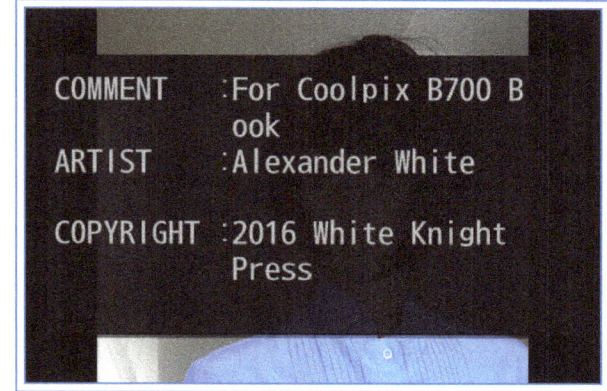

Figure 6-8. Playback Screen with Comment and Copyright Information

These four screens are available for still images. For movies there are only three screens; the screen with the histogram is not available. For a movie file, the camera displays the screen with blanks for copyright information, but does not display any information you have entered for those blanks, for some reason.

The detailed display screen, as noted above, includes a histogram. The histogram is a graph, or chart, representing the distribution of dark and bright areas in the image that is being displayed on the screen. The darkest blacks are represented by vertical bars on the left, and the brightest whites by vertical bars on the right, with continuous gradations in between.

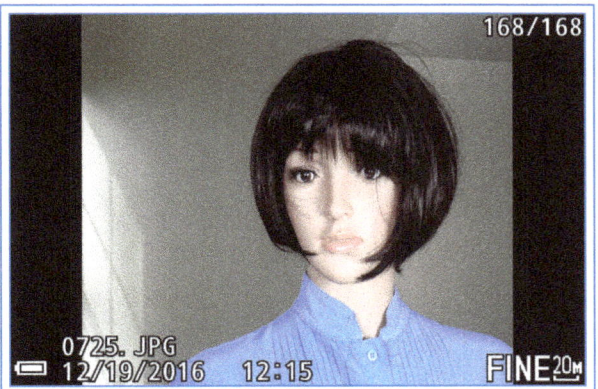

Figure 6-6. Playback Screen with Basic Information

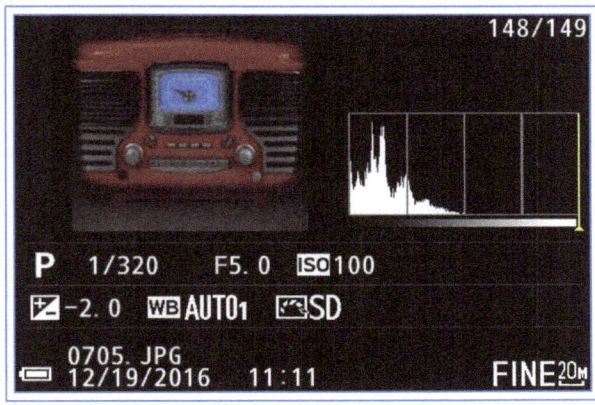

Figure 6-9. Histogram for Underexposed Image

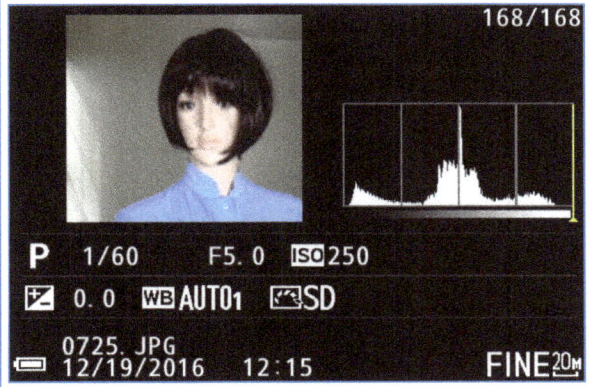

Figure 6-7. Playback Screen with Details and Histogram

If you have a histogram in which the pattern looks like a tall ski slope coming from the left of the screen down to ground level in the middle of the screen, that means there is an excessive amount of black and dark areas (high points on the left side of the histogram), and very few bright and white areas (no high points on the right), as shown in Figure 6-9.

A ski slope moving from the middle of the screen up to the top of the right side of the screen would mean just the opposite—too many bright and white areas, as illustrated in Figure 6-10.

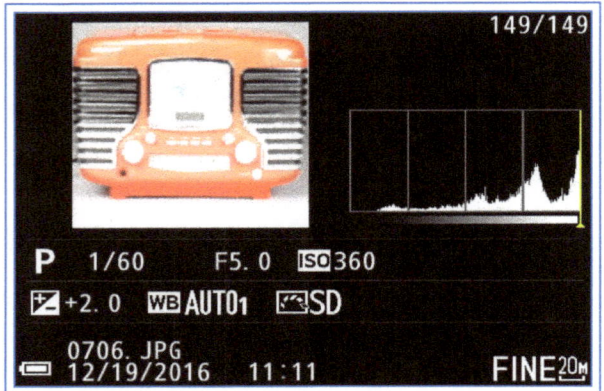

Figure 6-10. Histogram for Overexposed Image

A histogram that is "just right" would be one that starts low on the left, gradually rises to a medium peak in the middle of the screen, then moves gradually back down to ground level at the right. That pattern indicates a good balance of whites, blacks, and medium tones. Figure 6-11 shows a histogram of that sort.

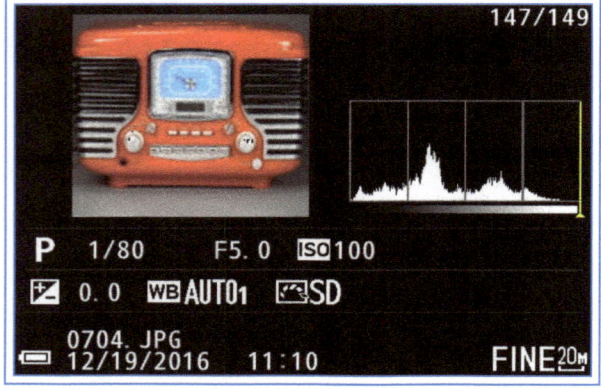

Figure 6-11. Histogram for Normally Exposed Image

The histogram is an approximation, and should not be relied on too heavily. It is useful to give some feedback on how evenly exposed your image is.

When the histogram is displayed in playback mode, you can see information about the tone levels in the image. To do this, look closely at the histogram, to the right of the thumbnail image, and you will see a vertical yellow line at the right edge of the histogram. If you press the Left and Right buttons or turn the multi selector dial, you can move that yellow line across the histogram. As you do this, you will see different parts of the image flash. What happens with this screen is that the camera gives you a way to select which parts of the tone map to check. As you move the yellow line over the chart, corresponding parts of the image will flash to show where in the image there are tones at that level. For example, as you start out with the line at the extreme right of the histogram, the whitest areas of the image will flash, and, as you move the line to the left, progressively darker areas of the image will flash. Using this tool, you may be able to gather some information about what parts of the image are too dark or too bright, so you can take another image that avoids having details swallowed up by blown highlights or excessively dark shadows.

When the histogram screen is displayed, you cannot use the Right and Left buttons to move through your images, because those buttons move through the tone level information. You can use the Up and Down buttons to navigate through the images in that situation.

Viewing Shots Taken in a Sequence

When you take photos with the Coolpix B700 in certain shooting modes or with certain functions, the images become part of what Nikon calls a "sequence." When you enter playback mode to view those images, you ordinarily will see only the "key" image of the sequence, usually the first one of the group that was taken. To see the rest of the images in the sequence, you have to take other steps. I'll discuss this process in some detail, because it can be a bit confusing at first.

Here is an example to illustrate the way the B700 handles sequences of still photos. Suppose you have placed the camera in Program mode by turning the mode dial to P, and then selected continuous high-speed shooting by selecting Continuous H from the Continuous item on the Shooting menu. Say you have selected Fine for the image quality and the maximum image size, 5184 x 3888 pixels, from the Shooting menu. Now, when you aim the camera at your subject and hold down the shutter button for a second or two, you will hear the sounds of the camera operating. The LCD screen (or viewfinder) will display the last captured image for several seconds and then return to the view of the live image.

When the camera settles back to the live view, you can press the Playback button to start viewing your images. If everything worked as expected, there will be as many as five new images to view, because that is the longest burst the camera can take using the Continuous H setting. However, when you press the Playback button, you will see only one image from this sequence. If you press any of the direction buttons or turn the multi selector dial, you will move to an entirely different image, assuming one exists; you will not see the other images from this sequence.

Look at the display on the screen, as shown in Figure 6-12, which has a few unusual aspects. (If you see a display with no information, press the Display button to show the screen with basic information.)

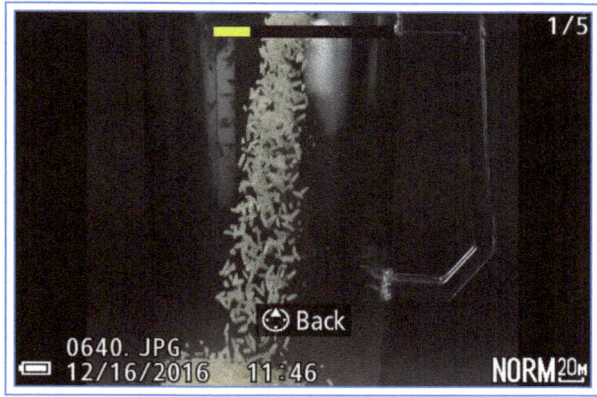

Figure 6-13. Group of Continuous Shots Viewed Individually

You will no longer see the OK notation; instead, you will see a yellow bar at the top center of the display, and the number 1 with a slash over the total number of images at the upper right. In Figure 6-13, the numbers are 1/5, because the sequence includes five images. The yellow bar will progress across the top of the screen as you navigate through the images in the sequence using the Left and Right (but not Up and Down) buttons or the multi selector dial, and the numbers will increase up to 5/5 as you move to the most recent images in the sequence.

You can magnify each individual image by moving the zoom lever toward the T position, but (naturally enough) you cannot call up an index screen by pressing the zoom lever in the other direction, because only the images in the single sequence are available for viewing at this point.

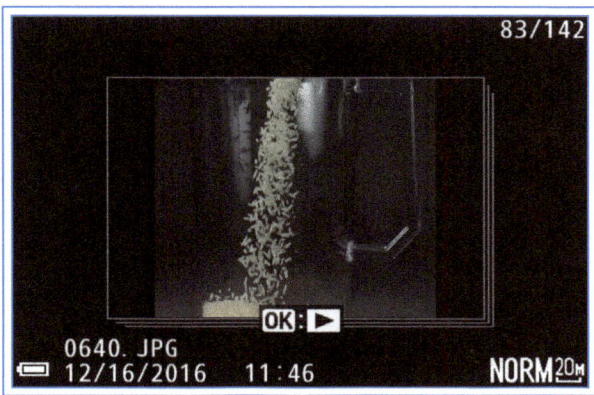

Figure 6-12. Initial Playback Display for Continuous Shots

For one thing, if you press the Display button, the basic image information will appear or disappear, but you cannot produce the detailed display with the histogram, because you are viewing the "key" image of a sequence rather than an individual picture. For another thing, you will see the notation OK at the bottom of the screen with a triangle, indicating the Play function, to its right. What this means is that, to view the other images in this sequence, you need to press the OK button.

If you now press the OK button, you will see the same image as before, but with a different appearance, as shown in Figure 6-13.

Now, pressing the Display button will cycle through the four possible views of each image, including the histogram and copyright information views, because you are viewing this photo as an individual image, not as a key image.

Once you have "entered" the sequence by pressing OK, you will be "stuck" inside it—you can keep navigating through these images, but you will continue to navigate through the same set of images, over and over, until you exit from the sequence and go back to viewing the key image. If you are viewing the display screen with basic information, there is a prompt on the screen that tells you how to do this: an icon highlighting the Up button with the notation Back, meaning you have to press the Up button to go back to the key-image view. That is, when you want to stop viewing these individual images and return to the key image so you can navigate through the rest of the images on your memory card, you have to press the Up button—the one marked with a lightning bolt, which controls flash functions when the camera is in shooting mode rather than playback mode.

If you would rather not have the camera display your continuous-mode shots in sequences, but would prefer to have them displayed as individual shots at all times, you can select that option using the Sequence Display Options item on the Playback menu, as discussed later in this chapter. However, if you take many sequences using a feature such as Continuous H: 120 fps, which takes 60 shots at a time, you may appreciate the ability to display just the key frame from the sequence when you browse through your images in playback mode.

The Playback Menu

Now it's time to discuss the options that are available on the two screens of the Playback menu. As you recall, to get access to this menu, you must put the camera into playback mode by pressing the Playback button (right-facing triangle). Then press the Menu button and, if necessary, move the yellow block on the screen to the left column and navigate to the triangle icon to select the Playback menu, as shown in Figure 6-14.

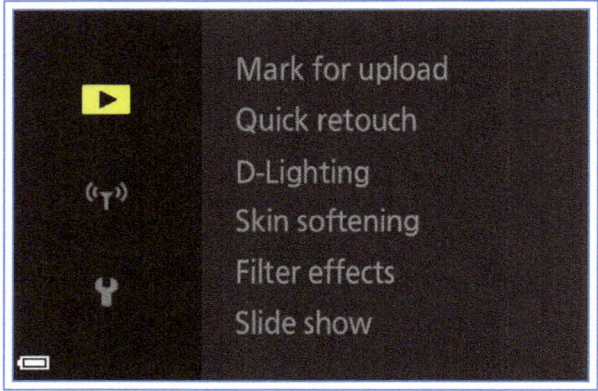

Figure 6-14. Icon for Playback Menu Highlighted at Left

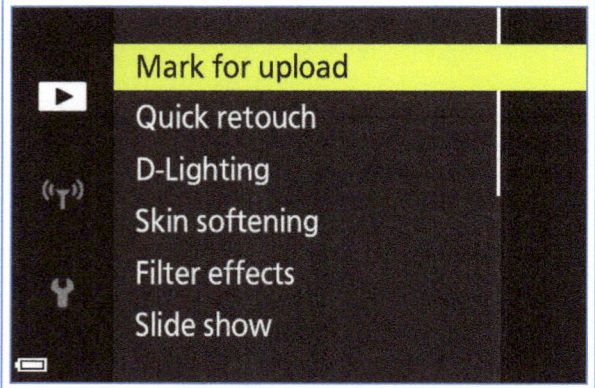

Figure 6-15. Screen 1 of Playback Menu

Then move the yellow block back to the right to highlight the various entries in the menu, whose first screen is shown in Figure 6-15. I'll discuss the options on the Playback menu one by one.

Mark for Upload

This first Playback menu option lets you mark images for later upload to a smartphone or tablet using the B700's built-in wireless capability using the SnapBridge app. To do this, highlight this option and press the OK or Right button to move to the next screen. The camera will display a screen with six thumbnail images. Scroll through those images with the Left and Right buttons or the multi selector dial. When you have highlighted an image you want to upload, press the Up or Down button to mark it; the camera will place a jagged, two-headed arrow below the thumbnail to indicate that that image is now marked for upload, as shown in Figure 6-16.

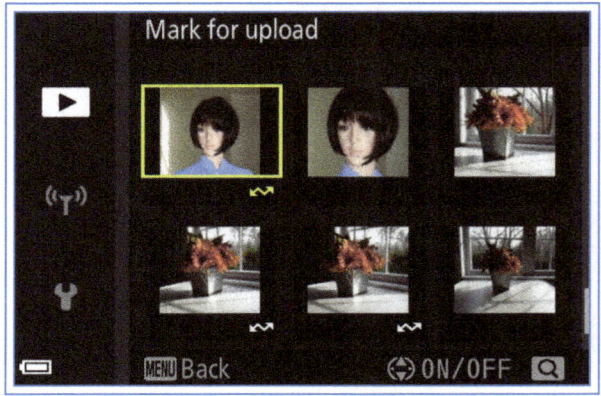

Figure 6-16. Images Marked for Upload

If you change your mind about an image, press the Up or Down button again, and the image will be unmarked. When you have finished, press the OK button to confirm your selections. Only still JPEG images, not videos or Raw images, can be marked with this option. In Chapter 9, I will discuss the procedure for uploading the marked images, along with other wireless operations.

Quick Retouch

The Quick Retouch option gives you a way to add "punch" to recorded images with in-camera processing. You can apply this enhancement to any individual JPEG image, but not to a Raw image or a panorama. If the image you want to enhance is displayed as part of a sequence, you first have to use the technique described earlier (pressing the OK button) to display the images individually. When the image you have selected is displayed, press the Menu button and choose Quick Retouch from the menu. You can then use the Up and

Down buttons or the multi selector dial to choose the desired amount of alteration—Low, Normal, or High, as shown in Figure 6-17.

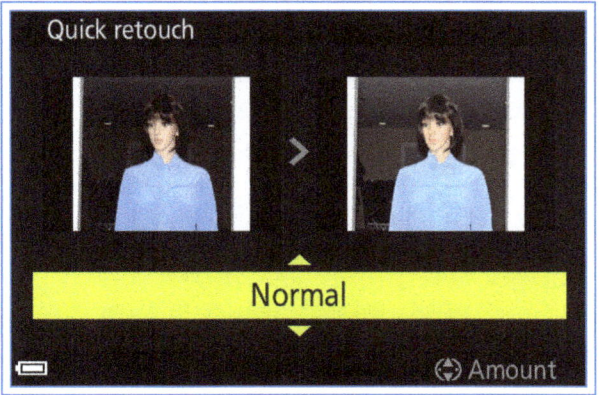

Figure 6-17. Quick Retouch Menu Options Screen

As you change the amount, you will see a preview of the finished image in a thumbnail on the right side of the screen, with the original on the left, for comparison. When you have selected the amount of change, press the OK button to confirm, and a new image will be saved with the retouched appearance and a new file number. It will have the Quick Retouch icon in the upper middle area of the image, as shown in Figure 6-18.

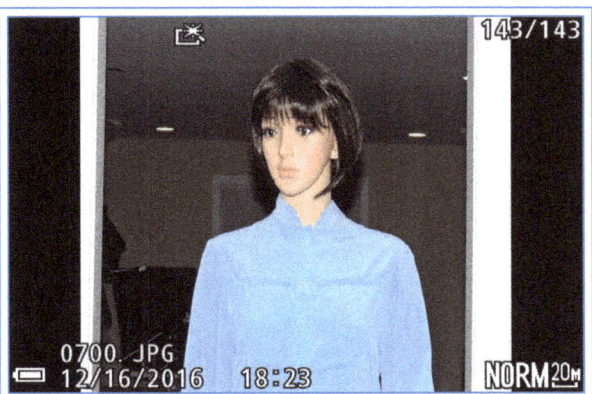

Figure 6-18. Image with Quick Retouch Applied

You cannot make any choices other than the level of the retouching. When it applies this processing, the camera alters the image's contrast (amount of difference between light and dark areas) and saturation (intensity of the colors).

This is a feature I don't use often, because I prefer to do my processing with software such as Photoshop. But there could be times when you take images at a party to display on a TV set during the party. You could use this function to brighten up some muddy images and make them livelier for the audience.

D-Lighting

The D-Lighting option works in the same way as the Quick Retouch feature. Select a still image that is being displayed individually (not as the key frame of a sequence), press the Menu button, select D-Lighting, press the OK button or the Right button to move to the next screen, and then choose Low, Normal, or High for the degree of enhancement, as shown in Figure 6-19.

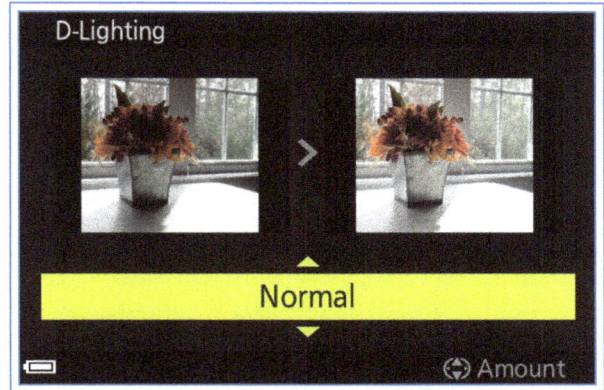

Figure 6-19. D-Lighting Menu Options Screen

In this case, the camera will attempt to add details in both the shadow and highlight areas, as it does when you use the Active D-Lighting option in shooting mode. (I discussed that option in Chapter 4.) This feature can be quite useful, because you can use it to recover details from an image that was taken in conditions with excessive contrast, or even one that was taken with inadequate lighting, at least to some extent.

Skin Softening

This next entry on the Playback menu gives you another way to modify your already-recorded images. In this case, you can add a softening effect to the areas in an image that the camera considers to be showing human faces. As with the previous two menu options, you select the image, press the Menu button, and then select how strong the effect should be. One difference with this feature from the other ones is that the camera will decide whether or not there are any faces in the image you have selected.

If the B700 does not detect any faces, it will display an error message saying the image cannot be modified and return you to the menu without doing any processing.

If it does detect a face, it will take you to a screen where you select the amount of processing. Once you select the amount and press OK, the camera will show you a larger view of the image with a preview of the effect, as shown in Figure 6-20; you can then press OK to save the processed image, or press Menu to go back and revise your setting.

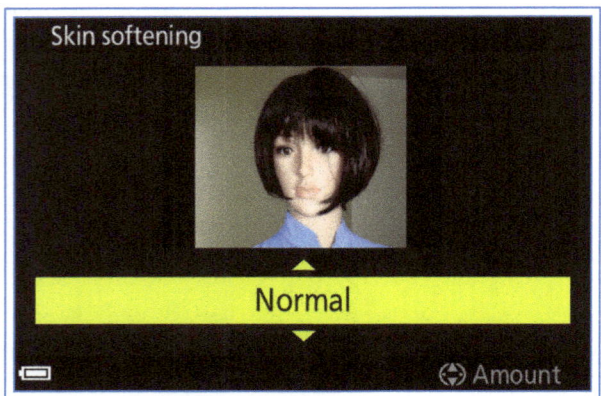

Figure 6-20. Skin Softening Preview Screen

Filter Effects

The Filter Effects menu selection has nine sub-options, the first six of which are shown in Figure 6-21. You can use these to make copies of your recorded images with altered aspects. Some of these are similar to settings in the Scene and Creative shooting modes, but some are available only through this menu option.

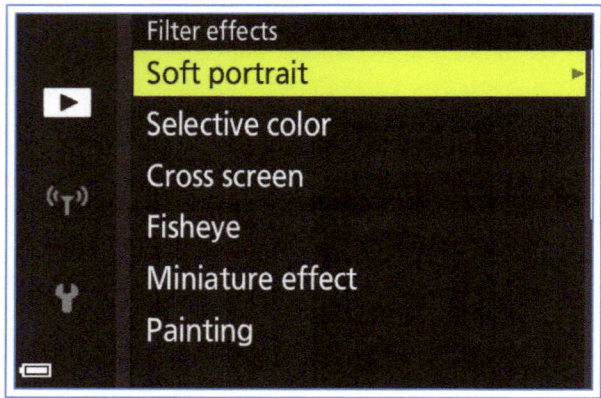

Figure 6-21. First Screen of Filter Effects Menu Options

The first entry on the list of effects is called Soft Portrait, which I used for Figure 6-22. With this option, the camera softens the focus of the image, leaving the center of the image, or a human face if one was detected, in sharp focus, and blurring the focus toward the edges of the image. As with the Skin Softening effect, after you have selected the effect, the camera displays a screen with a preview of the image with the effect applied; you then can press the OK button to apply the effect.

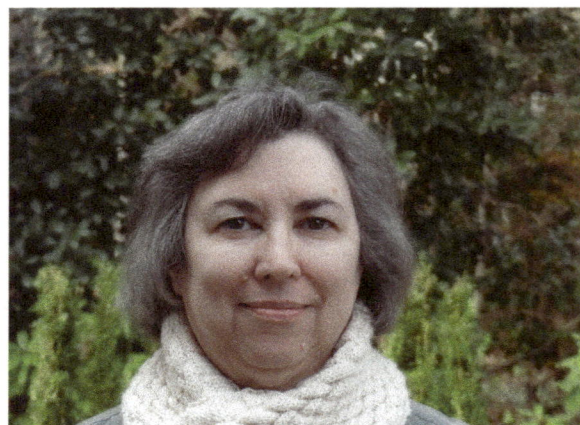

Figure 6-22. Soft Portrait Example

Second on the list is the Selective Color effect, similar to the Scene mode setting of that name, discussed in Chapter 3. As you can see in Figure 6-23, when you display an image on the screen, the camera places a vertical spectrum of colors to the right of the image, with a pointer that you can move up and down the scale using the multi selector dial or the Up and Down buttons.

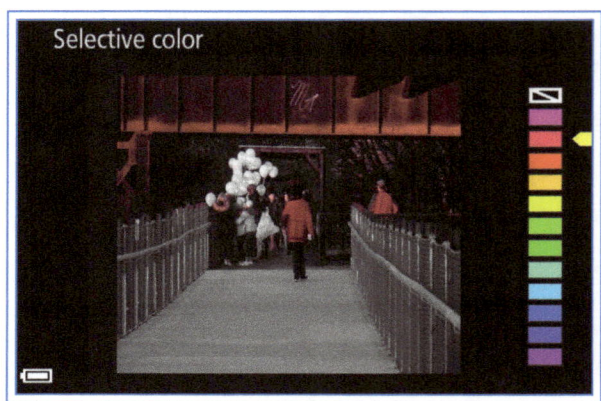

Figure 6-23. Selective Color Options Screen

As you move the pointer next to a color on the scale, the image changes to preserve only the portions that are approximately that color. If no parts of the image are that color, the image turns completely black-and-white. If you use this effect carefully, you can take an image with one area of bright color, remove all other colors using this effect, and end up with a photo that dramatically highlights the single colored object or area that remains, surrounded by a monochrome environment, as shown in Figure 6-24.

Chapter 6: Playback | 103

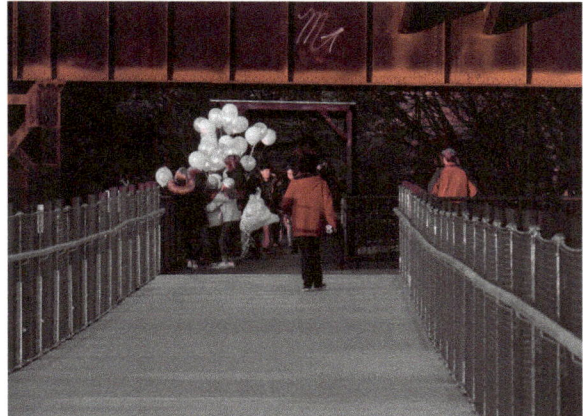

Figure 6-24. Selective Color Example

As with earlier effects, the camera presents you with a large preview screen before you decide whether to apply the effect.

The third choice on this menu is Cross Screen, illustrated in Figure 6-25.

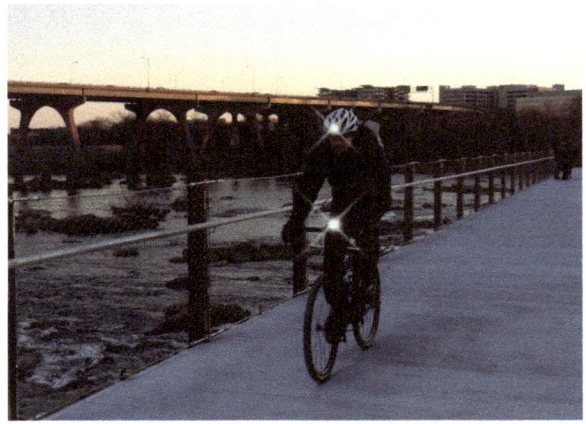

Figure 6-25. Cross Screen Example

With this option, as with Soft Portrait, there are no adjustments to make; you either save a copy of the image using this processing feature, or you cancel out of the selection. If you choose to go ahead, the camera makes a copy of your selected image with streaks radiating outward from bright objects such as the bicycle lights in this image. If there are no objects of that nature in the image, this option will not produce any changes at all, though the camera will still produce the new image. Figure 6-25 shows the final product, after the camera saved a new copy.

Next on the list is the Fisheye effect, which simulates photographs taken with a fisheye lens—a super wide-angle lens that distorts the image, making it look spherical as if seen through a fishbowl. Figure 6-26 is an example of this effect, showing the final product after the camera produced its new image. In this case, the effect distorted the view of a straight bridge over the river, transforming it with a wavy look like that of a fun house mirror.

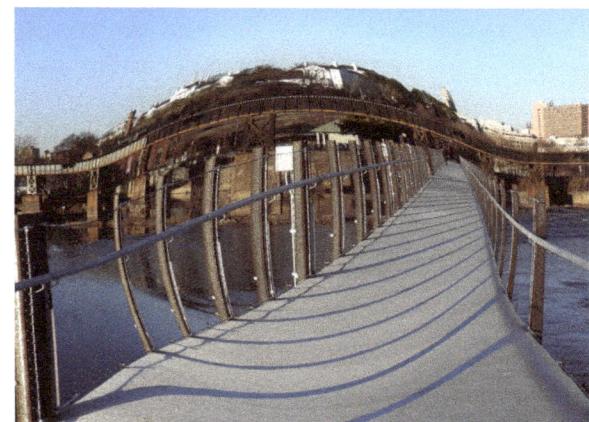

Figure 6-26. Fisheye Example

The result of using this effect will not look good unless you choose your subject carefully. I find this effect works best with a single, clearly identifiable subject, like the bridge in Figure 6-26. If you use the Fisheye effect on a busy or cluttered scene, it may be difficult to make out the subject at all, because of the distortion.

The next option is called the Miniature effect. With this feature, the camera adds blurring at the edges of the image to simulate a photograph of a tabletop model or miniature. Such images often appear blurred at the edges, either because of the shallow depth of field of close-up photos, or because of the use of a tilt-and-shift lens, which causes blurring at the edges.

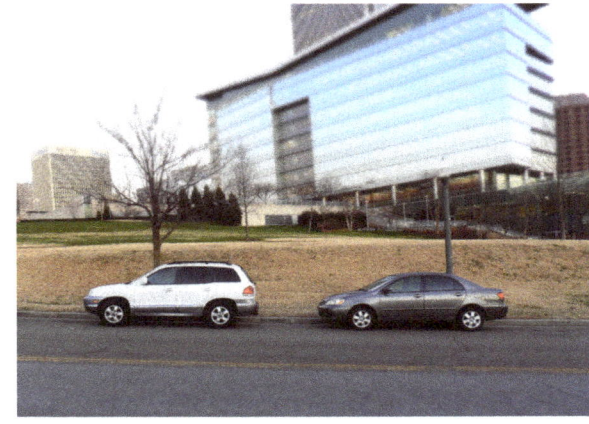

Figure 6-27. Miniature Effect Example

Here, again, you need to choose an appropriate subject. I have found that it works well with something like a street scene or a house, which might actually be reproduced in a tabletop model. For example, if you are able to get a photo from a vantage point above a parking lot or a street with cars driving on it, you can use this processing to make it look as if you had photographed a tabletop display with model cars. Try to keep the main subject in the center of the image. For Figure 6-27, I photographed a group of cars in front of a building, to try to establish a contrast between the cars and the surrounding scenery.

With the next option on the Filter Effects menu, Painting, the camera applies a distinctive form of processing that results in heightened emphasis on colors and imbues the image with a pastel-like look, as shown in Figure 6-28.

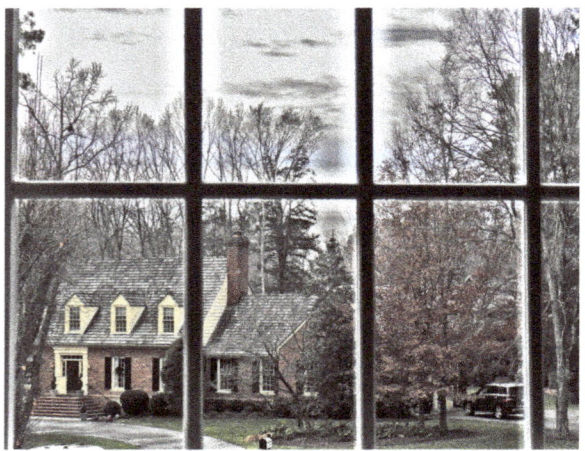

Figure 6-28. Painting Example

This is one of the more dramatic of the Filter Effects settings. The camera increases the intensity of colors and uses processing to give an appearance like that of an HDR image, as discussed in Chapter 4, with shadowed areas brightened. This setting is not appropriate if you are looking for a realistic representation of your subject; it is useful when you want a stylized, vibrant image, possibly for a poster or illustration. Also, I have found that this effect has a tendency to make the final image too bright, so it helps to start with an image that is darker than usual, perhaps through the use of negative exposure compensation.

The next option, Vignette, the first choice on the second screen of this menu item, darkens the image toward its edges to create the appearance of an old-fashioned vignette, with the center highlighted but fading out on the edges. An example is shown in Figure 6-29.

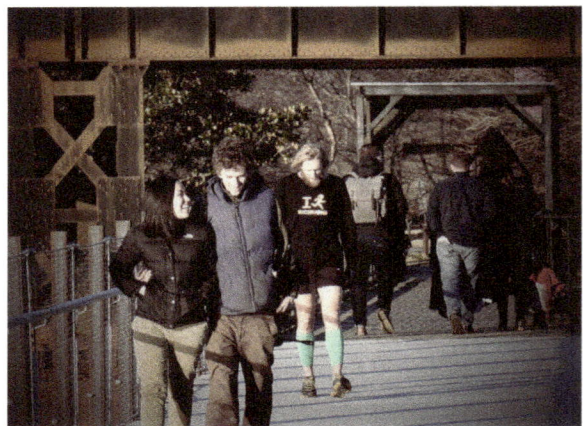

Figure 6-29. Vignette Example

The next option, Photo Illustration, alters the image by darkening the outlines of objects and reducing the number of colors, to make the image appear like a pen-and-ink illustration that has been colored in with poster paints. This option can create a very pleasant effect if used with an appropriate subject. I have enjoyed the results of using it with fairly large, colorful subjects like the river scene with blue sky in Figure 6-30.

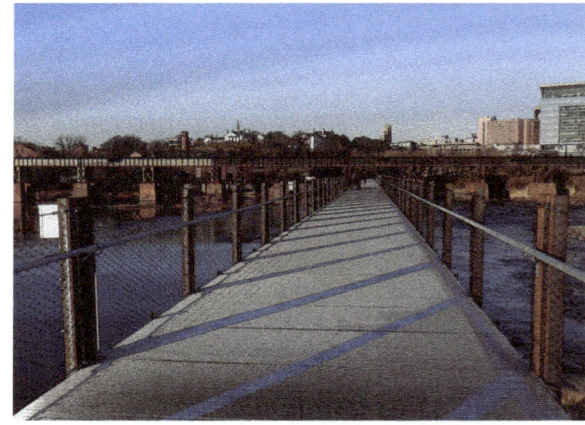

Figure 6-30. Photo Illustration Example

The final choice for the Filter Effects option, Portrait (Color+B&W), is somewhat like the Selective Color option, discussed above, but it is specially designed for images of people. If the camera detects a human face, it will leave the face in color and convert the background to black-and-white, as in Figure 6-31. If the camera does not detect a human subject, it will leave the central part of the image in color and convert the rest of the image to black-and-white. This can be an effective way to highlight a portrait subject and isolate him or her from the background.

Chapter 6: Playback

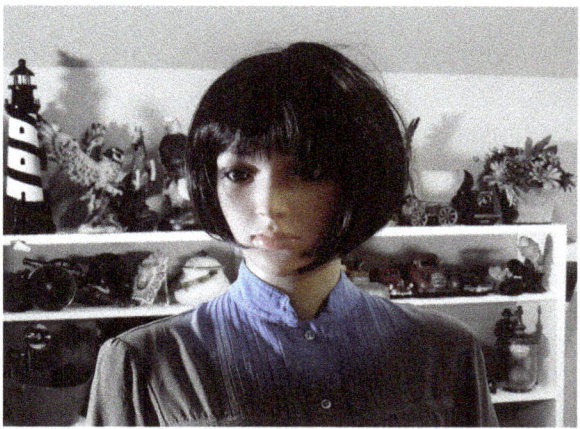

Figure 6-31. Portrait (Color+B&W) Example

Slide Show

Like most modern digital cameras, the Coolpix B700 can display the images on your memory card in a slide show that plays back on the camera's display or on a connected HDTV. The B700 does not offer elaborate options such as music or a variety of transitions; your pictures are played back with straight cuts between them and in silence.

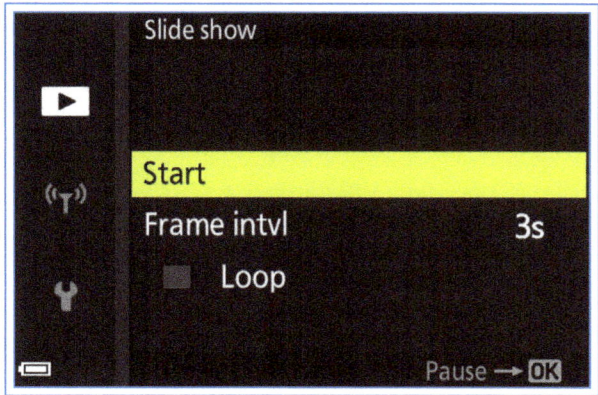

Figure 6-32. Slide Show Menu Options Screen

The only choices you can make from the Slide Show menu options screen, shown in Figure 6-32, are the length of time between images and whether or not the show will repeat in a loop. (The loop is not endless; the show will repeat for a maximum of 30 minutes.)

To start a slide show, go to the Playback menu and choose Slide Show. You can then navigate to the option for Frame Interval and select 2, 3, 5, or 10 seconds for the time between images. Next, press the Left button to go to the previous screen and press the OK button while the Loop option is highlighted, if you want the show to repeat. After selecting these options, highlight the Start option and press OK to start the show. To pause the show, press OK again. To restart it, highlight the playback triangle that appears on the screen and press the OK button. To stop the show, after pausing it, highlight the square "stop" icon and press the OK button. You also can stop the show at any time by pressing the Playback button. To skip forward or backward to the next image, you can press the Left or Right button at any time; hold either of those buttons down to move more rapidly through the images.

There is no way to select the images that will be played; all images on the memory card will be played. For movies, only the first frame will be played. For sequences of continuous shots, only the key images will be displayed, if the Sequence Display Options item on the Playback menu is set to Key Picture Only.

The remaining menu options are on screen 2 of the Playback menu, shown in Figure 6-33.

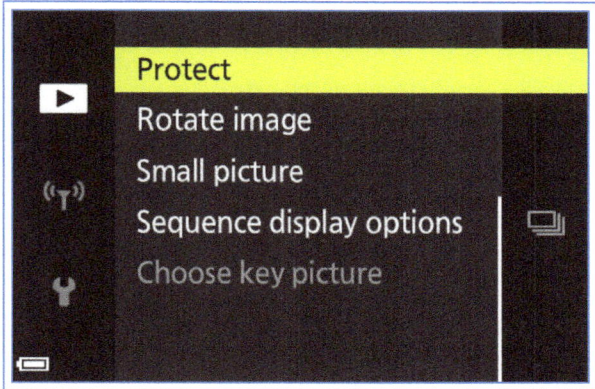

Figure 6-33. Screen 2 of Playback Menu

Protect

With the Protect feature, you can "lock" selected images so they cannot be erased with the normal erase functions using the Trash button. However, if you format the memory card using the Format command, all data will be erased, including protected images.

To protect images, after selecting this menu option, navigate through your images using the Left and Right buttons or the multi selector dial, and use the Up and Down buttons to mark or unmark any image you want to protect.

Press the Up or Down button to mark or unmark an image. A yellow key icon will appear beneath the thumbnail of each marked image, as shown in Figure

6-34. You can use the zoom lever to enlarge an image before deciding whether to apply protection to it.

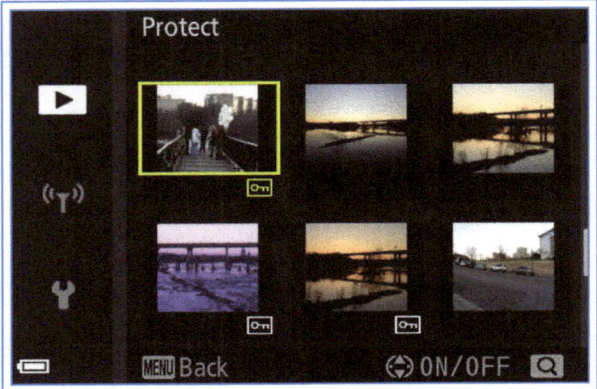

Figure 6-34. Selection Screen for Protect Menu Option

When you have marked all images as you want them, press the OK button to apply the protection. An image that is protected will have a key icon in the upper left corner, as shown in Figure 6-35.

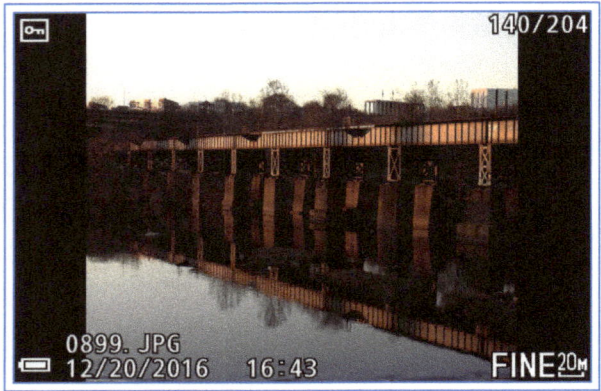

Figure 6-35. Protected Image with Key Icon

That icon will be visible when the image is viewed with the basic information screen; the icon will not appear in the image-only view, the copyright information view, or the detailed view with the histogram.

Rotate Image

Using this first option on the Playback menu's second screen, you can rotate still photos 90 degrees clockwise or counter-clockwise. You cannot rotate the key image of a sequence when it is displayed in sequence mode; you have to display the pictures from the sequence individually in order to rotate them.

After you select the Rotate Image option from the Playback menu, the camera displays the Select Image screen. Navigate with the multi selector dial or the Left and Right buttons until you have highlighted with a yellow frame the image you want to rotate, then press the OK button to select it.

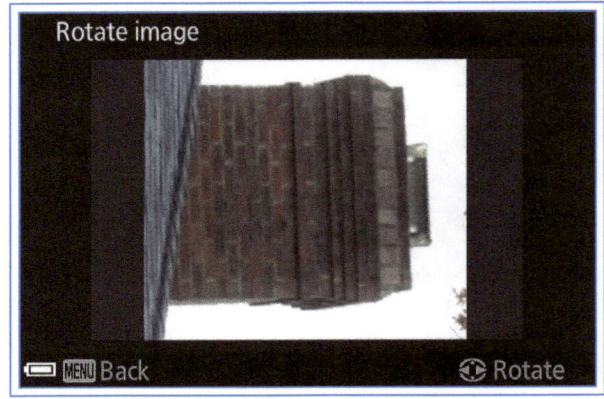

Figure 6-36. Screen for Rotating Image

On the next screen, as shown in Figure 6-36, use the Left or Right button to rotate the image counter-clockwise or clockwise. (You can also do the rotation by turning the multi selector dial.) Press OK when the image is rotated to the orientation you wish, then press the Menu button to exit from the Rotate screen.

If an image was taken with the camera turned sideways, so the camera ordinarily has to be turned in order to view the image right-side-up, the camera will let you rotate it as much as 180 degrees; other images can be rotated only 90 degrees.

Small Picture

This option lets you do a simple form of in-camera editing. This feature allows you to take any saved JPEG image that was captured in the aspect ratio of 4:3 or 16:9 and create a new version in a file small enough to send by e-mail or post on the internet. This feature does not work with Raw images or images in the aspect ratio of 3:2 or 1:1.

This function is useful if you need to take a quick photo and e-mail it to a friend or colleague. If you don't have software available on your computer or other device to edit the image down to a smaller size, you can let the camera take over this task. Of course, you could take the image in the small size to begin with, but you might want to have a higher-resolution version available for later editing or printing, and also be able to create a small version for e-mailing after you have already recorded the original version.

Chapter 6: Playback | 107

To use this feature, navigate to the JPEG image you want to alter. Once it is displayed, in either full-frame or thumbnail view, press the Menu button, then select the Small Picture option. On the next screen, as shown in Figure 6-37, for an image captured in the 4:3 aspect ratio, you can choose from three options: 640 x 480 pixels, 320 x 240 pixels, or 160 x 120 pixels. For an image captured in the 16:9 aspect ratio, there is only one option: 640 x 360 pixels. Each of these choices produces a low-resolution image, well under one megapixel in size.

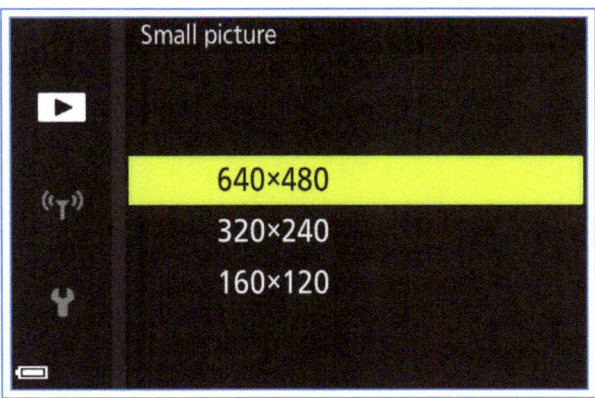

Figure 6-37. Small Picture Menu Options Screen

If you confirm the operation on the next screen, the camera will copy the selected image at the chosen size and save it at the end of the images on the memory card.

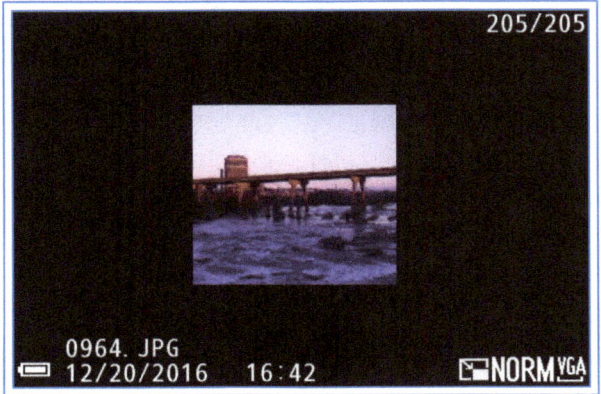

Figure 6-38. Image After Small Picture Option Applied

The image will be displayed in the camera with a large, black border area around the image itself, as shown in Figure 6-38, to show that this is a "Small Picture" copy. This border does not become part of the actual image; it displays only in the camera.

SEQUENCE DISPLAY OPTIONS

This menu option controls how the camera displays images taken in one of the continuous-shooting modes such as Continuous H, Continuous L, Pre-shooting Cache, and others, which normally are displayed as "sequences." This option is straightforward: You have just two choices—Individual Pictures or Key Picture Only, as shown in Figure 6-39.

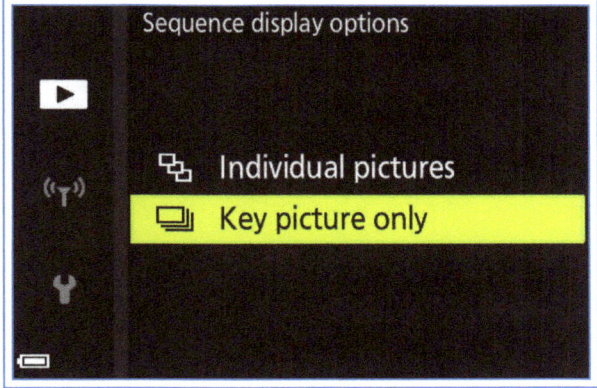

Figure 6-39. Sequence Display Options Menu Screen

If you choose Key Picture Only, then, as you navigate through your images, when you come to a sequence, only the key image will display; it will be displayed in a frame that appears like a stack of images, indicating that it is the key frame of a sequence, and there will be a prompt to press the OK button to view the individual images from the sequence.

You cannot call up the detailed information for the key image with the Display button or use the Playback menu options to manipulate the image; you first have to press the OK button to "enter" the sequence and display the individual images. If you choose the Individual Pictures option, all sequences will automatically be opened up, so the images from the sequences all display as you scroll through your saved images; you will not see any key images or have to "enter" into the sequences.

CHOOSE KEY PICTURE

This final option on the Playback menu lets you change the key picture that displays for a sequence. Ordinarily, the first image in a sequence is used as the key picture. If you would prefer to display one of the other images when the shots are displayed in sequence mode, you can use this feature. First, you have to set the previous menu option, Sequence Display Options, to Key Picture

Only. Then display the sequence whose key picture you want to change. Select this menu option and press the OK button or the Right button to activate it.

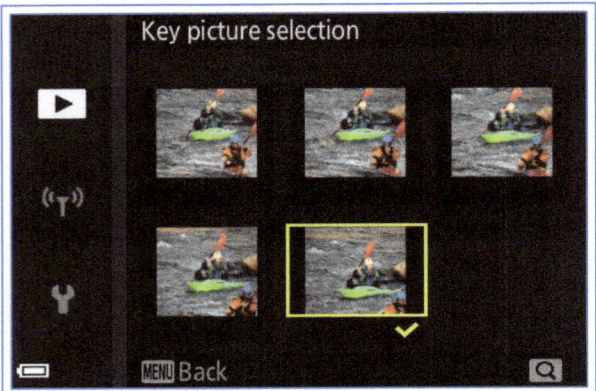

Figure 6-40. Selection Screen for Choose Key Picture Option

The camera will display all of the images from the sequence, as shown in Figure 6-40.

Navigate through those images using the multi selector dial or the Left and Right buttons. Press the OK button when the picture you want to use is highlighted.

Because of the nature of continuous shooting, which captures a stream of images rapidly, in many cases the images will be quite similar. However, there may be occasions when one image stands out above the others in quality and you will want to have it display as the representative of its sequence, so it will show up in slide shows, for example.

Printing Images

There are various ways to produce copies of digital photographs on paper. You can import the photographs into a program such as Adobe Photoshop or Photoshop Elements, or use the ViewNX-i software supplied by Nikon for the Coolpix B700, or any of many other programs that are available for photo editing. Once you have edited the images to your satisfaction, you can print the finished products from that software.

However, in some cases you may not be willing or able to spend the time to manipulate the pictures in software before printing them out. You may have access to a printer that will connect directly to the camera, and you may need or want to print out copies on photo paper without going through the time-consuming process of transferring the images to a computer first. The following discussion covers the high points of this procedure.

Printing Directly from the Camera

The Coolpix B700 uses the PictBridge printing protocol, which lets the camera communicate directly with a wide variety of printers. The basic procedure is quite simple: First, make sure the Charge by Computer option on the Setup menu is set to Off. Then, plug the black USB cable that came with the camera into the micro-USB port inside the door on the right side of the camera. (This is the upper of the two ports in that location.)

Then plug the other end of the cable into the USB port of a PictBridge-compatible printer. (This USB port is different from the one for the cable that connects the printer to a computer; this one is rectangular; the port for the cable to the computer has more of a square shape.) The printer does not have to be made by any particular company; I plugged the camera directly into a Brother laser printer, and the two devices communicated with no problems.

Once the connection is made and the printer is turned on, the camera should turn on automatically and display the PictBridge logo, as shown in Figure 6-41.

Figure 6-41. PictBridge Logo on Camera's Screen

The PictBridge logo is followed quickly by a special screen that appears only when the B700 is connected to a PictBridge printer, as shown in Figure 6-42.

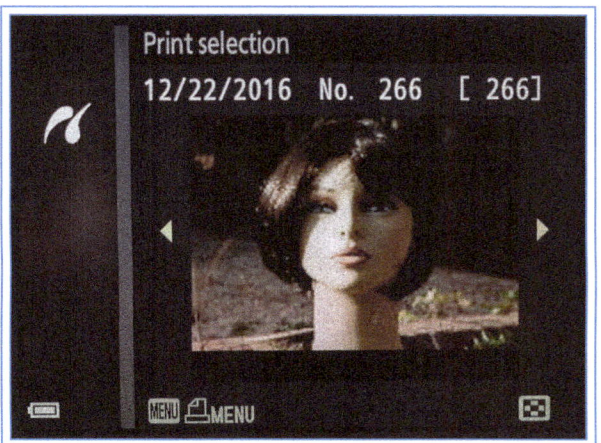

Figure 6-42. Special Screen When Camera Connected to Printer

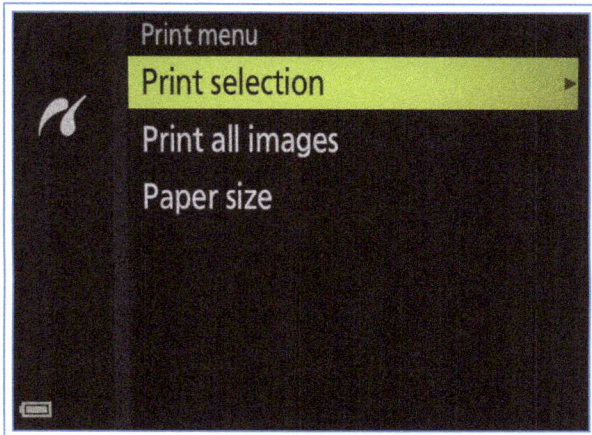

Figure 6-43. Print Menu

To print an individual image, navigate to it and press the OK button; the camera will prompt you for the number of prints and the paper size. To print multiple images, when the print display screen initially displays, press the Menu button to bring up the Print menu, shown in Figure 6-43.

From that screen, you can select images to print or print all images.

Once you have all of the settings as you want them, press the OK button on the camera to print out the photograph or photographs. For further details about these procedures, see the Nikon Coolpix B700 Reference Manual at pages 103-106.

Chapter 7: The Setup Menu

The next menu system to discuss is the Setup menu. (I'll discuss the Movie menu along with the various options for motion picture recording in Chapter 8 and the Network menu in Chapter 9.)

The Setup menu gives you options for housekeeping matters such as settings for date and time, screen brightness, and operational sounds, but it also includes some settings that affect how you take your images, including Vibration Reduction, Peaking, and Digital Zoom. In addition, this menu is where you perform the important task of formatting a memory card.

As a reminder, you enter the menu system by pressing the Menu button. The available menus change depending on whether the camera is set to shooting mode or playback mode, and, in shooting mode, which exposure mode is selected (Program, Auto, or Scene, for example). However, no matter what mode the camera is set to, you can always enter into the Setup menu. After you press the Menu button, use the Left button to move the yellow highlight to the far left column and move the highlight down the line of icons to the wrench icon that indicates the Setup menu, as shown in Figure 7-1.

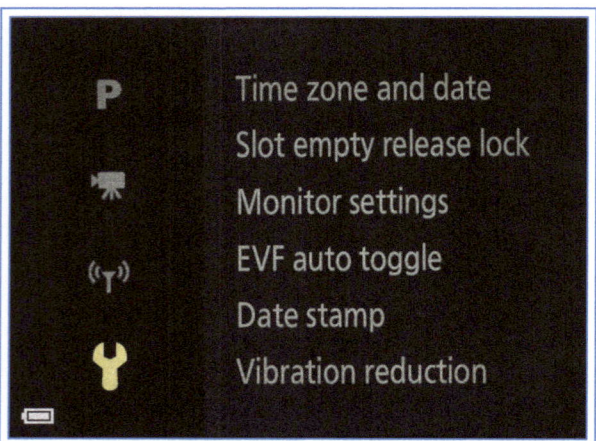

Figure 7-1. Wrench Icon for Setup Menu Highlighted at Left

Once that icon is highlighted, use the Right button to move the selection block back into the list of menu items, and then use the multi selector dial or the Up and Down buttons to navigate through the various options on the menu. The first screen of the Setup menu is shown in Figure 7-2. I'll discuss all of the choices on the menu in the order in which they appear.

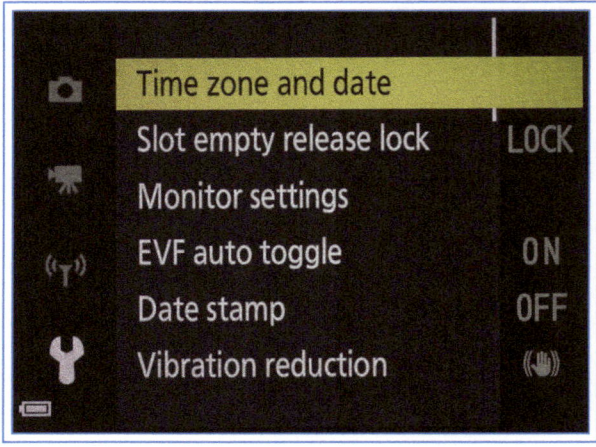

Figure 7-2. Screen 1 of Setup Menu

Time Zone and Date

Chances are you set the date, time, and time zone when you first turned on the camera. If you haven't done so or need to change these settings, use this menu option.

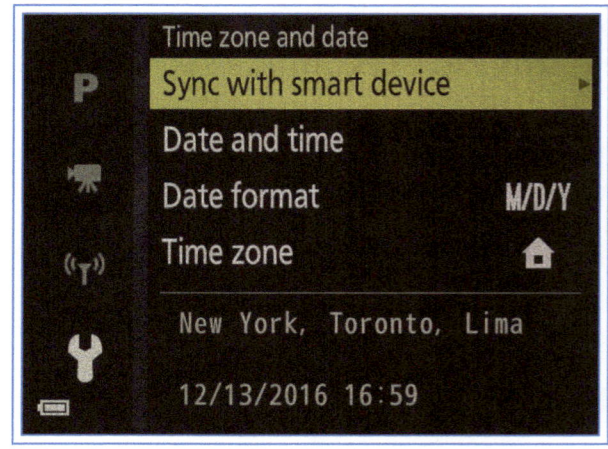

Figure 7-3. Time Zone and Date Menu Options Screen

Chapter 7: The Setup Menu | 111

First, select Time Zone and Date, and press the Right button or the OK button to move to the Time Zone and Date options screen, shown in Figure 7-3.

Select Date and Time and press the Right or OK button to get to the Date and Time Settings screen, shown in Figure 7-4. Navigate through the selections for month, day, years, hour, and minute using the Left and Right buttons; change the values using the multi selector dial, the command dial, or the Up and Down buttons.

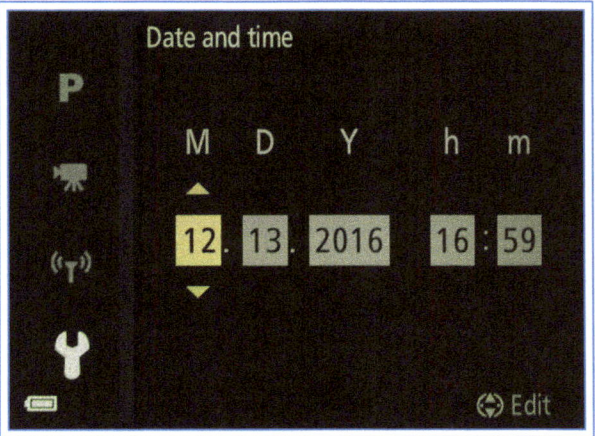

Figure 7-4. Date and Time Settings Screen

When you have finished with the minutes setting, press the OK button to confirm all of the values. You can then return to the Time Zone and Date screen, select Date Format, and choose your preferred order for displaying the month, day, and year.

Then, return to the Time Zone and Date page and select Time Zone. For this option, you have two choices—the home time zone and the travel destination, as shown in Figure 7-5.

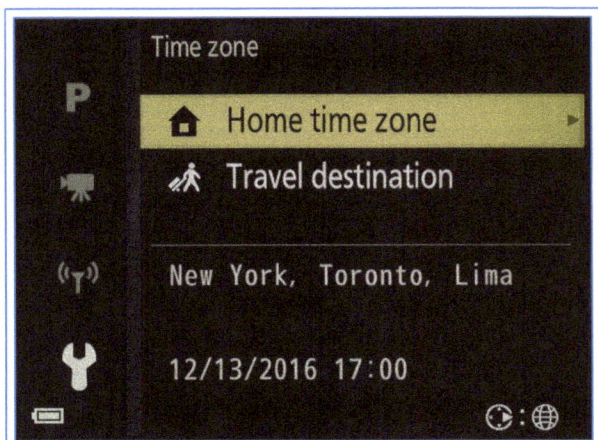

Figure 7-5. Time Zone Menu Options Screen

Set the home zone to the location where you spend most of your time, and, if you want, set the travel zone for an area you are most likely to travel to. Or, you can wait until you travel to set this zone.

When you take a trip, select the travel time zone from this menu option and the camera's time and date will change as required, so your images will have the correct dates and times when you take pictures in your destination time zone.

There is one other option on the Time Zone and Date menu screen, called Sync with Smart Device, which was shown above in Figure 7-3. If you turn on that option, the camera will synchronize its date and time settings with the information provided by a smartphone or tablet the camera is connected to using the SnapBridge app, as discussed in Chapter 9.

Slot Empty Release Lock

This second option on the Setup menu, whose choices are shown in Figure 7-6, determines how the camera reacts if you press the shutter button when there is no memory card installed in the camera.

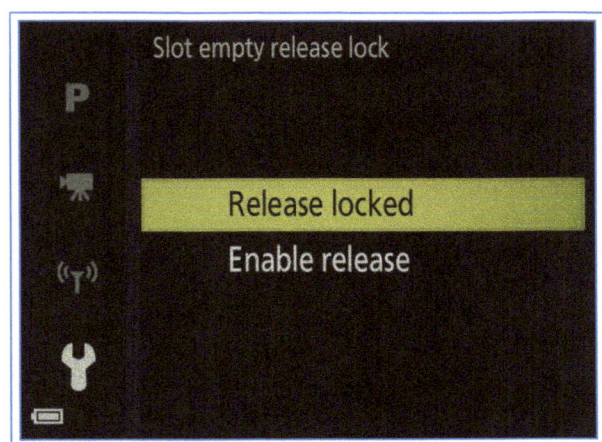

Figure 7-6. Slot Empty Release Lock Menu Options Screen

If you choose the default option, Release Locked, the camera will display the message "No card present" and will not let you operate some of the controls, including the shutter button and zoom lever. You can still use the Menu button, however, to get access to the menu system.

The Release Locked option is the one I prefer, because it ensures that I will not press the shutter button, thinking I am recording images, when there is no card in the camera.

If you choose the second option, Enable Release, then, even when no card is in the camera, you can press the shutter button and an image will appear on the display screen. You can capture one or two large images or about 10 small ones and play them back, but they will have the words Demo Mode marked in the lower right, and they will not be saved if you turn off the camera. There is no easy way to transfer these images to a computer or print them on a printer. (In an emergency, you could use a video capture device to transfer them through the HDMI port.) In other words, these images can be taken only for the purpose of demonstrating how the camera works, such as in a camera store.

Monitor Settings

With this menu item, whose main screen is shown in Figure 7-7, you can control several aspects of the way your camera's monitor (LCD display screen or viewfinder) displays your images.

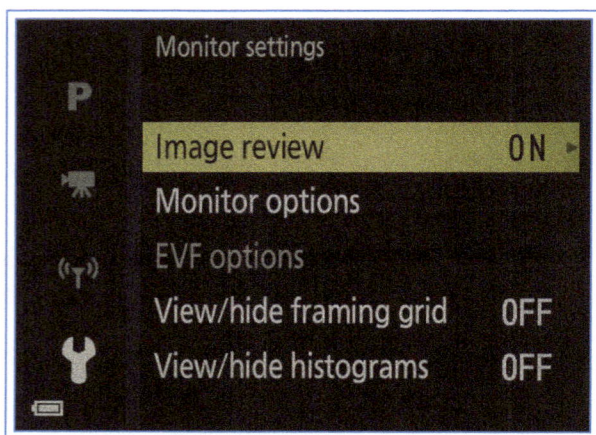

Figure 7-7. Monitor Settings Menu Options Screen

Image Review

First, you can turn the Image Review feature on or off. If it is turned on, a new image shows up on the display for about one second when you first take the picture. If this option is turned off, the display immediately goes back to the shooting screen when you take a picture. There is no way to control the length of time the image displays; this feature is either on or off. If you want to view a new image for a longer period of time, press the Playback button and use the normal playback procedures.

Monitor Options

The next option, Monitor Options, lets you select from six levels of brightness for the LCD screen, including levels 1-5 and a final level called Hi, as shown in Figure 7-8. This option is available only when the LCD screen is active.

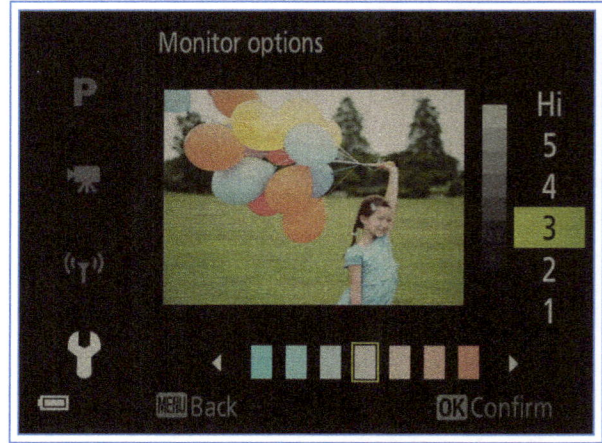

Figure 7-8. Monitor Options Screen

Select the brightness level you want using the Up and Down buttons. The default value is level 3. The Hi setting is extremely bright, and can be used when you are shooting in bright conditions but still want to use the LCD screen rather than the viewfinder. If you use the higher settings, the camera's battery will run down faster than usual, so take that factor into account when increasing the brightness.

On the same screen, you can also adjust the color of the display on the screen, using the scale at the bottom of the display. Use the Left and Right buttons to change this aspect of the display to be more bluish or reddish in order to calibrate the display for the conditions in which you are shooting. Press the OK button to confirm the settings.

EVF Options

When you are using the viewfinder instead of the LCD screen, the next menu option, EVF Options, becomes available for selection. With this menu item, you can make similar adjustments that affect only the display in the viewfinder. There are only three levels of brightness for the viewfinder, because its view is shaded and it should not need large changes in brightness. You can make the same color adjustments as for the LCD display.

View/Hide Framing Grid

Next, the View/Hide Framing Grid option, shown in Figure 7-9, lets you turn on or off a grid of vertical and horizontal lines that divide the screen into nine blocks.

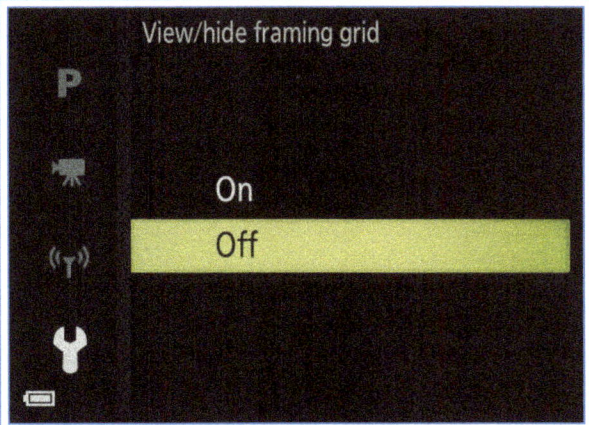

Figure 7-9. View/Hide Framing Grid Menu Options Screen

The grid itself is shown in Figure 7-10, as it is displayed on the shooting screen.

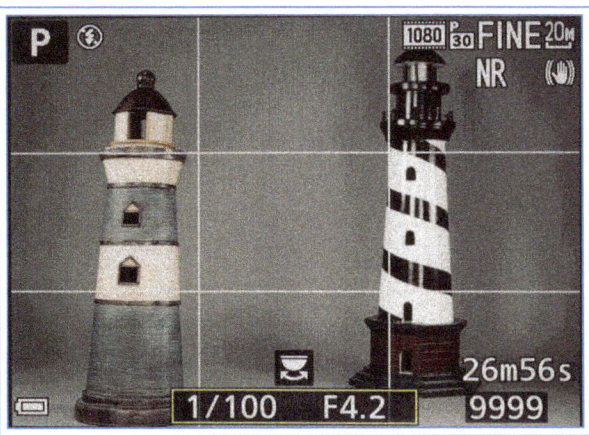

Figure 7-10. Framing Grid on Shooting Screen

You may appreciate having this grid available to help you compose your images according to the Rule of Thirds, which calls for placing the most important subject close to the intersections of these lines, to increase visual interest in the photo.

The grid also may help you keep the horizon or your subject, such as a building, properly horizontal or vertical by lining it up against one of the lines on the screen. If turned on, the grid will appear on all three shooting screens, including the screen with no shooting information. In addition, the camera will add a fourth screen with no information and no grid. The grid does not appear when a video is being recorded. If you don't find the grid useful, just leave it turned off.

View/Hide Histograms

The final option under Monitor Settings, View/Hide Histograms, controls whether or not the histogram is displayed when the camera is in shooting mode. If you turn this option on, the histogram appears in the left half of the screen when the camera is set to shooting mode, as shown in Figure 7-11.

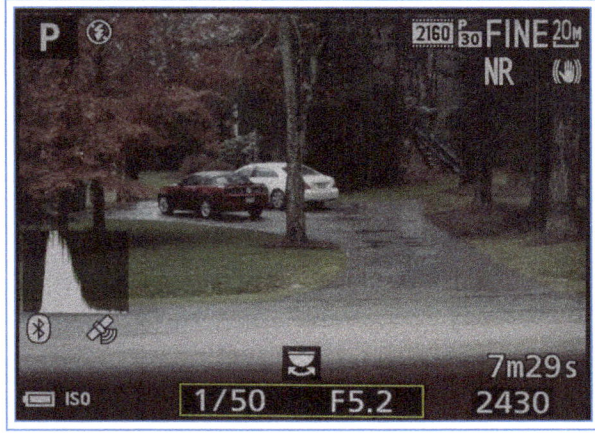

Figure 7-11. Histogram on Shooting Screen

Like the framing grid, the histogram displays on all shooting screens except for a fourth screen that the camera adds, with no shooting information and no histogram. The histogram is not displayed when AF Area Mode is set to Target Finding, when the camera is in Auto mode, with some Scene mode types, or when a movie is being recorded.

Even with this option turned on and the detailed information screen activated, the histogram does not display in some cases, including when you use the enlarged view for manual focus or the pop-up menu to select focus mode, the self-timer, or flash mode.

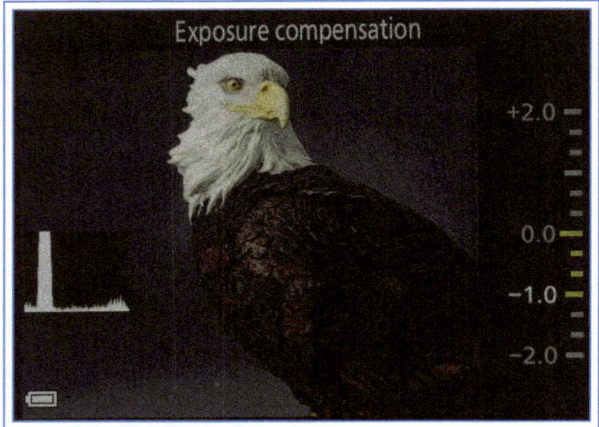

Figure 7-12. Histogram Displayed with Exposure Compensation Scale

If this option is turned off, you can still display the histogram in shooting mode by pressing the exposure compensation button (Right button) to adjust the exposure. The histogram will turn on in that situation to help you gauge how much exposure compensation to apply, as shown in Figure 7-12.

I prefer to display histograms in shooting mode using the exposure compensation control, because I can call up the histogram when I want it and it will disappear quickly after I am finished using it. However, exposure compensation is not available with Manual exposure mode, so this menu option is the only way to display a histogram when shooting in that mode. If you want to display a reasonably accurate histogram in Manual mode, you need to turn on the Manual Exposure Preview option on the last screen of the Shooting menu; otherwise, the histogram will not reflect the settings you have made for your manual exposure.

The histogram will always display for still images in playback mode when you have selected the histogram display screen by pressing the Display button, as discussed in Chapter 6.

As also discussed in Chapter 6, the histogram is useful for indicating whether your image will be underexposed or overexposed. In most cases, it's a good idea to adjust exposure settings so the histogram looks roughly like a mountain with gradual slopes from left and right to a moderate peak in the center.

EVF Auto Toggle

This next option on the Setup menu, if turned on, causes the view of shooting displays and playback displays to switch from the LCD screen to the viewfinder when your head (or another object) gets close to the viewfinder. If this option is turned off, the view does not switch when an object approaches the viewfinder. In that case, you can use the Monitor switch, directly to the right of the viewfinder, to switch the view. You also can switch the view to the viewfinder by folding the LCD screen in against the camera, with the viewing screen concealed.

I prefer to leave this option turned on most of the time, because it is convenient to have the view of the scene switched to the viewfinder when I lift the camera to my eye. However, if I have the camera on a tripod and am examining the LCD display closely, I may turn this option off so the view does not switch unexpectedly as I get close to the screen to look at the details of the shot.

Date Stamp

With this option, you can control whether the camera places the current date, or date and time, on the image when the image is recorded, as shown in Figure 7-13.

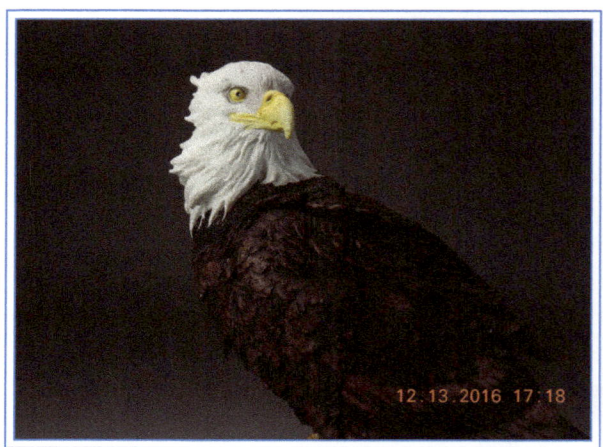

Figure 7-13. Date Stamp Option in Use

This option places the information permanently on the image, and the information cannot be deleted (unless you use Photoshop or similar software to edit it out). You might want to use this feature if you are taking images as part of a scientific experiment in which you need to record this information as part of your data, but you probably would not want to use it for general picture-taking, because the date (or date and time) information will mar the image. For ordinary images, you can always use editing software to retrieve the date and time information, which is recorded invisibly with the images (assuming the camera is set to the correct date and time).

To use this feature, go to the Date Stamp option on the menu, then select either Date, Date and Time, or turn the option off altogether. This setting does not operate when you have selected Pre-shooting Cache or the 60 fps or 120 fps continuous shooting options. It also does not function with some Scene mode settings or with Raw images.

Vibration Reduction

This is one of the more important settings for the camera, especially because of the extreme telephoto range of the B700's lens. When you activate Vibration Reduction (VR), the camera uses its lens-shift system to stabilize the image. When you are handholding the camera, there is bound to be a slight amount of camera motion or shake. At slow shutter speeds, this motion can cause blurring of your images. Any such blurring is magnified at higher telephoto levels, as you can see if you look through the viewfinder or at the LCD screen at a zoomed-in level. The slightest motion can make the image appear to jiggle uncontrollably.

There are three available settings for the VR system: Off, Active, and Normal, as shown in Figure 7-14.

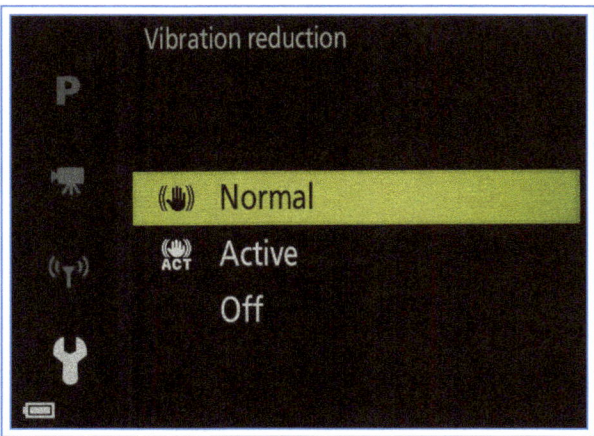

Figure 7-14. Vibration Reduction Menu Options Screen

These settings are applicable when you are shooting either still pictures or movies. When you have the camera on a tripod, you should make sure the VR setting is turned off, because the camera's circuitry can get confused and attempt to correct for camera shake when there is none, thereby degrading the image.

The Normal setting is intended to correct relatively mild camera movement, such as when you are handholding the camera and panning it to record a movie (moving side to side smoothly). The camera can detect the smooth horizontal motion, and will correct only for vertical motion caused by the handholding. It also can detect smooth vertical motion, such as when you are tilting the camera vertically to take a movie of a tall building. When the Normal setting is turned on, the camera places a hand icon in the upper right corner of the screen, as shown in Figure 7-15.

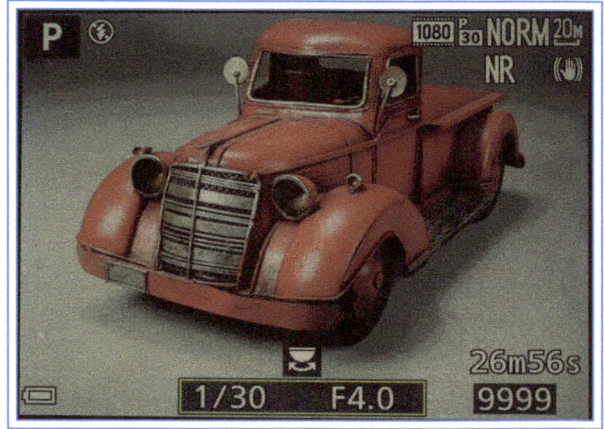

Figure 7-15. Icon on Shooting Screen for Normal Vibration Reduction

The Active setting is intended for situations in which the camera motion is more erratic and dramatic, such as when you are shooting out of a car window, or from a boat or helicopter, and the camera may be jerked in various directions unpredictably. In this case, the camera does not allow for horizontal motion, but corrects for motion in any direction as well as it can. Of course, the camera cannot erase the effects of violent motion, but it can reduce the blurring effects of milder motion to a fair extent. When the Active setting is turned on, the camera displays the icon shown in Figure 7-16 instead of the plain hand icon.

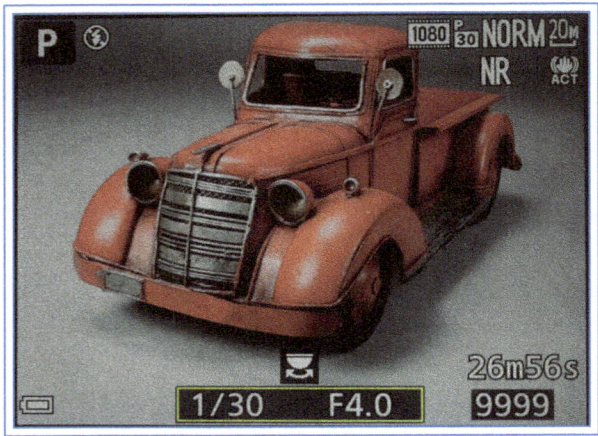

Figure 7-16. Icon on Shooting Screen for Active Vibration Reduction

I recommend that you use the Normal setting when handholding the camera in most situations, if you are able to hold the camera quite steady. However, if you are a passenger in a vehicle on a bumpy road or otherwise encountering random motion, you may want to switch to the Active setting to see if it can help even out the jolts to the camera.

The next items are on screen 2 of the Setup menu, shown in Figure 7-17.

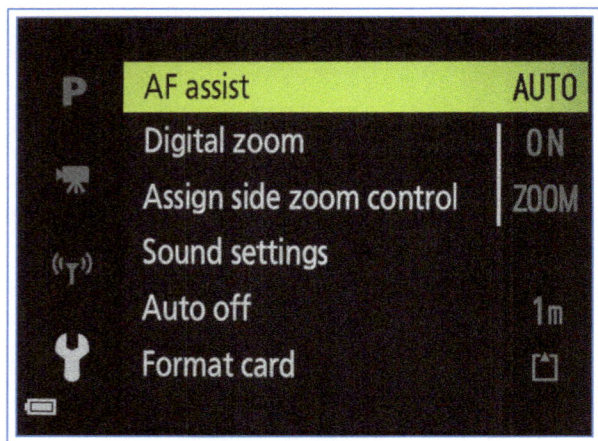

Figure 7-17. Screen 2 of Setup Menu

AF Assist

This first option on the second screen of the Setup menu lets you turn on or off the reddish light beam that shines from the autofocus assist lamp on the front of the camera. When the camera is trying to focus in a dark area, this light helps the autofocus mechanism find the patterns and shapes it needs to evaluate in order to achieve proper focus. You should usually leave this setting turned on, but you may want to turn it off when you're taking pictures in a place where the beam could be distracting or annoying to others, or where it might alert the subjects of your candid photography. If the camera then has difficulty in focusing, you can switch to manual focus and adjust the focus yourself. Also, if you are shooting subjects beyond the illuminator's range of about 13 feet (four meters), it is of little or no use and can be turned off.

The choices for this setting are Auto or Off. With the Auto setting, the lamp will fire when needed, except with some focus settings and some scene settings in which it is disabled and cannot activate. The lamp will always light up when the self-timer or Smile Timer is used; there is no way to disable the self-timer lamp, though you can cover it with black tape if you need to suppress it. The lamp also serves as the Red-eye Reduction lamp, and lights up when a flash mode with red-eye reduction is selected. That function also is not affected by the setting of the AF Assist menu option.

Digital Zoom

This next item on screen 2 of the Setup menu lets you zoom in on a scene electronically, beyond the magnifying power of the camera's optical zoom. Because it is an electronic zoom and not an optical one, it does not increase the optical information received by the camera; instead, it enlarges the image digitally, which can result in a blocky, pixellated look.

That's not to say that digital zoom is useless. It can help you to compose a scene the way you want to, or to measure the exposure on a small part of the scene before you zoom back out to take the picture without the digital zoom effect, for example. And, in some cases, the use of digital zoom does not actually degrade the quality of the image; it just uses a smaller portion of the image sensor's surface, resulting in a lower-resolution image, but without the pixellation of an artificially magnified image. In addition, Nikon uses a feature called Dynamic Fine Zoom that improves the quality of digitally zoomed images.

There are two settings available on the Setup menu for this feature: On and Off. If you choose Off, the camera will be limited to using its optical zoom, which is really not much of a limitation, since the B700's lens has the impressive range of 24mm to 1440mm.

If you choose On, the lens will "zoom" electronically beyond the optical limit of 1440mm, to a maximum of an amazing (though somewhat illusory) 5760mm for still images. (With 4K movies, the digital zoom range is only up to 2880mm.) With digital zoom, beyond a certain level of magnification the camera will use its circuitry to "interpolate" pixels—that is, it will create new pixels in between the ones produced by the sensor, in order to expand the image to a greater magnification. When the camera is using interpolation, the image deteriorates to some extent because the camera is using pixels that are not part of the original image.

If you use an Image Size setting smaller than the maximum, then you can use digital zoom to a certain extent without image deterioration. This is because the camera needs a smaller number of pixels in order to create the desired image. Therefore, instead of interpolating new pixels among the existing ones, the camera crops out the actual pixels that appear on a portion of the image sensor, and enlarges that portion to fill the entire area of the sensor. In this way, the camera uses only the actual pixels captured through the lens to the image sensor; it does not have to interpolate any new pixels.

Whenever you move the zoom lever, a scale appears in the top center of the image to show how far the lens has been zoomed.

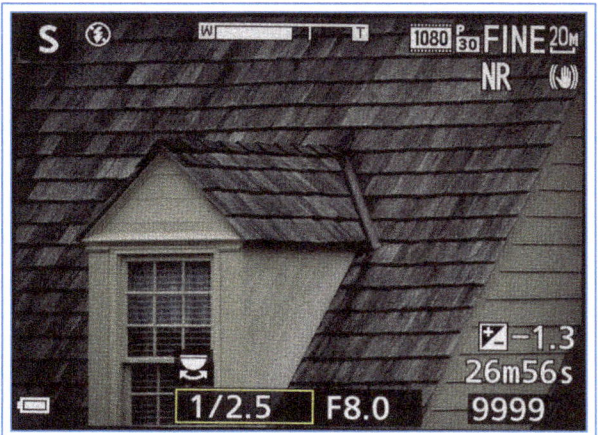

Figure 7-18. Zoom Scale with Optical Zoom Only

The scale has a small vertical line past the halfway point; that line shows the point where the camera starts using digital zoom instead of optical zoom. As shown in Figure 7-18, when only optical zoom is in use, the zoom scale is white, and the bar stays to the left of that small line.

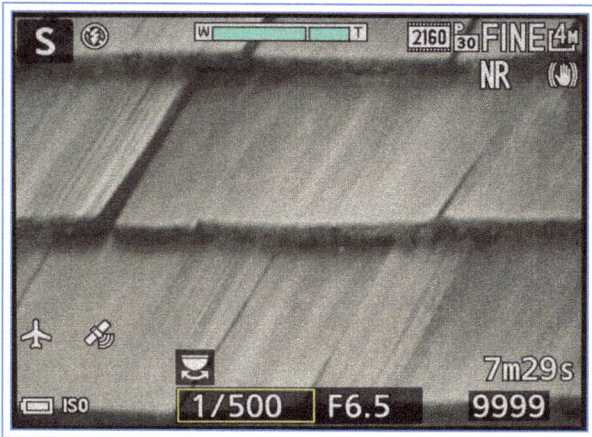

Figure 7-19. Zoom Scale Turned Blue Past Optical Zoom Range

If you zoom the lens in so the zoom bar moves past the vertical line, the bar may turn blue, as shown in Figure 7-19, indicating that zoom is being used beyond the normal optical zoom range, but that the picture quality should not deteriorate too much.

When the zoom bar turns blue, that can mean one of two things. Either the Image Size is set to a value smaller than the maximum of 5184 x 3888 pixels, so the camera can use the extra pixels to magnify the image without loss of quality, or the camera is using what Nikon calls Dynamic Fine Zoom, special processing that increases magnification without compromising image quality in the way that normal digital zoom does. According to Nikon, you can zoom the lens to a focal length of 2880mm, twice the optical zoom range, before image deterioration occurs, even when using the largest Image Size setting.

If you set Image Size to a value smaller than the maximum, the zoom bar will stay blue for longer ranges. If you use one of the smaller options, such as 2272 x 1704 (4M), the bar may stay blue for the entire digital zoom range, because the camera can use the unneeded extra pixels to increase magnification. An example of this effect is seen in Figure 7-19, where digital zoom was used with Image Size set to the 4M size and the zoom bar is blue to the end of its range.

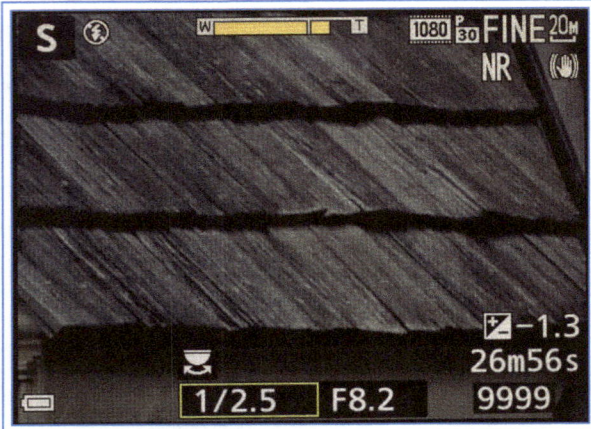

Figure 7-20. Zoom Scale Turned Yellow with Digital Zoom

If you continue to zoom in with the largest Image Size settings, the bar may turn yellow, indicating that image deterioration will occur because the camera is adding extra pixels. For example, Figure 7-20 shows the zoom scale with Image Size set to its maximum value, and with the lens zoomed well beyond the optical zoom range.

I rarely use Digital Zoom. The maximum optical zoom range of the B700's lens is so phenomenal (1440mm) that I seldom find a need to zoom beyond that. It becomes difficult to maintain a completely steady image, even with a tripod, at magnifications greater than that. However, if you are trying to capture an elusive bird or other creature with your lens, or have some other special photographic need, I recommend you use the Digital Zoom menu option if it will help you, bearing in mind that you may want to use some of the smaller image sizes in order to preserve image quality. I suggest you keep the zoom scale in the white range if possible, and let it go into the blue range in some cases. I recommend that you avoid the full digital zoom, with its yellow zoom bar. There is usually no advantage to be gained from pushing the camera to this limit.

Digital zoom is not always available, depending on other settings in use. It cannot be used along with any of the following settings: Image Quality of Raw, Smile Timer, Zoom Memory, or AF Area Mode set to Subject Tracking. It also cannot be used with certain scene settings (Scene Auto Selector, Portrait, Night Portrait, Backlighting with HDR turned on, Pet Portrait, Easy Panorama, Time-lapse Movie, and Superlapse Movie). Focus will always be in the center of the frame when digital zoom is used. (The focus can be in another part of the frame while the lens is being zoomed through its optical range, but the focus will shift to the center when the Digital Zoom range is reached.)

Assign Side Zoom Control

I mentioned this menu option briefly in Chapter 5, in discussing the side zoom control. This switch, on the left side of the lens, ordinarily serves as an alternative zoom control. When you leave it set to that function, you may find you can hold the camera steadier by using the side control rather than the zoom lever on top of the camera. (Personally, I don't notice much difference; when using the zoom lever I can hold the camera quite steady.)

You also might want to use the side control for zoom when you have turned on the Zoom Memory menu option, because, with that option, the zoom lever can zoom only to specific focal lengths. You might want to use the side control to zoom continuously to all focal lengths when needed. If you want to use the side zoom control to handle manual focus instead, choose that option with this menu item, as shown in Figure 7-21.

If you choose manual focus, the side control can be used to adjust the focus, but you can still use the multi selector dial or the command dial to focus. I find I can focus more rapidly using the side zoom control, so, if you use manual focus often, it may be useful to use this control for initial focus adjustment and the multi selector dial or command dial to fine-tune the focus.

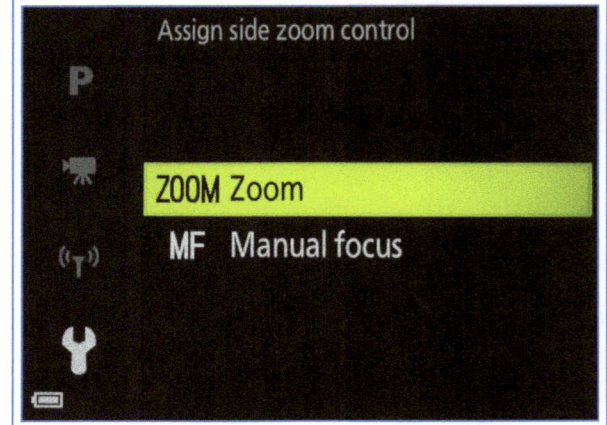

Figure 7-21. Assign Side Zoom Control Menu Options Screen

Sound Settings

This next option on the Setup menu, shown in Figure 7-22, gives you a quick way to silence all of the electronic beeps and chirps that sound off when the camera performs certain actions, such as turning on, achieving focus and exposure, or having the shutter pressed to take a picture.

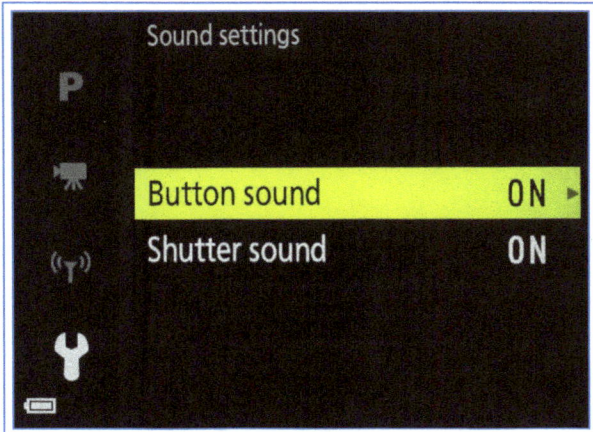

Figure 7-22. Sound Settings Menu Options Screen

There are not many options. First, you can turn the "Button sound" on or off. This option controls whether the camera beeps when it starts up, when settings are made, when it achieves focus, and when an error occurs. With the Bird-watching or Pet Portrait setting of Scene mode, sounds are automatically disabled, regardless of this setting.

The other option is to turn on or off the shutter sound, which ordinarily is heard when you press the shutter button all the way to take a picture. Again, this sound is automatically disabled with the Bird-watching and Pet Portrait settings, as well as with the Easy Panorama scene type. It also is disabled with the highest-speed settings for continuous shooting (H:60 and H:120) as well as with the Pre-shooting Cache setting.

Auto Off

This option controls the length of time before the camera enters standby mode to save power. By default, the camera will stay fully powered on for one minute when you are not touching the controls; after that time, it enters standby mode, in which the display goes blank and the green light around the power button blinks about twice per second. After about three minutes in that mode, the camera turns completely off. During standby mode, you can bring the camera back to full-power mode by pressing the power switch, the shutter release button, the Playback button, or the Movie button.

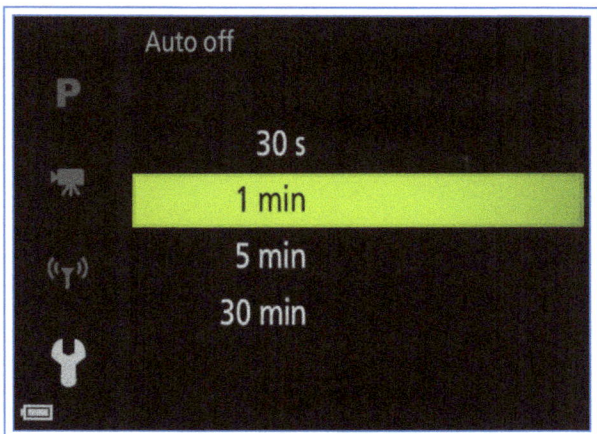

Figure 7-23. Auto Off Menu Options Screen

If you want a different interval before the camera enters standby mode, you can choose 30 seconds, five minutes, or 30 minutes with this menu item, as shown in Figure 7-23.

Note, however, that those times apply only when the camera is in shooting mode, displaying the shooting screen. When menu screens are displayed, the camera will enter standby mode in three minutes, if a shorter setting than that is chosen for Auto Off. Also, during slide show playback, the camera will stay active for up to 30 minutes, and when the optional AC adapter is connected, the time before entering standby mode will always be 30 minutes.

Format Card

This is one of the most important menu options. Choose this process only when you want or need to completely wipe all of the data from a memory storage card.

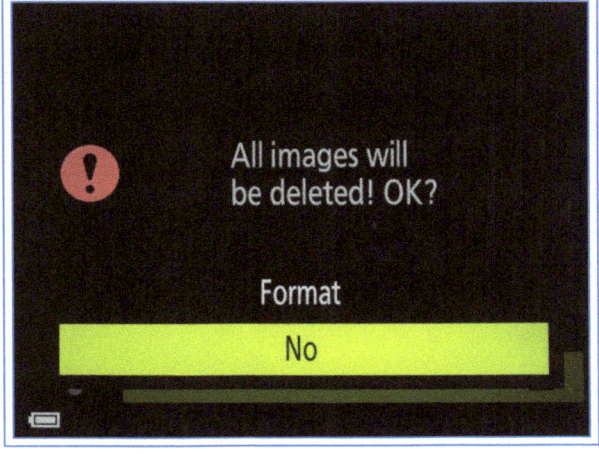

Figure 7-24. Format Card Confirmation Screen

When you select the Format Card option, as shown in Figure 7-24, the camera will warn you that all images currently on the card will be deleted if you proceed. Because of the seriousness of this step, the camera displays the selection bar in red if you highlight Format; the bar is yellow if you highlight No.

If you highlight Format with the red selection bar and press the OK button to confirm, the camera will format the card that is in the camera, and the result will be a card that is empty of images and properly formatted to store new images from the camera.

With this procedure, the camera will erase all images, including those that have been protected from accidental erasure with the Protect function on the Playback menu. It's a good idea to periodically save your good images and videos to your computer or other storage device and then re-format your memory card, to make sure it is properly set up to start recording new images and videos. It's also a good idea to use the Format Card command on any new memory card when you first insert it in the camera. Even though it likely will work without that procedure, it's best to make sure the card is set up with Nikon's method of formatting.

Here is one important note: The Format Card option cannot be selected while a wireless connection is active with the camera. This situation confused me at first, because I needed to format a card, and forgot that I had not ended a wireless session with the camera. You can select Airplane Mode from the Network menu to disable all wireless connections quickly.

The next items to discuss are on screen 3 of the Setup menu, shown in Figure 7-25.

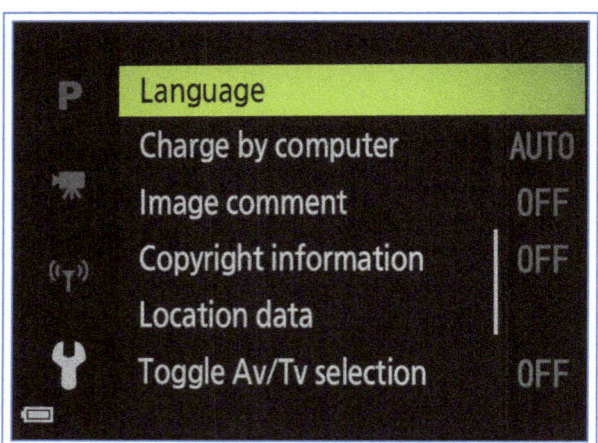

Figure 7-25. Screen 3 of Setup Menu

Language

This option gives you a choice of 36 languages for the display of commands and information on the camera's display.

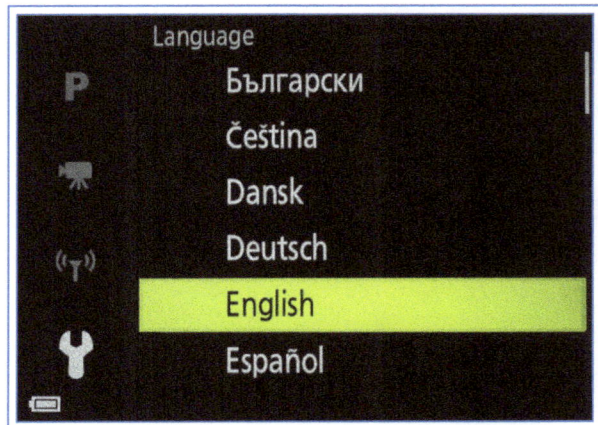

Figure 7-26. Language Selection Screen

Once you have selected this menu item, scroll through the language choices using the multi selector dial or the direction buttons and press the OK button when your chosen language is highlighted, as seen in Figure 7-26.

Charge by Computer

By default, when you connect the B700 to a computer using the USB cable, the battery is gradually charged by power from the computer through the cable. To disable this capability, choose Off instead of Auto from this menu item, whose screen is shown in Figure 7-27, and the camera will not get power from the computer.

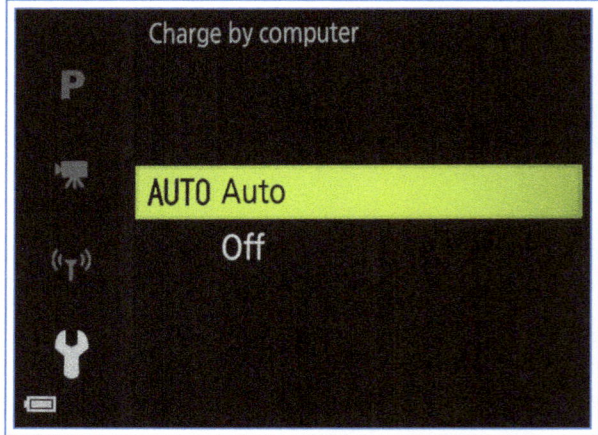

Figure 7-27. Charge by Computer Menu Options Screen

You may want to turn this feature off when you are using a laptop computer and you don't want to run down the computer's battery unnecessarily. You also should turn this option off if you are going to connect the camera to a printer for direct printing of images. Unless you have some particular need to charge your battery by this method, I recommend that you turn this option off.

Image Comment

This menu option gives you a mechanism for creating a comment to be displayed in the metadata for your still images, and also to attach that comment to all still images you take with this camera after the comment is created. The menu option has two sub-options, as shown in Figure 7-28: Attach Comment and Input Comment.

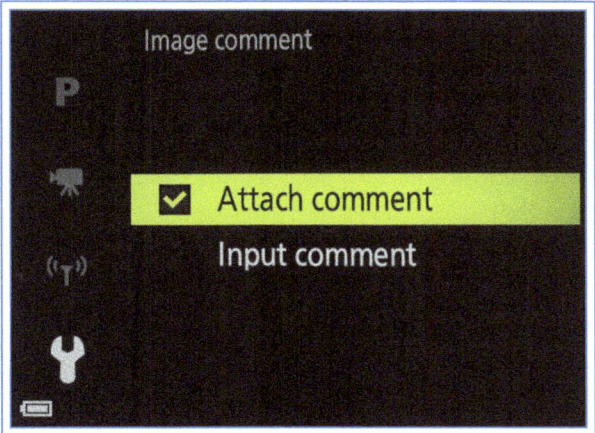

Figure 7-28. Image Comment Menu Options Screen

To create a comment that will be attached to your images, use the Input Comment option. When you select that option, the camera displays the screen shown in Figure 7-29, with a virtual keyboard.

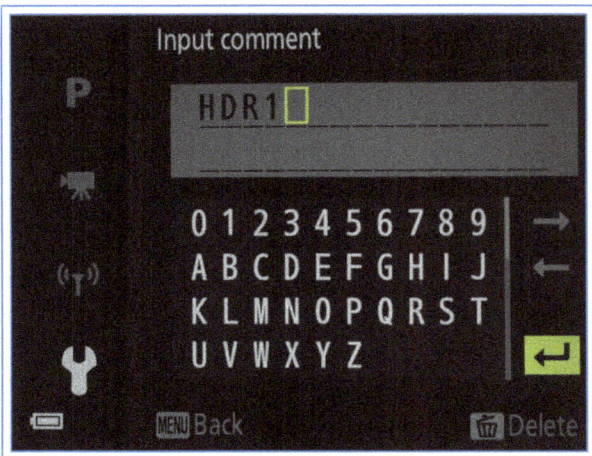

Figure 7-29. Virtual Keyboard for Input Comment Option

To use the keyboard, navigate through the letters, numbers, and symbols using the four direction buttons and the multi selector dial. Scroll down past the capital letters to reach the small letters and symbols. The space is located after the small letter z. You can scroll back and forth through the text you are entering using the top two arrows at the right side of the screen. You can enter up to 36 characters. When you have finished inputting the characters, select the Return/Enter arrow at the bottom right of the screen to accept the comment.

Once your comment has been created, highlight the Attach Comment option on the menu screen and press the OK button, to enter a check box beside that option, as shown in Figure 7-28. The comment then will appear on the copyright information screen for all still images that you enter in the future. It also will be readable in software that reads metadata, including the Nikon ViewNX-i software.

Copyright Information

This option works the same way as the previous one. You can input two items, as shown on its menu screen in Figure 7-30: Artist and Copyright.

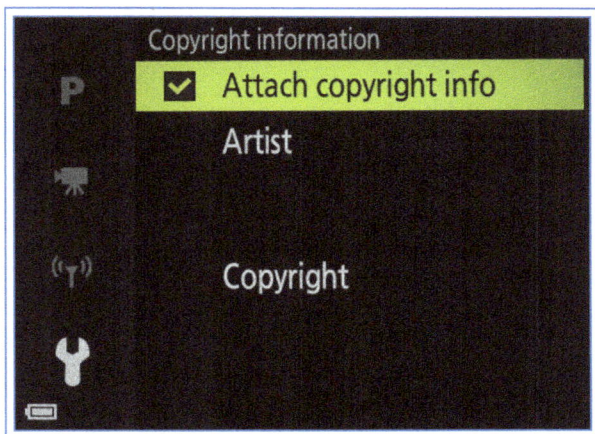

Figure 7-30. Copyright Information Menu Options Screen

With these fields, you can type in the name of the photographer and the name of the company or person who owns the copyright for the still images to be taken in the future with this camera. You can use up to 36 characters for the artist and up to 54 characters for the copyright holder. Once either or both of those fields have been filled in, select the Attach Copyright Info item to check its box and attach the information to future images.

Location Data

This menu item, whose options screen is shown in Figure 7-31, enables the camera to download location information from a smartphone or tablet that is connected to the camera using the SnapBridge app. I will discuss that process in Chapter 9. Once the information has been downloaded to the camera, you need to select the Position option to cause the camera to display the latest, updated data from the device. If the location changes, you need to select Position again to display the updated location information. Figure 7-32 shows this option in use.

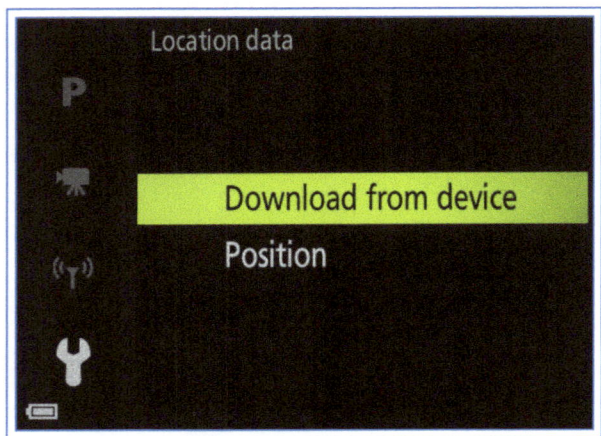

Figure 7-31. Location Data Menu Options Screen

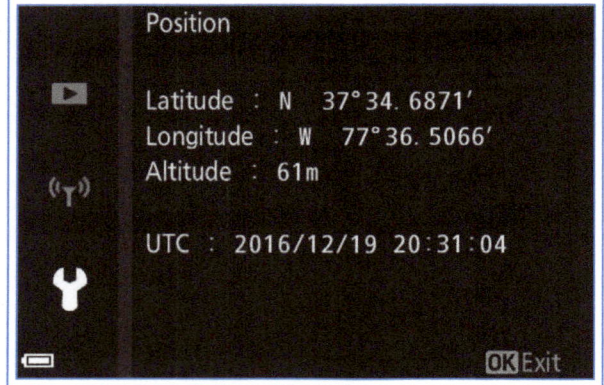

Figure 7-32. Location Data Information Displayed from Position Option

Toggle Av/Tv Selection

This menu option has just two possible settings: On or Off. The purpose of this setting is to determine whether the command dial is used to set shutter speed and the multi selector dial is used to set the aperture, as is normally the case, or not.

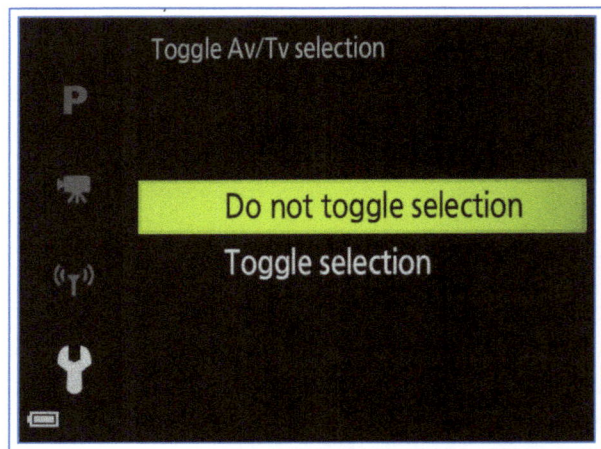

Figure 7-33. Toggle Av/Tv Selection Menu Options Screen

On the menu screen, as shown in Figure 7-33, the two choices are labeled as Do Not Toggle Selection and Toggle Selection.

If you select Do Not Toggle Selection, the dials' assignments are not changed. If you change the setting to Toggle Selection, the camera switches the functions of the command dial and the multi selector dial for setting aperture and shutter speed in Aperture Priority, Shutter Priority, and Manual shooting modes. Ordinarily, the command dial sets shutter speed and the multi selector dial sets aperture. If Toggle Selection is active, the command dial sets aperture and the multi selector dial sets shutter speed.

Using this option also affects the functions of those dials for controlling the Flexible Program option. Normally, the command dial has that function; you can re-assign that duty to the multi selector dial by setting this menu option to Toggle Selection.

Figure 7-34 shows the final screen of the Setup menu.

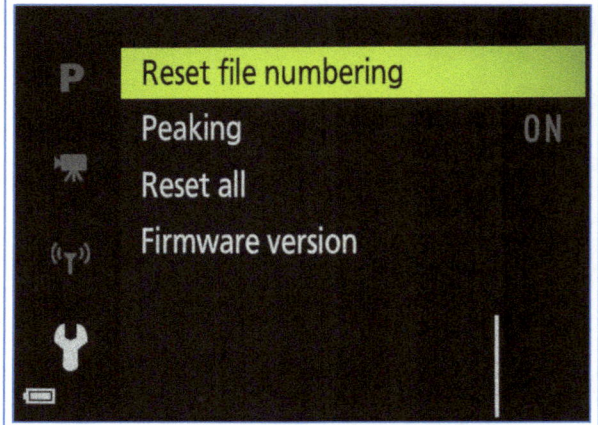

Figure 7-34. Screen 4 of Setup Menu

Reset File Numbering

This option resets the file numbering system back to 0001. Ordinarily, the camera assigns increasing file numbers to your images, even when they occupy many folders on the memory card.

For example, when you start out with a new camera and memory card, your first images will be stored in a folder named 100NIKON. The first image will be named 100NIKON-0001.jpg (or 100NIKON-0001.nrw for a Raw file). (On your computer, there will be a prefix such as DSCN, so the name may be DSCN0001.jpg or DSCN0001.nrw.) Once the 100NIKON folder has 999 files stored in it, the camera will automatically create a new folder called 101NIKON. If none of your files has been deleted, which would interrupt the numbering scheme, the first file in the new folder will be numbered 101NIKON-1000.jpg. That is, each folder can hold only 999 files before a new folder is created, but the individual files' numbers will keep increasing, even over multiple folders, until the individual file numbers reach 9999. Thus, after roughly 10 folders are filled with files, the individual numbers start over again at 0001.

If you don't like the idea of your folder and file numbers continuously increasing, you can use this menu option at any time to reset the file numbering back to the beginning. That is, if the file numbers have increased to a number such as 0476.jpg, and you don't want to wait until the numbers reach 9999 before they start over, you can invoke this procedure, select Yes when prompted by the menu, and the camera will start numbering your next image back at 0001.jpg. The folder numbers will continue to increase, however. Whenever the newest folder contains 999 files, a new folder will be created.

Peaking

The Peaking feature, which is turned on by default, provides additional assistance for accurate manual focus. When Peaking is active and you are using manual focus, the camera displays white pixels in areas that are in sharp focus, as illustrated in Figure 7-35.

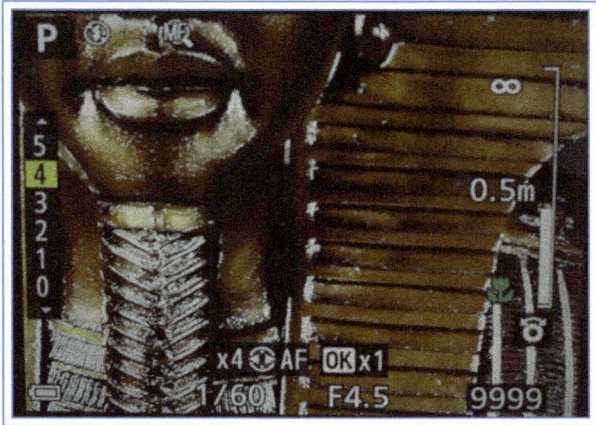

Figure 7-35. Peaking Option in Use on Shooting Screen

As the focus gets sharper, you will see thicker areas of white. Try to adjust the focus to get the white areas to be as thick and obvious as possible.

You can adjust the intensity of the Peaking feature by using the Up and Down buttons to choose a value on the scale that appears at the left of the image. The higher the value, the greater the intensity of the feature. If your subject is one with naturally high contrast, such as one with straight lines and clear differences in color or brightness, you may see the Peaking pixels more clearly with a lower value. If the subject has low contrast, you may do better with a higher Peaking level. If you find the white pixels distracting, you can, of course, just turn the Peaking feature off using this menu item.

Reset All

Choose this menu option when you want to reset all of the camera's settings to their original (default) values. This action can be useful if you have been experimenting with different settings and you find that something is not working as expected. It will give you a fresh start with known values for all of the major settings on the menus and for shooting. There are a few settings that will not be reset, including items such as date and time, time zone, and language. In addition, user settings that you selected for the User Settings shooting mode are not reset by this menu item. To reset those settings, use the Reset User Settings option on screen 3 of the Shooting menu. This option cannot be selected when a wireless connection is active. You can disable wireless connections by turning on the Airplane Mode option on the Network menu.

Firmware Version

This final option on the Setup menu lets you see the current version of the firmware that is installed in your camera. The Coolpix B700, like other digital cameras, is programmed at the factory with firmware, which is a semi-permanent set of computer instructions that are electronically implanted in the camera. These instructions control all aspects of the camera's operation, including the menu system, functioning of the controls, and in-camera processing of your images.

The reason you might want to check to see what version is installed is that, in many cases, the manufacturer will release an updated version of the firmware that may fix problems or bugs in the system, provide minor enhancements, or, in some cases, even provide major improvements, such as including new shooting modes or menu options.

In fact, with the Coolpix B700, Nikon released an upgrade of the firmware to Version 1.1 on December 27, 2016, that addressed a few issues with the SnapBridge app and with the proper recording of images in color. This upgrade can be downloaded at http://downloadcenter.nikonimglib.com/en/download/fw/213.html.

To see what firmware version is currently installed in your camera, highlight this menu option, then press the OK button or the Right button, and the camera will display the version number, as shown in Figure 7-36.

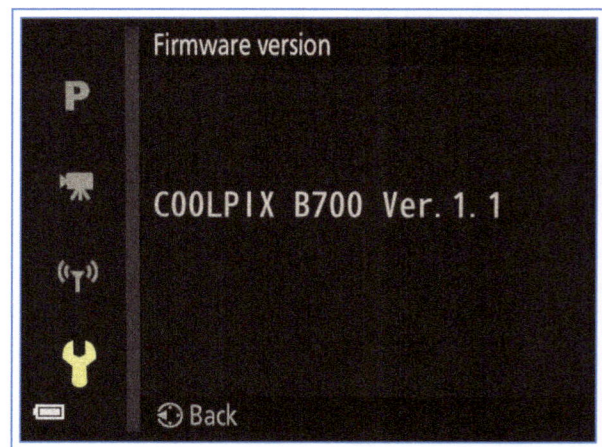

Figure 7-36. Firmware Version Displayed for Coolpix B700

To check whether any later firmware upgrades have been released, you can visit Nikon's support web site. For United States customers, the address is http://support.nikonusa.com; for Europe, the site can be found by starting at http://www.europe-nikon.com. Find the Download Center, and look for the link to current firmware versions. The site will provide detailed instructions for downloading and installing the new firmware.

Chapter 8: Motion Pictures

The Nikon Coolpix B700, like most advanced compact cameras, includes high-definition (HD) video recording among its capabilities. It also provides a limited ability to record using the higher-quality 4K/UHD format, and offers high-speed video recording, which results in slow-motion footage when you view it. I will explain the various options for movie-making in this chapter. Before I discuss the specific settings you can make for your movies, I'll begin with a brief overview of the process.

Movie-making Overview

In one sense, the fundamentals of making videos with the Coolpix B700 can be reduced to four words: "Press the red button." Having a dedicated motion picture recording button makes things easy for the user of the B700, because anytime you see a reason to take some video footage, you can just press the Movie button while aiming at your subject, and you will get results that are very likely to be usable. If you're more of a still photographer and not that interested in movie making, you don't need to read any further. Be aware that the red button sits on the camera's back just under the mode dial, and if interesting action starts to happen before your eyes, you can press that button and record the events on video with a minimum of effort.

But for B700 users who would like to delve further into the camera's motion picture capabilities, there is more information to discuss. Even though you can press the red button at any time to start recording a video sequence, there are several settings that can have a significant impact on your footage. The shooting mode the camera is set to for still images, and the menu and control-button settings you make, all have some effect on your movie recordings. So, it is helpful to be aware of the current settings, even if you just want to capture a brief clip of a scene during your vacation.

Quick Guide to Recording a Movie Clip

I will discuss the details of movie-related settings later in this chapter. For now, here are suggested guidelines for quick settings when you want to record the action and you don't care about fine-tuning menu options and other settings. I'll discuss these steps with extra detail, in case you have turned to this section before reading about the camera's various controls and menus.

1. Turn the mode dial on top of the camera, to the right of the viewfinder, so the green camera icon is at the white indicator mark, putting the camera into the Auto shooting mode, as shown in Figure 8-1.

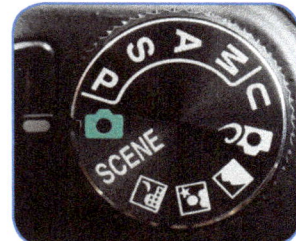

Figure 8-1. Mode Dial at Auto Mode

2. Remove the lens cap and turn on the camera with the power switch.

3. Press the Menu button at the bottom left of the control area on the right side of the camera's back, and then press the Left button (left edge of the ridged dial that surrounds the OK button), which will move the yellow selection highlight to the far-left column of menu icons.

4. Press the Down button to move the yellow highlight down to the movie camera icon.

5. Press the Right button to move the yellow selection block into the list of menu options. Using the direction buttons or the multi selector dial (the ridged dial that surrounds the OK button), highlight the top line of the menu, Movie Options.

6. Press the OK button or the Right button to get to the next menu screen, and make sure the second line is highlighted, as shown in Figure 8-2. It should say 1080/30p. (If it says 1080/25p, see the discussion of Frame Rate later in this chapter.) Press the OK button to select this option.

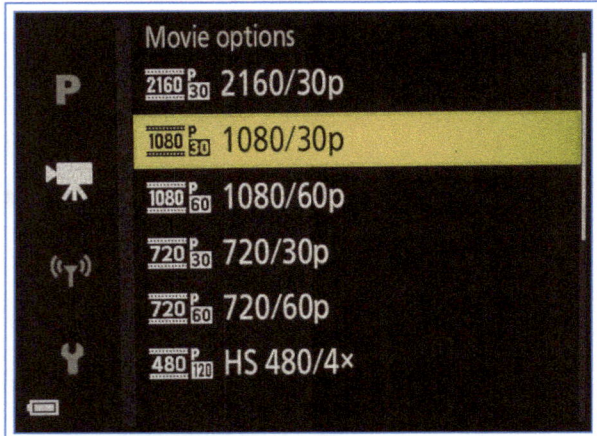

Figure 8-2. 1080/30p Highlighted for Movie Options

7. Press the Menu button to go back to shooting mode.

8. Press the Display button on the back of the camera, just below the mode dial, until you see black lines outlining the area for movie recording.

9. Aim at your subject and use the zoom lever on top of the camera to frame the scene as you want within the black lines on the display, zooming in or out as needed. (The video frame, designated by the black lines, captures only part of the overall view of the scene on the camera's display screen.)

10. Press the shutter button halfway down until you hear a beep, to have the camera evaluate exposure and focus.

11. When the action starts or you're ready to begin, press the red Movie button to start the recording. Then take your finger off the button and hold the camera as steady as possible.

12. Continue to hold the camera steady, and pan (move the camera from side to side in one direction) slowly and smoothly if appropriate to take in the scene before you. When the scene has ended, press the red Movie button again to end the recording.

Other Settings for Movies

The above steps will get you started recording video with the Coolpix B700 using one of the highest quality recording formats and standard settings for white balance, autofocus, and other options. Once you are familiar with the basic steps for movie-making, you may want to experiment with some of the other available settings. There are quite a few items that can be adjusted for recording videos with this camera.

Still Photo Settings Available for Movies

When you record movies with the Coolpix B700, several settings you make for still photos, either through controls or through the Shooting menu, carry over to your movies, as long as the camera remains set to a shooting mode in which that setting remains in effect. For example, if, as discussed above, the camera is in the Auto shooting mode, the Auto mode settings will be in effect, including autofocus and Auto White Balance.

If the camera is set to Program mode, the settings for focus and white balance will be whatever you have set through the Shooting menu. In Program mode you can select manual focus, any white balance setting, and several other options, though not all items on the Shooting menu will carry over to affect video recordings.

You also can shoot movies in Creative mode, and the movie will take on the appearance of the setting you select, such as Melancholic, Sepia, Charcoal, Somber, etc. You can record a movie when the camera is set to any Scene mode setting, though some of the settings, such as Fireworks Show or Easy Panorama, will not have their normal effect, while others, such as Dusk/Dawn, will use their normal attributes, such as altered colors.

Focus Mode

The first setting that carries over to video shooting is the focus mode, set by pressing the Down button, as shown in Figure 8-3. However, the only choices that operate for movie recording are autofocus and manual focus. Whichever setting you make will remain in effect for video shooting, as long as the camera stays set in the mode in which you made the setting. Note that you have to make the setting before you press the red Movie button to start the recording; you cannot change the focus mode once the recording has started.

Chapter 8: Motion Pictures

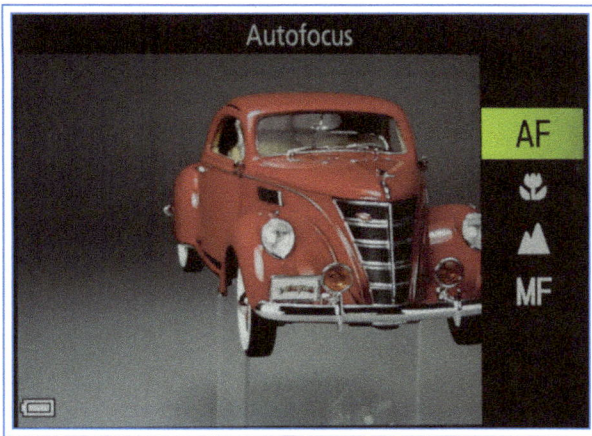

Figure 8-3. Focus Mode Menu

You can select manual focus in the P, A, S, M, or Creative shooting mode, and with the Sports, Fireworks Show, Bird-watching, Soft, Selective Color, Multiple Exposure Lighten, Time-lapse Movie (Night Sky or Star Trails), and Superlapse Movie settings of Scene mode. (Of course, if you record a movie while the camera is set for Time-lapse Movie, the Time-lapse functions will not operate.)

In those cases, you can adjust the focus manually while recording the movie, using the multi selector dial, the command dial, or the side zoom control, if you assigned it to adjust manual focus using the Setup menu. The Peaking feature does not work for movie recording, though. Later in this chapter I will discuss other focus settings you can make for movie recording.

Exposure, Exposure Compensation, and Exposure Lock

You cannot change the exposure mode used by the camera for recording movies; it will use automatic exposure adjustment, as if it were in Auto mode, no matter what mode is set on the mode dial. Even if you select Manual exposure on the mode dial and make fairly extreme settings for shutter speed and aperture, such as 1/500 second and f/7.6, the B700 will use its autoexposure programming to expose the footage as normally as possible, subject only to whatever exposure compensation and exposure lock functions you employ.

One adjustment you can make that stays in effect during video recording is exposure compensation, available in all modes except Manual exposure and the Fireworks Show, Multiple Exposure Lighten, and Time-lapse (Night Sky and Star Trails) settings of Scene mode. Whatever adjustment you make before pressing the red Movie button, to brighten or darken the image, will stay

in effect during video recording; you will see the effect on the screen in the brightness of the image. You cannot make changes to this setting during the recording.

There is one other exposure adjustment you can make for movies, and this one can be made during video recording. You will see a prompt at the bottom of the display stating that you can press the Right button for AE-L, or autoexposure lock, as shown in Figure 8-4.

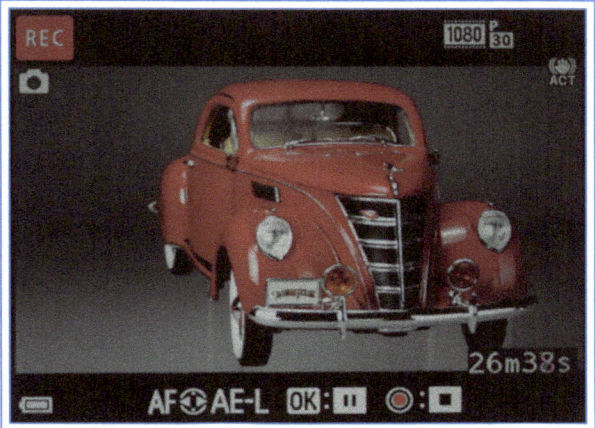

Figure 8-4. Prompt to Press Right Button for AE-L

If you press the Right button to lock exposure, the message will change to indicate that you can turn AE-L off by pressing the same button again, and a message at the top of the display will state that AE-L is in effect, as shown in Figure 8-5.

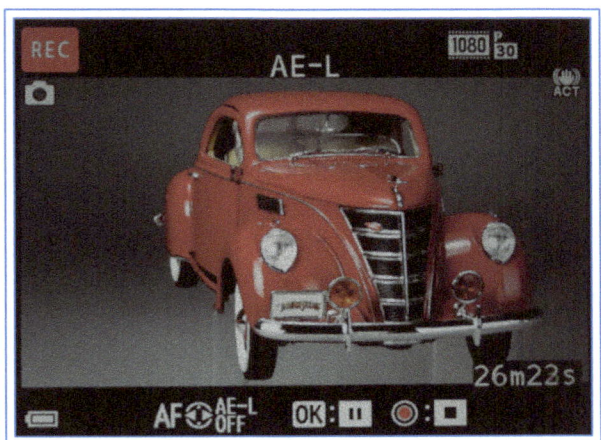

Figure 8-5. Message Showing AE-L is in Effect

Although this button is the same one used to adjust exposure compensation, in this situation the button will not affect exposure compensation, only exposure lock. However, any exposure compensation that was in effect before the recording started will remain in effect, even if you use this button to lock the exposure.

The exposure will be locked, taking into account any exposure compensation value that has been selected.

Self-timer

The self-timer also will function for movie recording. Just set it as you normally do, by using the Left button and selecting either two seconds or 10 seconds for the delay. Then, when you press the red Movie button, the camera will delay the designated length of time before starting to record.

Picture Control

Turning to menu options, the first item on the Shooting menu that works for movie recording as well as stills is the Picture Control feature, as shown in Figure 8-6.

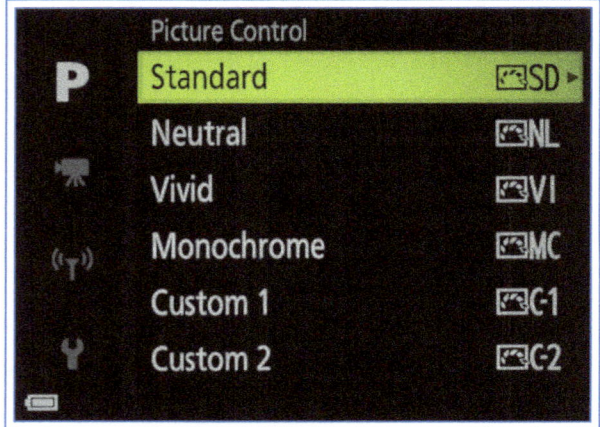

Figure 8-6. Picture Control Menu Options Screen

This menu option lets you add a distinctive style to your movie footage. You can shoot in black-and-white, for example, or you can use the Vivid setting to enhance colors. In most cases, though, you are likely to be satisfied with the Standard setting. Of course, you can only use this option for movie recording if you set it while the camera is in the P, A, S, or M mode, and if the camera remains in that mode during the video recording.

White Balance

The next Shooting menu option that carries over to video shooting is white balance. Generally, Auto White Balance is adequate, so you can usually use the Auto shooting mode with no problem in this respect. However, if you are shooting your movie indoors, perhaps with non-standard artificial lighting, you may want to turn to the P, A, S, or M mode and use one of the specific white balance settings or even the Preset Manual option. Being able to set white balance however you want it gives you the option of purposely setting a "wrong" white balance to achieve an unusual color cast. For example, if you set manual white balance using a blue surface as the standard rather than a gray or white one, your footage will take on an eerie reddish appearance, suitable for some science fiction or horror scenes, perhaps.

Metering

You can set Metering however you want it for movies, if you set it while the camera is in one of the advanced shooting modes (P, A, S, or M). Of course, normal limitations apply, so you cannot change the metering setting if Active D-Lighting is turned on.

Here, as with white balance, the standard setting (Matrix) is likely to be satisfactory for most of your shooting. However, having the option to use Center-weighted or Spot metering may be of use in some specific lighting situations. You need to be careful, though, because using Spot metering can result in a dramatic, and potentially unwelcome, shift in the lighting of a scene if the center spot is pointed at a particularly bright or dark object. The camera does not display any circle or other indicator showing the area that Spot or Center-weighted metering uses, so you will not have any reminder that a different metering mode is in effect until you see a sudden change in the scene's exposure.

Vibration Reduction

The Vibration Reduction setting on the Setup menu carries over for your movie recording. I recommend that you usually leave this setting at Normal for movies unless you have the camera on a tripod. When you are handholding the camera, this setting can reduce the shakiness that may result from an unsteady hand. You don't need to worry when you are purposely moving the camera in a panning motion, because the camera will detect that motion and ignore it for purposes of vibration reduction; it will attempt to counteract only the vertical shakiness that may be caused by your hand motions. I don't recommend using the Active setting for movie recording unless you are in an unusual situation, such as handholding the camera in a moving car on a bumpy road.

Note that there is another option on the Movie menu, Electronic Vibration Reduction, discussed later in this chapter, that can provide additional stabilization for your video footage, though at the cost of cropping away the edges of the video frame.

Zoom

The optical and digital zoom both will function during video recording. However, when the Movie Options menu item is set to 2160/30p or 2160/25p, the digital zoom operates only up to a focal length of 2880mm, rather than the 5760mm for still images. In addition, to use digital zoom while recording a movie, you have to stop at the limit of optical zoom, release the zoom control, and then start zooming again to cause the digital zoom to keep moving, whether zooming in or out.

I don't have much of a problem with this limitation of the B700. My own preference is to zoom in, if necessary, before starting the recording. I try to avoid zooming while shooting a video if at all possible, because I find that the zooming motion can be unsettling to the viewer. Also, as Nikon points out in its reference manual, the noise of the zoom mechanism is likely to be audible on your recording. So, unless you're following action that requires you to adjust your zoom range, you're probably better off adjusting it before you start recording. (Of course, if you don't mind using editing software, you can later edit out any parts of the movie where you adjust the zoom range.)

TAKING STILL IMAGES DURING MOVIE RECORDING

When you are shooting movies with the Coolpix B700, you can capture still images at a limited resolution. When a video sequence is being recorded, the camera will display a camera icon in the upper left corner of the display, as seen in Figure 8-7, meaning that still images can be captured while the recording is in progress.

All you have to do is press the shutter button while the movie is being recorded. This is a useful feature when you want to record an entire event on video, such as a school graduation ceremony, but you also want to capture still images at critical moments (such as when the graduate you are most interested in receives his or her diploma). There are several limitations to this feature, as you might expect. First, the sound of the shutter being activated will be recorded. The sound is not too loud and does not interfere with the audio track of the video too severely, but it is a clearly noticeable clicking sound.

A more significant limitation is that all of the images will be recorded at a small size and in the 16:9 aspect ratio. For example, when Movie Options is set to 1080/30p, the resulting still images are about two megapixels in size.

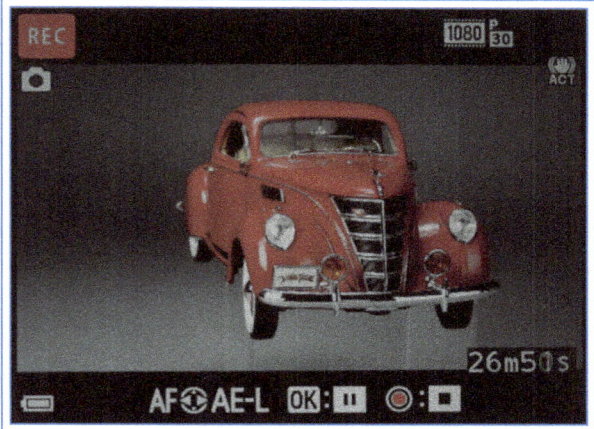

Figure 8-7. Camera Icon Showing Still Images Can be Captured

With Movie Options set to a 2160 setting, the still images are about eight megapixels in size. These images may not be useful for making very large prints, but they will provide a record of the event, and you will have them immediately available for viewing, e-mailing, or other uses.

This feature is not fully available for all video recording formats. When the Movie Options menu item is set to 2160/30p (or 2160/25p), the camera can take only 20 still images while recording a movie. (With other settings, there is no limit other than the capacity of the memory card.) The camera cannot record any still images when Movie Options is set to any of the HS (High Speed) settings. If one of those formats is selected, the B700 will display a camera icon with a diagonal line drawn through it during video recording, indicating that still pictures cannot be recorded for the current video format.

You also cannot capture a still image while a video recording is paused, or when there are fewer than five seconds of video recording time remaining.

It also is worth pointing out another option for creating still images from video recording. As is discussed at the end of this chapter, when you are playing a movie in the camera, you can save a still frame from it using the on-screen editing controls. With ordinary HD footage, this action generates a small image that may not suit your purposes. However, if you extract a frame from 4K footage (using the 2160/30p setting for Movie Options), the saved frame will have about 8 megapixels of resolution, and may serve as a more useful still image.

Pausing Recording with the OK Button

When you are shooting a video, you can pause the recording by pressing the OK button and resume it by pressing that button again. The camera indicates this capability by displaying at the bottom of the screen two vertical lines for the pause function next to a symbol for the OK button. When the recording is paused, the camera displays the vertical lines at the upper left of the display and places at the bottom of the frame a pair of icons indicating that you can press the OK button to resume recording, as shown in Figure 8-8.

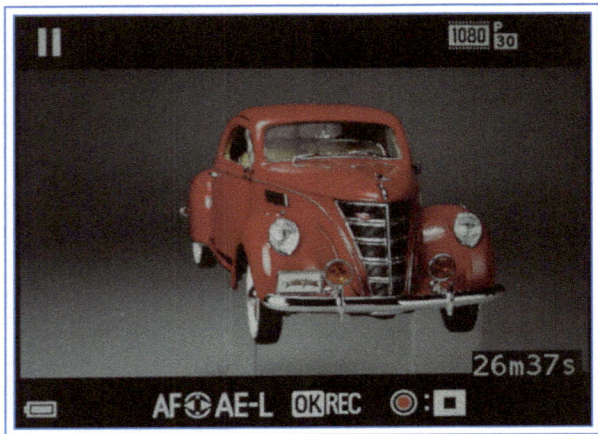

Figure 8-8. Pause Icon on Screen When Recording Paused

This function can be useful if, for example, you are recording action at a sporting event and you want to pause during a lull in the action, but you don't want to completely stop and re-start the video each time the action stops. The pause cannot last more than five minutes; after that time, the recording will terminate. This feature is not available with the HS settings of Movie Options.

Fn Buttons: Do Not Operate During Video Recording

Several of the settings you can program for the two Function buttons (Fn1 and Fn2) are not available for movie recording, such as Image Size, continuous shooting, ISO, and AF Area Mode. However, a few of the settings you can assign to these buttons do affect video recording: Picture Control, White Balance, Metering, and Vibration Reduction. The Function buttons do not operate while a movie is being recorded. However, you can, of course, press either button before pressing the Movie button, to activate whatever setting is programmed into that Function button.

Settings That Are Not Adjustable for Video Recording

Although, as discussed above, several settings for still photography are available for use during video recording, several others are not. You cannot adjust the ISO, aperture, shutter speed, Autofocus Area Mode, or Active D-Lighting. Several other settings from the Shooting menu do not apply for shooting movies, either because there are specific settings for movies (Image Quality, Image Size, and Autofocus Mode) or because they apply only to still photography (exposure bracketing, flash exposure compensation, noise reduction filter, multiple exposure, and Manual Exposure Preview).

The Movie Menu

When the camera is in shooting mode, you can get access to the Movie menu, represented by the movie camera icon, as shown in Figure 8-9.

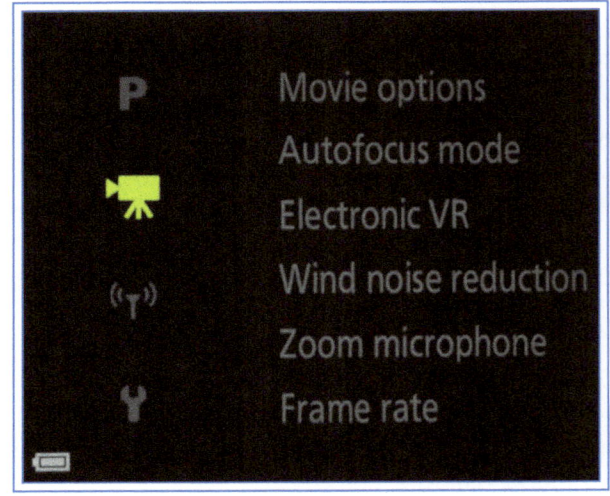

Figure 8-9. Icon for Movie Menu Highlighted at Left

To reach this menu, press the Menu button, then use the Left button to move the selection highlight to the left column and use the Up or Down button or the multi selector dial to move to the movie camera icon. Then use the Right button to move the highlight back into the list of menu options on the main part of the screen, as shown in Figure 8-10. There are six main options on the single screen of this menu.

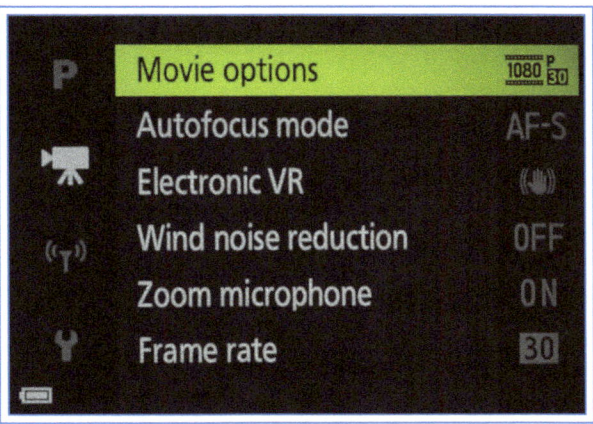

Figure 8-10. Single Screen of Movie Menu

Before I discuss the specifics of these options, I will point out some limitations of the Coolpix B700 with respect to the length of its video sequences. Like many modern digital cameras that are not primarily video cameras, the B700 is limited to recording only about 29 minutes or four GB of video, whichever comes first, in any single sequence. The duration is slightly less, between 26 and 28 minutes for some formats; 29 minutes is available only with Movie Options set to the 720/30p or 720/25p setting. With Movie Options set to the 1080/60p or 1080/50p setting, the recording time is only 13 minutes. The camera can record only about seven minutes of video in a single sequence when Movie Options is set to 2160/30p or 2160/25p.

You can store considerably more than 29 minutes of high-quality video on a large SD card, but you cannot record more than about 29 minutes (or less, as noted above) at a time; you have to then stop and re-start your recording. And, as noted, any one sequence cannot exceed four GB in size. So, choosing a lower-quality video format that lets you store a great deal of video may not mean as much as it would if you could store a very long single sequence.

Frame Rate

As you can see in Figure 8-10, the six Movie menu options are Movie Options, Autofocus Mode, Electronic VR, Wind Noise Reduction, Zoom Microphone, and Frame Rate. Ordinarily, I would discuss these options in that order. However, the selection you make for Frame Rate determines what choices you can make for the first item, Movie Options, so I will discuss Frame Rate first.

There are two major video standards used in various countries—NTSC and PAL. NTSC is used in the United States, Canada, much of South America, South Korea, and other areas; PAL is used in Europe and some other areas. With the NTSC system, the standard rate for video playback is 30 frames per second; with the PAL system, it is 25 frames per second. These differences are reflected in the Movie Options menu choices.

If you are using your camera with the NTSC system, the selections for the HD formats will include the number 30 or 60, such as 1080/30p or 1080/60p. If you are using the B700 with the PAL system, the options will include the number 25 or 50, as in 1080/25p or 1080/50p. Similarly, the primary choices for the HS (high-speed) options will include numbers that are multiples of 30 for NTSC and 25 for PAL. The final HS option, for recording video at half-speed, will include the number 15 for NTSC or 12.5 for PAL.

Some camera makers sell different versions of a camera for the North American (NTSC) and European (PAL) markets. With the Coolpix B700, though, Nikon provides the Frame Rate menu option to let you set your camera to whichever system you prefer to use. The choices for this menu option are shown in Figure 8-11.

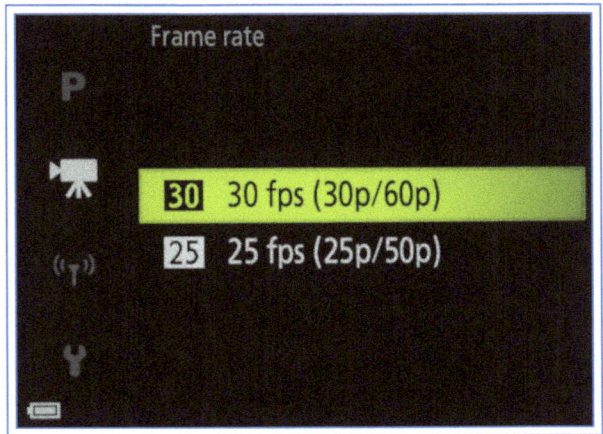

Figure 8-11. Frame Rate Menu Options Screen

If you are located in North America or another area that uses the NTSC system, you should ordinarily set Frame Rate to 30 fps; if you are in Europe or otherwise using the PAL system, you should ordinarily set Frame Rate to 25 fps. However, if you prefer using one of these systems over the other, it is not critical to use the system intended for your geographical area.

Now that I have discussed Frame Rate, I will discuss the first item on the Movie menu, Movie Options. I will assume in this discussion that the NTSC/30 fps system is being used, because that is what is used in the United States, where I am located. If you are using the PAL system, you should substitute 25 or 50 for the number 30 or 60 in this discussion.

Movie Options

The first choice on the Movie menu, whose main options screen is shown in Figure 8-12, lets you set the aspect ratio, quality, and speed of your video footage from eight possibilities.

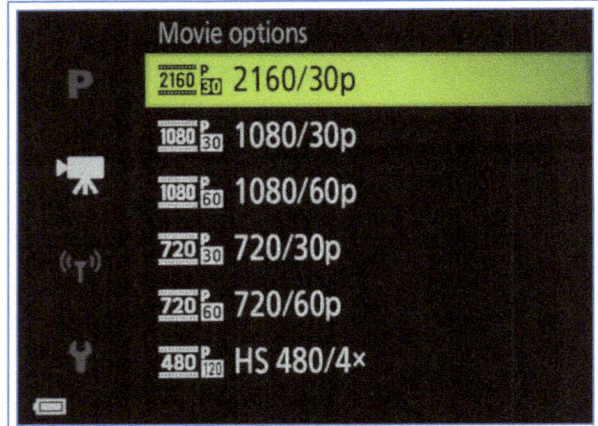

Figure 8-12. Movie Options Menu Screen

The top option is designated as 2160/30p if Frame Rate is set to 30, or 2160/25p if Frame Rate is set to 25. This option sets the camera to record video in a 4K/UHD format, meaning the format's frames have a horizontal count of roughly 4,000 (4K) pixels, also sometimes designated as UHD for ultra-high definition. (The actual frame size is 3840 by 2160 pixels.) The letter "p" following the number 30 stands for "progressive," which provides higher quality than formats designated with the letter "i," where the "i" stands for "interlaced." (There are no interlaced formats available with the B700 camera.)

This format provides the highest quality of video footage for the B700 camera, but naturally requires the most processing power and memory card capability to record and manipulate. If you use this format, Nikon recommends that you use a memory card rated in UHS Speed Class 3, the fastest currently available. If you use a slower card, the camera will attempt to record the video, but recording may stutter or halt. The camera can record only about seven minutes of continuous video in this format. With this setting, Electronic VR cannot be turned on, and digital zoom will not function beyond 2880mm. Also, only up to 20 still images can be captured while recording with this option.

To take full advantage of this format, the footage should be displayed on a TV set that is capable of showing 4K video. However, even if you don't have access to a 4K TV, video captured in this format can be viewed on any HDTV set. It also can be edited and down-sampled to a normal HD format such as 1080/30p.

The second selection is designated as 1080/30p, which means it provides high-definition video with 1080 vertical pixels and 1920 horizontal pixels, the same amount as on many HDTV sets. This standard is sometimes called "Full HD" to distinguish it from the lesser-quality HD that provides only 720 vertical pixels and 1280 horizontal pixels.

The third option on the Movie menu screen, 1080/60p, provides the same quality as the top choice on the menu, but, instead of 30 frames per second, it provides 60 frames per second. This option provides higher quality than the 30p choice, because the camera records more information. The camera records 60 full frames of video each second, but converts the footage internally so it will play back at 30 frames per second without looking slowed down.

However, if you import your 1080/60p footage into video-editing software such as iMovie, Windows Movie Maker, or any other up-to-date program, you can slow it down to one-half normal speed, producing a smooth slow-motion effect. So, if you want higher quality for your HD video or you want to slow it down to half-speed on your computer, you can choose this option.

The next option on the menu, 720/30p, produces footage with 720 vertical and 1280 horizontal pixels, which still provides high-definition video in the 16:9 aspect ratio, but with fewer pixels and somewhat reduced quality. Choose this option if you want HD, but with less-taxing storage and speed requirements for your memory card and computer.

Next, you can choose the 720/60p option, which also records your footage with 720 vertical pixels and 1280 horizontal ones. This choice is similar to the 1080/60p option discussed above, except that it has the reduced resolution of regular HD as opposed to Full HD. Here again, you can slow down the footage to half-speed using editing software.

HS (High Speed) Movie Options

The next choice on the Movie Options menu, HS 480/4x, is the first one for recording silent HS (high-speed) movies with the Coolpix B700. I have not discussed the HS capabilities of the camera before now, so I will take this opportunity to explain the use of this interesting feature.

The abbreviation HS, for high-speed, is really a misnomer, because the HS choices include both high-speed and low-speed options. It would be more accurate to use a term such as "non-standard-speed." HS is a convenient shorthand, but bear in mind that it is not precisely accurate.

Here is a brief explanation of how the HS feature works. The standard rate for recording and playing back video (in the United States) is 30 frames per second. That is, the camera takes 30 individual images each second and then plays them back at that same rate. When your eyes see those images, the pictures follow each other so rapidly that it seems as if the motion in the scene is continuous, rather than 30 separate still photos, which is the actual situation.

If you have seen old silent movies, they sometimes seem unnaturally fast and jerky. That happens because those movies were recorded at a slower speed than movies of today, but sometimes are played back on modern projectors at a faster rate. If a movie was recorded at, say, 15 frames per second, and then played back at 30 frames per second, the action in the movie would appear to be twice as fast as normal, resulting in a jumpy, jerky, speeded-up appearance. Similarly, if you set a camera to record at 60 frames per second and then play back the footage at 30 frames per second, the action will appear to be slowed down to one-half its normal rate.

The Coolpix B700 gives you the ability to either increase or decrease the frame rate at which it records video footage. It's important to note that the video will always play back in the camera at the standard 30 frames per second; the only factor you can change is the speed at which the video is recorded.

With that background, I will discuss each of the HS options on the Movie Options menu screen.

The first choice, HS 480/4x, sets the camera to record video at 120 frames per second, four times faster than the normal 30 fps. (If you have set Frame Rate to 25 fps, the recording speed will be 100 frames per second.) When this footage is played back in the camera, any movement will appear to be at one-fourth normal speed. This setting provides a capability for slow-motion video, which you can use to analyze a golf swing, slow down the beating of a hummingbird's wings, or for any of a myriad of sports and nature applications. Or, you might just like slow-motion video for its dreamlike, underwater-style appearance.

One major caveat with this setting is that, not surprisingly, the use of this setting requires a sharp trade-off of speed against quality. When you set the camera to record 120 frames per second, it automatically reduces the quality to VGA, which provides noticeably lower quality than HD. The aspect ratio is 4:3. There also is another limitation: The camera can only record for seven minutes and 15 seconds at this rate, which results in a playback time of 29 minutes, the limit for video playback. However, for many applications, that amount of time should be sufficient.

The other choices for Movie Options appear on the second screen of the Movie Options menu item, as shown in Figure 8-13.

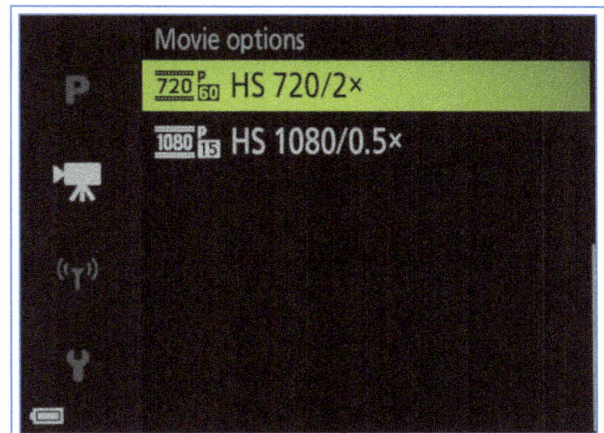

Figure 8-13. Screen 2 of Movie Options Menu Item

The next setting, HS 720/2x, gives you half as much slowing of motion as the first one, with higher image quality. In this case, your footage is recorded at 60 frames per second (50 frames per second if Frame Rate is set to 25 fps), so it will play back at one-half normal speed, and at 720 HD quality. You can record at this speed for 14 minutes and 30 seconds, again resulting in the full 29 minutes of playback time.

The last option for this menu item, HS 1080/0.5x, turns the whole exercise in a different direction. This is the only setting with which the camera records video at a slower than normal speed. In this case, in which the footage is recorded at 15 frames per second (12.5 if Frame Rate is set to 25 fps) and played back at the normal speed (30 or 25 frames per second), the footage will appear to be speeded up to twice the normal speed. And, as a bonus, because the camera is actually doing less work in terms of speed, it can provide higher-quality video: full HD, at 1920 by 1080 pixels, recording for 29 minutes, with a playback time of 14 minutes and 30 seconds.

This option could be used to create a movie with the speeded-up look of old silent films, maybe to inject a light touch into a business presentation, or just for fun with footage of the family at the beach. You also could use this mode for some situations in which you need a video record, but you don't need (or want) to have a real-time recording. For example, to record patterns of vehicle traffic at a street intersection, you could set up the camera on a tripod, set it to the HS 15 fps mode, and you would then have a video that would reveal the patterns at twice the normal speed, which actually might be easier to interpret than a real-time movie.

Of course, shooting half-speed movies (to show them at double speed) is a mild form of time-lapse photography. If you want to do time-lapse photography for applications such as showing a flower opening up or recording the progress of a construction project, you probably would be better off using the Time-lapse Movie setting of Scene mode, discussed in Chapter 3, or the interval timer function of the Continuous shooting item on the Shooting menu, discussed in Chapter 4.

With all of the HS options, the video is recorded with no sound. The zoom position of the lens, the focus, exposure, and white balance are all set when the Movie button is first pressed; no further adjustments to those settings can be made while the video is being recorded.

Autofocus Mode

This second choice on the Movie menu controls how the camera focuses when recording videos.

Your focus options for movie-making with the B700 are a bit tricky, so I'll discuss them again here. First, it's important to remember that the shooting mode you select has an impact on focus options for video recording.

That is, if you choose the Auto shooting mode, the camera will use autofocus for video shooting. However, if you choose the Program, Aperture Priority, Shutter Priority, Manual, or Creative exposure mode, or the Sports, Fireworks Show, Bird-watching, Soft, Selective Color, Multiple Exposure Lighten, Time-lapse Movie (Night Sky or Star Trails), or Superlapse Movie setting of Scene mode, you can choose manual focus if you wish, and that choice will carry over to video shooting.

If you want to use the Autofocus Mode menu option while shooting a movie, make sure you have the camera set for autofocus, not manual focus. If the camera is set for manual focus, you will still be able to set the Autofocus Mode option on the Movie menu, but it will have no effect. When you press the red Movie button to start recording, you will see the MF indication on the screen, indicating that manual focus is in effect, as shown in Figure 8-14.

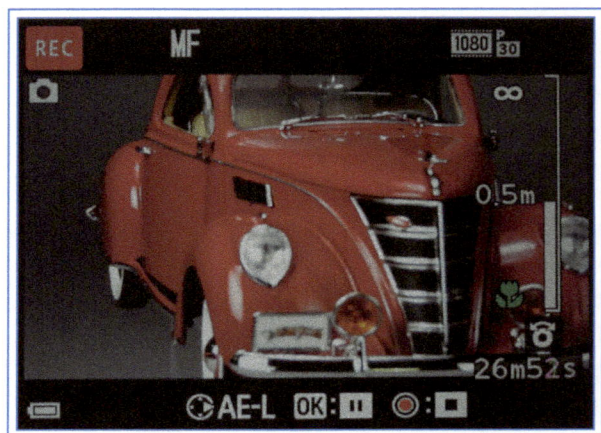

Figure 8-14. MF Icon on Screen During Video Recording

At that point, you cannot switch the camera into autofocus mode; you will have to focus manually using the multi selector dial, the command dial, or the side zoom control, if you have programmed that control to handle manual focus.

Assuming you have activated autofocus, you have the choice of two options for Autofocus Mode on the Movie menu, as shown in Figure 8-15: AF-S for Single AF, or AF-F for Full-time AF.

These labels are self-explanatory. If you select AF-F, the camera will continually adjust its focus as the scene changes. The camera will focus on any object in the center of the frame. The advantage of this mode is that the focus will remain sharp (with some blurring during focus adjustments) throughout the duration of the scene; the disadvantage is that the battery will be drained more quickly because of the demands on the autofocus mechanism.

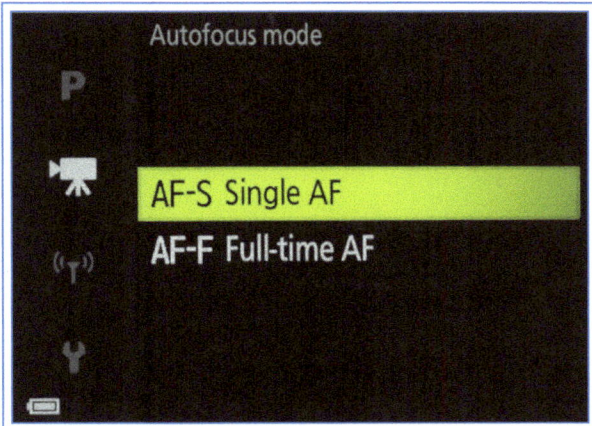

Figure 8-15. Autofocus Mode Menu Options Screen

The Nikon user's guide says that, with this setting, the camera is likely to record the sounds of the continuous focus adjustments, but I have not found any such sounds to cause problems. The zoom mechanism makes sounds that can be distracting, but the continuous autofocus sounds have been moderate, in my experience. Therefore, if you are in a situation in which the distance to the subject may change during the recording, I recommend that you use the AF-F setting.

If you select the AF-S setting, the camera will initially focus just once, when you start recording. However, while the recording is in progress, you can force the camera to re-focus at any time by pressing the Left button. You will see a prompt at the bottom of the screen with an icon that shows that button next to the letters AF, as shown in Figure 8-16.

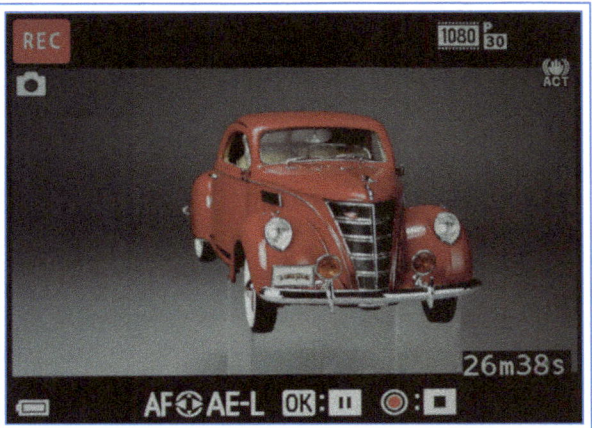

Figure 8-16. Prompt to Press Left Button for AF

So, you don't need to worry that the scene will stay out of focus if the distance to the subject changes; just press the Left button and the lens will focus again.

I recommend that you choose your autofocus setting according to the situation. For example, if you are recording in an environment where the focus distance is likely to keep changing and you have sufficient battery power to last for the entire recording session, I recommend using the AF-F setting so the camera continues to adjust its focus as needed.

However, if you are recording a school play or concert, where the focus distance should not change that dramatically and the battery needs to last for a long time, you may be better off using the AF-S setting, with the knowledge that you can adjust the focus at any time by pressing the Left button if necessary. Also, if you want the focus to remain sharp on a particular subject at a fixed distance, you may want to choose AF-S so the focus is not changed if another subject moves temporarily into the center of the scene.

Electronic VR (Vibration Reduction)

This third option on the Movie menu controls whether the camera uses electronic vibration reduction. This option is another type of image stabilization, in addition to and independent of the Vibration Reduction setting on the Setup menu. The VR setting on the Setup menu uses lens-shift technology, which means that, when the camera detects image blur from movement, it moves the lens in the opposite direction to counteract the blur. The Electronic VR setting uses a different method. When the camera detects blur, it uses image processing to eliminate the blur as much as possible. With this method, the resulting image will lose some pixels at the edges as the camera processes the information to remove blur. In other words, the video image will be slightly cropped at the edges when Electronic VR is used.

It is possible to use both Vibration Reduction from the Setup menu and Electronic VR from the Movie menu when shooting movies. My recommendation is to turn off both types of VR when you are using a tripod. When you are shooting handheld in normal conditions, I would use VR from the Setup menu. When you are shooting in more unsettled conditions, where more stabilization is needed, you may want to use Electronic

VR in addition to the Normal or Active setting for VR on the Setup menu.

Electronic VR is not available when recording movies in any of the HS formats or with Movie Options set to 2160/30p or 2160/25p. Electronic VR is always on when the Superlapse Movie setting of Scene mode is in effect.

WIND NOISE REDUCTION

This next option controls the use of a feature designed to reduce extraneous sounds that may be recorded in windy conditions. I recommend leaving this option turned off unless the wind noise appears to be a problem, because this feature reduces the sounds that are recorded at certain frequencies. I prefer to record all of the available sounds and adjust them later using editing software if necessary. If you won't be using editing software, you may want to use this feature on a windy day. This option is not available with the HS formats for movie recording.

ZOOM MICROPHONE

When this option is turned on, as it is by default, the camera adjusts the angle of the microphone's sensitivity to sound according to the zoom position of the lens. So, when the lens is zoomed back to its wide-angle position, the microphone picks up sounds in a wide pattern. As the lens is zoomed in, the angle of sound recording is narrowed in accordance with the angle of view of the lens.

I recommend you leave this feature off unless you are sure you want to have it operate. For example, if you are recording a stage play and have zoomed in on the actors, you might want to have this option turned on so the camera will pick up the sounds of the play more clearly than the sounds from the audience. However, if you are recording general scenery, you may want the camera to pick up sounds from every direction, even if you have zoomed in on a distant mountaintop or ship at sea.

FRAME RATE

I discussed the Frame Rate option earlier, because of its importance in determining what movie formats you can choose for your recordings.

Movie Playback and Editing

In Chapter 2, I discussed the fundamentals of movie playback. Now it's time to go into more detail about that topic and to discuss how to edit your video footage in the camera.

PLAYBACK

When the camera is in full-screen playback mode, you can recognize a movie by the movie format icon in the lower right corner of the display, as seen in Figure 8-17.

On index screens of four, nine, or 16 images, you can recognize a movie by the sets of small gray blocks that look like movie film sprocket holes on the sides of the images, as in Figure 8-18.

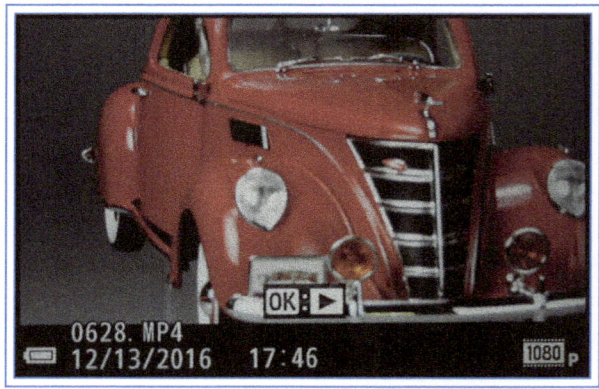

Figure 8-17. Movie Ready to Play in Camera

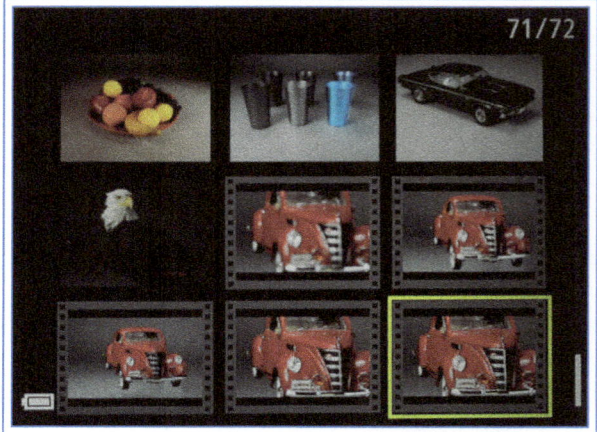

Figure 8-18. Movie Files Shown on Index Screen

(With screens of 72 thumbnails, the sprocket holes do not appear, but, when a movie is highlighted, a movie camera icon appears at the top left of the display.)

With a movie's frame on the display, press the OK button to start it playing. You will see a line of DVR-like icons at the bottom left, as seen in Figure 8-19.

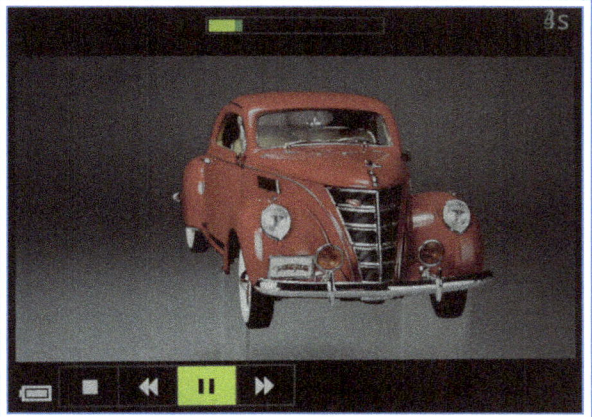

Figure 8-19. Initial Movie Playback Controls

Use the Left and Right buttons to move to any of those icons, then press OK to choose that function. The controls are, from left, stop, rewind, pause, and fast-forward. You can also turn the command dial or multi selector dial to the left or right to rewind or fast-forward the movie. You can press the zoom lever to the right or left at any time to raise or lower the audio volume.

While the movie is paused, there will be a somewhat different group of icons, as shown in Figure 8-20.

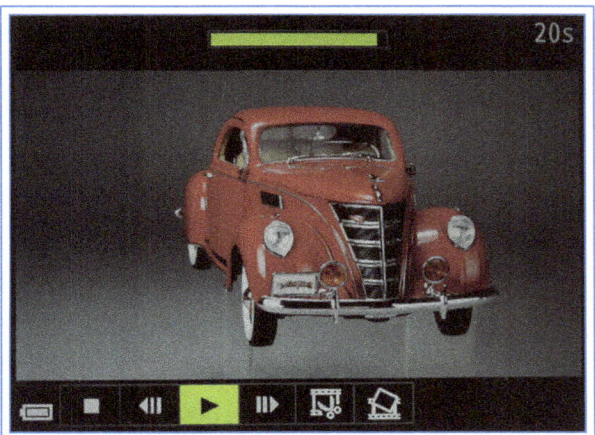

Figure 8-20. Playback Controls When Movie Paused

From the left, those icons represent stop, single-frame back, play, single-frame forward, edit, and extract and save a single frame. If you hold down the OK button while highlighting the single-frame forward or back icon, the frames will advance continuously, one at a time, at a slow rate. You can also use the command dial or the multi selector dial to advance and rewind the frames at that rate. If the battery has very little power remaining, movie editing functions are not available and the edit icon will not even appear.

EDITING

Of course, you cannot do full video editing in the camera; if you want to get really involved in editing, you need to import your footage into a program that has serious editing capabilities, like Adobe's Premiere Pro or Premiere Elements, or Apple's Final Cut or Final Cut Express. If you don't want to purchase a dedicated editing program, if you're a PC user you may already have Windows Movie Maker; Mac users often have iMovie available. Finally, you can use Nikon's software suite, which includes Nikon Movie Editor software.

However, if you're on a camping trip away from your computer or you need to put together a quick video to play on a hotel's TV screen, you can do some basic trimming of your B700 video files in the camera. (As noted earlier, you cannot perform in-camera editing when the battery is low.) Here is what you can do.

First, you can save a portion of an original video to a new file by trimming away footage at the beginning and/or end of a clip. To do this, start by playing the video to the approximate location where you want the new, shorter clip to start. Then pause the clip by pressing the OK button, use the Right button to highlight the scissors icon, and press OK to select that icon.

You will see a vertical menu of icons, from top to bottom: Choose Start Point; Choose End Point; Preview; Save; and Back, seen in Figure 8-21. There will be a yellow bar at the bottom of the screen, with a yellow pointer at the left end and a gray pointer at the right end.

The top icon, Choose Start Point, will be highlighted. Use the Left and Right buttons (or the command dial or multi selector dial) to move the pointer to the starting point for the new clip. If you have to move the pointer more than a few seconds, it may take quite a while, because each press of a button or turn of a dial moves the pointer only a fraction of a second in the clip. You can hold down the Left or Right button to move more quickly through the footage.

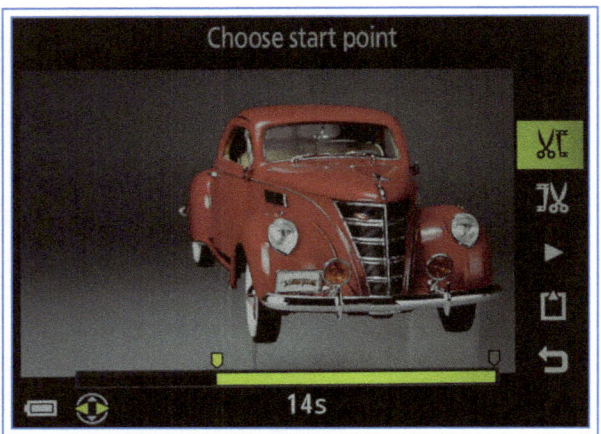

Figure 8-21. Choose Start Point Highlighted on Menu of Movie Editing Icons

When you have finished moving the left (start) point, press the Down button to highlight the second icon, for Choose End Point, as shown in Figure 8-22.

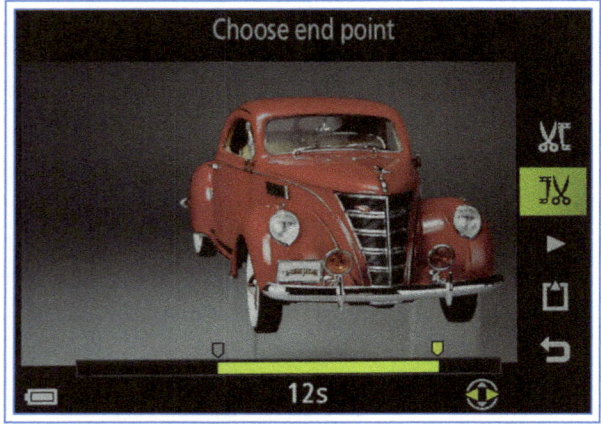

Figure 8-22. Choose End Point Highlighted on Menu of Movie Editing Icon

Repeat the previous procedure, but move the right (end) point in toward the center of the yellow bar. When both the start and end points are set as you want them, highlight the third icon, which looks like a Play button; this control lets you preview the adjusted clip. If the preview looks okay, press the OK button to stop it if necessary, and move down to the next icon, a rectangle with a small triangle at its top. When that icon is highlighted, as shown in Figure 8-23, press the OK button, and the camera will display a message saying Save OK? asking if you want to save the clip in its new length. The camera will place Yes and No bars at the bottom of this screen. If you want to save the clip, highlight the Yes bar and press the OK button. The camera may take quite a while to save the new, shorter version of the clip; the original will remain untouched.

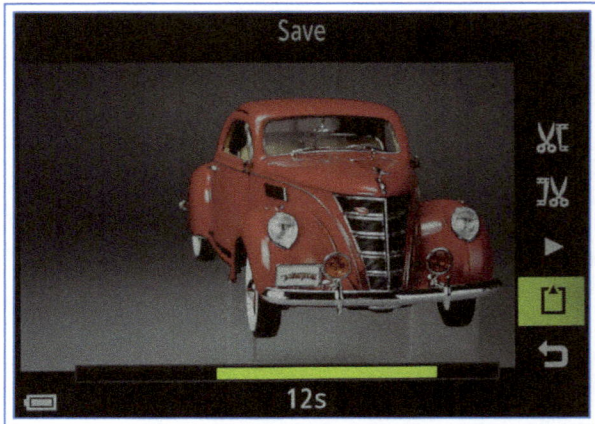

Figure 8-23. Icon to Save Movie Clip Highlighted

Finally, you can save a single frame from any video clip, except for clips made using any of the HS formats. To do this, start playing the movie to the approximate point where you want to extract a frame, then press the OK button to pause the movie. You will then see the icons for playing, advancing, or reversing by single frames, as well as editing and saving a single frame.

Use the advance and reverse controls to move to the exact frame you want to save. Then, highlight the icon at the far right that looks like a frame next to some movie footage, and press the OK button.

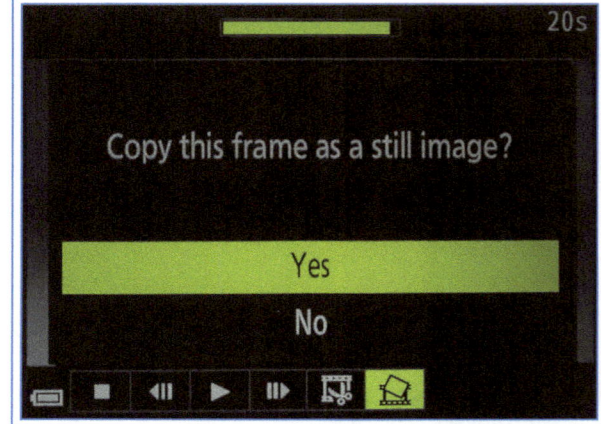

Figure 8-24. Confirmation Screen to Save Still Frame

When the camera displays a message asking if you want to copy that frame as a still image, as shown in Figure 8-24, highlight the Yes bar and press the OK button to confirm.

The camera will display the new still image with a .jpg file name. The still picture will be saved with Normal quality, at the same image size as that of the format of the movie it was extracted from.

Chapter 9: SnapBridge App, Superzoom Lens, and Other Topics

SnapBridge App

Although earlier Coolpix models, such as the P610 and the P900, have the ability to transfer images and to control the camera remotely over a Wi-Fi network from a smartphone or tablet, the Coolpix B700 takes an updated approach, using an app called SnapBridge. The SnapBridge app lets the B700 connect to devices that use iOS (such as iPhone, iPad, and iPod Touch) or Android (such as Samsung Galaxy phones and tablets) using a combination of Bluetooth and Wi-Fi. I will not discuss every feature of the app in this section, but I will describe the basic procedures for transferring images from the camera to a smart device and for controlling the camera remotely from a smart device. For more information about the app, see the online help at http://nikonimglib.com/snbr/onlinehelp/en/index.html.

Initial Connection and General Overview

As noted above, the SnapBridge app lets a smartphone or tablet connect to the camera through a Wi-Fi network, as similar apps do for many other cameras. The notable difference with SnapBridge is that it also lets a smart device connect to the camera by Bluetooth, a low-power local network commonly used to connect speakers, headphones, or other devices to a phone or tablet. The Bluetooth connection makes it possible for the camera to upload new images to your phone or tablet as soon as they are taken, automatically. This connection can remain active even when the camera is turned off, so images stored on the camera's memory card can continue to upload to your phone or tablet.

When you first set up a new B700 camera, the camera will prompt you to set up a connection using the SnapBridge app. When it does, follow the prompts if you wish. For now, I will assume that you decide to set up the connection later. Following is a summary of the steps to take to set up the connection, which involves pairing the camera with your smartphone or tablet over Bluetooth, using the SnapBridge app. I will illustrate the discussion using an iPhone, though the steps are similar for an Android device.

1. First, download the SnapBridge app from the App Store for Apple devices or from Google Play for Android devices. The icon for the app on an iPhone is shown in Figure 9-1.

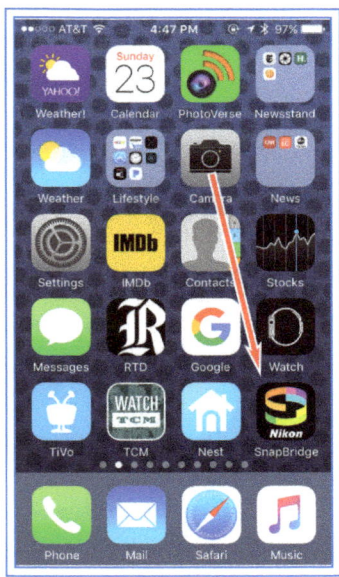

Figure 9-1. Snap Bridge App Icon on iPhone

2. On the camera, go to the Network menu (marked by a Wi-Fi icon) and highlight Connect to Smart Device, as shown in Figure 9-2. Press the OK button or the Right button to activate this option.

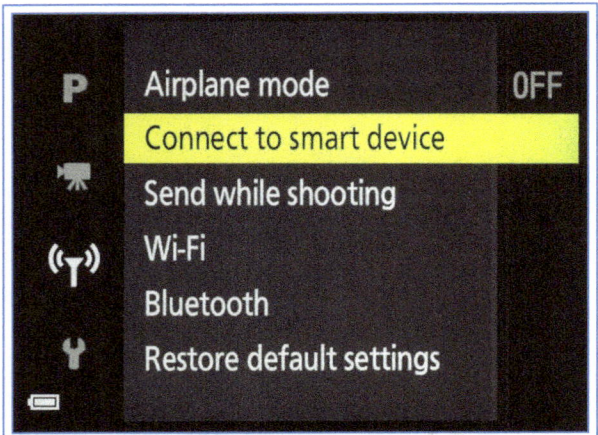

Figure 9-2. Connect to Smart Device Highlighted on Network Menu

3. When you see additional screens with text, press OK once or twice until you see a screen like that shown in Figure 9-3, prompting you to obtain and use the SnapBridge app.

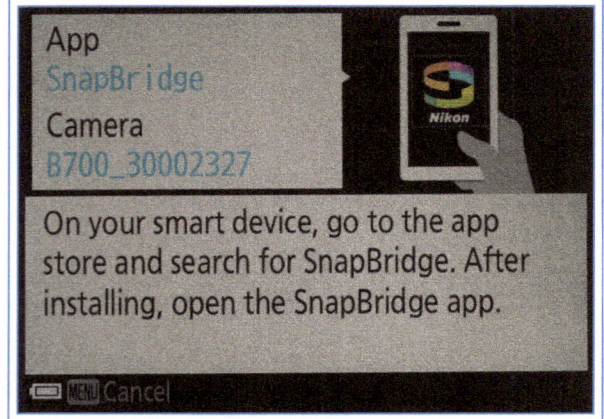

Figure 9-3. Camera Screen with Prompt for Using SnapBridge

4. On your smart device, open the SnapBridge app.

5. Follow the prompts on the camera's screen. On the SnapBridge app, you should see a Connect screen like that shown in Figure 9-4.

6. On that screen, tap the exclamation point at the top. On the next screen, you should see the name of the camera, which should be something like B700_30002327, as shown in Figure 9-5.

7. Press on that name, and another screen should appear, asking you to select the Bluetooth accessory to connect to, as in Figure 9-6.

8. Press the name of the camera on that screen, and you should see a screen prompting you to approve the pairing of the two devices. Verify that the authentication code shown on the camera matches that on the smart device, and select Pair on the smart device and OK on the camera. The camera and smart device are now paired.

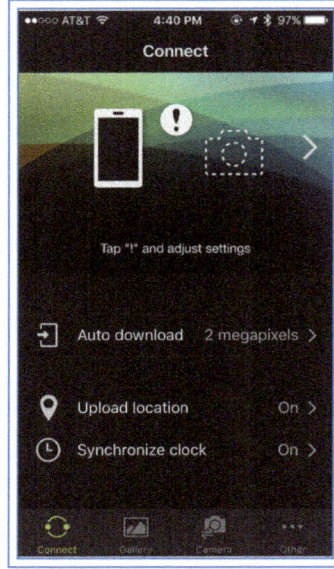

Figure 9-4. Connect Screen on SnapBridge App

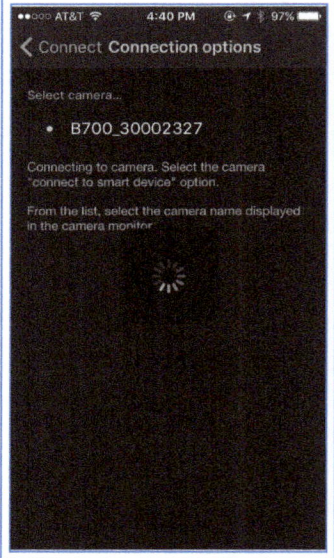

Figure 9-5. Name of Camera on SnapBridge App Screen

Chapter 9: SnapBridge App, Superzoom Lens, and Other Topics | 141

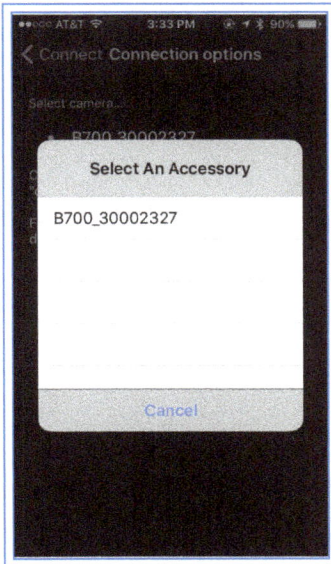

Figure 9-6. Screen in SnapBridge App to Select Bluetooth Accessory

9. In the SnapBridge app, select the Connect tab at the bottom of the screen (if necessary), then choose Auto Download, shown in Figure 9-7, and select the settings you want for automatic downloads of images from the camera to the smart device.

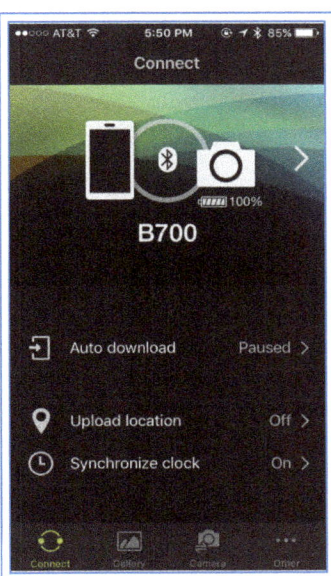

Figure 9-7. SnapBridge Connect Screen After Connection Made

10. The choices are 2 megapixels or original size, as shown in Figure 9-8. The camera cannot automatically download Raw files or movies to a smart device, although you can download movies using a Wi-Fi connection, as discussed later. You also can choose whether to automatically upload images to the Nikon Image Space. You need to sign up with that service separately, through the Other tab in the app.

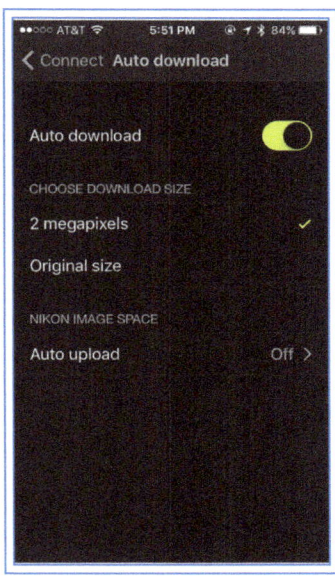

Figure 9-8. Auto Download Screen for SnapBridge App

11. If you want to control the camera remotely from a smartphone or tablet, select the Camera tab at the bottom of the SnapBridge app, as shown in Figure 9-7. You will then see the screen shown in Figure 9-9, with choices for Remote Photography and Download Selected Pictures.

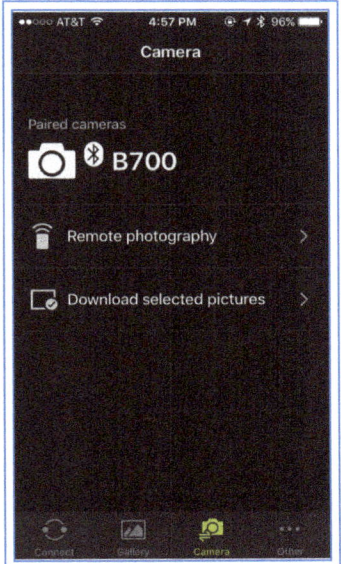

Figure 9-9. Options on Camera Screen in SnapBridge App

12. Select Remote Photography, and the smart device will display a message saying Wi-Fi has been enabled on the camera, and directing you to select the listed name of the camera's Wi-Fi network in the Settings area on the phone.

13. On the iPhone, the camera's network should show up as seen in Figure 9-10. The first time you connect, you will have to enter the password for the camera's Wi-Fi network. The password can be found (or changed) on the camera's Network menu, under Wi-Fi/Network Settings. On my camera, the default password was Nikon_B700.

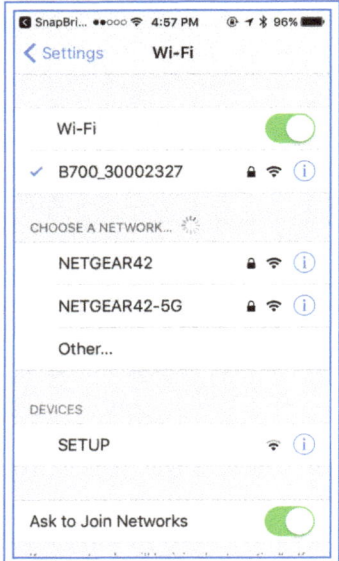

Figure 9-10. Camera's Network Shown on iPhone Screen

14. Once the phone has connected to the camera's Wi-Fi network, return to SnapBridge. You should see the Camera screen, as shown in Figure 9-9, above. Select Remote Photography, and you should see a screen like Figure 9-11 on the phone. The camera will display a blank screen saying Connected.

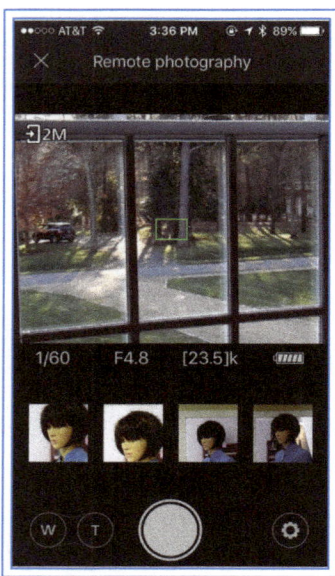

Figure 9-11. Remote Photography Screen in SnapBridge App

15. On the phone's remote control screen, you can use the controls to take still images. You can zoom the lens in and out with the W and T buttons at the lower left, and take a picture with the large white shutter button icon in the lower middle.

16. Tap the gear icon at the lower right to open the Settings screen, as shown in Figure 9-12. On that screen, you can choose whether to download images automatically to the smart device after shooting; the download size; and the self-timer. You also can turn off the live view through the camera's lens, though I can't imagine why you would do that.

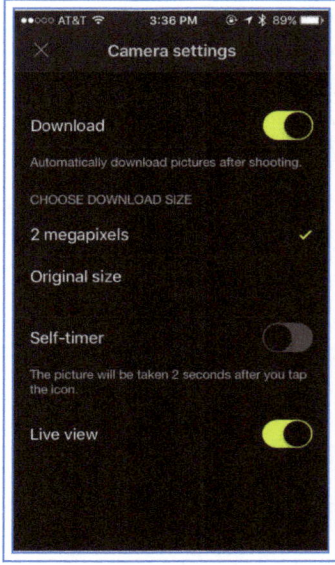

Figure 9-12. Settings Screen for Remote Photography

17. You can cause a connected smart device to upload location data for your images to the camera. To enable this function, choose the Upload Location option on the Connect screen of the app. To synchronize date and time information between the camera and the smart device, choose the Synchronize Clock option on the Connect screen.

18. If you want to download selected images and movies from the camera to a connected smart device, you can do that using the Camera tab in the SnapBridge app. First, disable the Auto Download option on the Connect tab in the app. Then, on the app's Camera tab, choose the Download Selected Pictures option. You can then view the images in the camera and select the ones to be transferred to the smart device by tapping Select at the upper right of the app's screen, as shown in Figure 9-13. Then tap each image you want to download, and

Chapter 9: SnapBridge App, Superzoom Lens, and Other Topics

tap Download Selected Images at the bottom of the app's screen when you have finished selecting images.

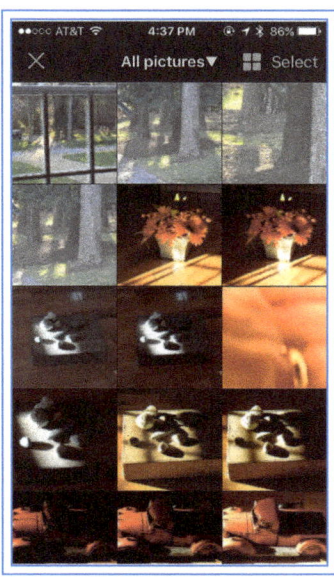

Figure 9-13. SnapBridge Screen to Select Images for Download

19. To display and download movies as well as still images, you have to establish a Wi-Fi connection. To do that, go to the Settings app on your smart device and select the camera's Wi-Fi network. Once that connection is made, the movies should be displayed on the Download Selected Pictures screen.

SUMMARY OF OPTIONS FOR TRANSFERRING IMAGES AND MOVIES TO A SMART DEVICE WIRELESSLY

I found it confusing at first to understand the various options for uploading images and movies to a smartphone or tablet wirelessly, so I have put together a brief summary:

- All of the options below assume that the camera has been paired with the smart device by Bluetooth using the SnapBridge app.

- To have JPEG still images (not Raw files or movies) sent automatically to a smart device while shooting, on the camera's Network menu, select the Send While Shooting option and choose Yes for Still Images. Using the Upload (Photos) option under that menu item, you can further specify whether to upload only single images, or to include all burst photos, or just the key frames for bursts of photos.

- If you have not turned on the Send While Shooting option, you can still have the camera upload JPEG still images automatically to a smart device through the paired connection. To do that, go to the Mark for Upload option on the Playback menu, and select the images you want to have uploaded, as discussed in Chapter 6. The selected images will then be uploaded to the smart device.

- If you have not turned on Send While Shooting and have not marked any images with the Mark for Upload option, you can still upload JPEG (not Raw) images, as well as movies, to a smart device. To do this, go to the SnapBridge app, select the Camera tab at the bottom of the screen, and, on the next screen, select Download Selected Pictures. In the dialog box that appears, asking if you want to switch to Wi-Fi, select Yes.

- Then make sure the smart device is connected to the Wi-Fi network generated by the camera. Once that connection is made, go back to the Download Selected Pictures option. At the top of the screen, choose the Select option. You should then see thumbnails of all images and movies on the camera's memory card. It may take a long time to load them.

- When they are loaded, tap on one or more thumbnails, then select Download Selected Pictures at the bottom of the app's screen. For a movie, the app may advise you that it will wait until more bandwidth is available, because the file is so large.

The Network Menu

The Coolpix B700 has a special menu with several options for controlling the camera's wireless functions. That menu is represented by the wireless network icon, which is highlighted in Figure 9-14.

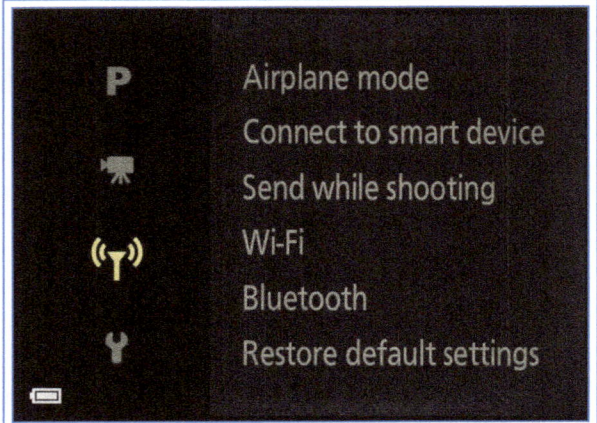

Figure 9-14. Network Menu Icon Highlighted at Left

Once you have highlighted that icon, press the Right button to move the highlight into the list of options on the single screen of the menu, as shown in figure 9-15. I will discuss these menu options in order.

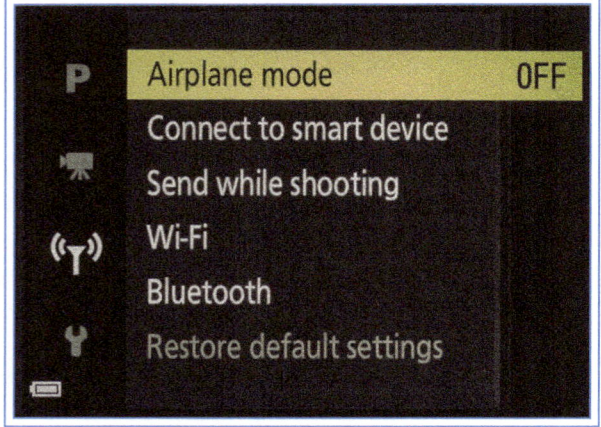

Figure 9-15. Single Screen of Network Menu

Airplane Mode

This first option gives you a quick way to disable all wireless connections with the camera, including both Bluetooth and Wi-Fi. As the name implies, you can use this menu item when you are on an airplane and need to disable wireless devices. The constant connection between the camera and a smart device will be interrupted, but will be automatically re-established when you later turn Airplane Mode back off. As noted in Chapter 7, you cannot use the Format Card option when the camera has a wireless connection active. In that case, you can use the Airplane Mode option to disable that connection temporarily if you need to format a memory card.

Connect to Smart Device

I discussed this option earlier, in describing the steps for setting up a connection between the camera and a smartphone or tablet. If you don't establish that connection during the initial setup of the camera, you can use this menu option to set up the connection, as discussed earlier in the chapter.

Send While Shooting

This option lets you specify what images will be automatically uploaded to a smartphone or tablet that is connected to the camera through a Bluetooth connection. This menu item has two sub-options: Still Images and Upload (Photos). The Still Images option can be set to either Yes or No. If you choose No, then no images will be uploaded automatically while you are shooting. In that case, you can still upload images using the Mark for Upload option on the Playback menu to mark images to be uploaded, and then later upload them from the camera using the SnapBridge app. (You also can use the Download Selected Pictures option on the Camera tab in the SnapBridge app to transfer pictures to the smart device later.)

If you choose Yes for the Still Images option under the Send While Shooting menu item, you can go to the Upload (Photos) item and choose what types of still images will be automatically uploaded to the connected device: only individual still images; all still images, including ones taken in a burst of continuous shooting; or only key frames from continuous bursts of shots.

Essentially, the Send While Shooting menu option gives you a way to limit the items that are automatically transferred to a smart device when the camera is connected by Bluetooth to such a device. Note that movies cannot be transferred using the Bluetooth connection; they can only be transferred by a Wi-Fi connection using the SnapBridge app.

Wi-Fi

This next option on the Network menu gives you the ability to control the settings the camera uses for establishing a Wi-Fi network that can be used to connect to a smart device. As discussed earlier, you can connect the camera to a smart device over a Bluetooth connection that can remain active as you shoot, letting the camera automatically upload still images to the smart device. However, if you want to upload movies or large volumes of still images, or if you want to control the camera remotely using the smart device, you need to have the camera establish a Wi-Fi network that the camera can connect to. Use the settings available with this menu option to assist in making that connection.

The first sub-option, Network Settings, is dimmed and unavailable when a wireless connection is active. To get access to this menu option, you can choose the next menu option, Bluetooth, and then select Connection/Disable to disable the current connection. (You can re-enable it once you have finished using the menu option.)

Using the Network Settings item's various sub-options, you can change the SSID (identifying name) of the camera's Wi-Fi network (the default is a name such as B700_30002327); the encryption standard (the default is WPA2-PSK-AES); the password of the network (the default is Nikon_B700); the channel (the default is 6); the subnet mask (the default is 255.255.255.0); and the DHCP server's IP address (the default is 192.168.0.10).

I have not found a need to change any of these items, but you may want to change the password or SSID of the network to enhance security in some circumstances.

The Current Settings option under the Wi-Fi item displays the settings that have been established with the Network Settings item.

Bluetooth

This next menu item lets you control the way in which the camera's Bluetooth connection operates. The three sub-options are Connection, Paired Devices, and Send While Off. With the first option, you can choose either Enable or Disable for the current connection. So, if you have paired the camera with a smartphone or tablet using the Bluetooth capability, you can temporarily disable it with this option. You can re-enable it at any time by selecting the Enable option.

The Paired Devices item displays the identification of any device that is currently paired with the camera via Bluetooth. You can use this option to verify that the Bluetooth connection was successfully made (or that a connection was successfully terminated).

The Send While Off option lets you specify whether or not the camera maintains its Bluetooth connection with a smart device while the camera is turned off or in standby mode. If you select On for this option, then the connection stays on when the camera is off, so images can continue to be uploaded. If you want to conserve battery power, you can set this option to Off. In that case, the camera will still upload images to the smart device, but only while the camera remains powered on and active.

Restore Default Settings

This final option on the Network menu lets you reset all wireless settings to their original factory values. This can be useful if you are having problems getting the camera to connect to a smartphone or tablet and you want to get a fresh start with the original settings. This option cannot be selected while a wireless connection is active. You can choose Bluetooth/Connection/Disable to disable a connection so this menu item can be used.

Using the Superzoom Lens

One of the most outstanding features of the Coolpix B700 is its lens, which has an amazing range of focal lengths, at both the wide-angle and telephoto ends of its zoom. The focal length of a lens is a measure of how wide is its view of a scene and how powerfully it enlarges the view.

A "normal" lens—used for everyday shooting of family scenes, portraits, and the like—is often considered to be a 50mm lens. A "wide-angle" lens is in the range of 35mm or lower, and a "telephoto" lens is one with a focal length of 100mm or greater. The B700, of course, has a zoom lens, which means it can change its focal length. The focal lengths of this lens range from a very wide 24mm to an astounding 1440mm at the telephoto end, for an overall range of 60 times optical zoom. (None of this discussion will involve digital zoom, which is not a "real" zoom capability, as discussed in Chapter 7.)

Of course, there are trade-offs for having this great zoom range. It is not possible to provide the same quality in a zoom lens of this type as in lenses used by professional photographers at major sporting events, for example. A good Nikon zoom lens for a DSLR can easily cost more than $1,000, and a top-quality Nikon telephoto (non-zoom) lens can cost more than $10,000. However, for everyday photography, the B700 provides you with the ability to capture scenes with an array of focal lengths far greater than for many consumer cameras.

The 24mm minimum focal length of the B700 is very useful when you need to photograph a large group of people without standing back a great distance. Also, if you need to photograph the interiors of rooms, this wide-angle focal length is a great bonus, because you will likely be able to capture an excellent view of the entire room by standing in one corner.

But the telephoto power of its lens is one of the B700's most dramatic features, so I will concentrate on its use.

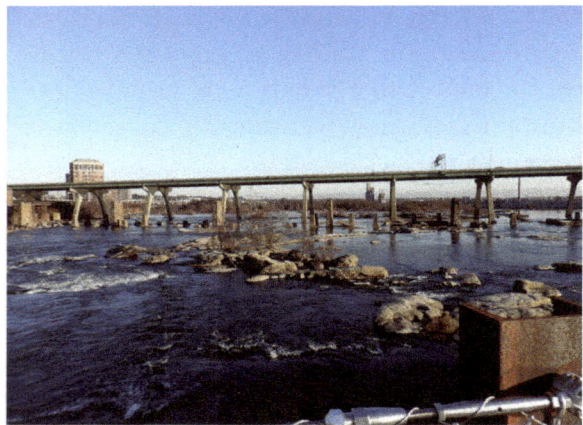

Figure 9-16. Image Taken at Focal Length of 24mm

First, the B700's zoom lens is so powerful it can capture details you cannot see with the naked eye. For example, I took the shot of the James River in Figure 9-16 with the lens zoomed out to its 24mm wide-angle setting. In this image, you cannot see many details in the distance.

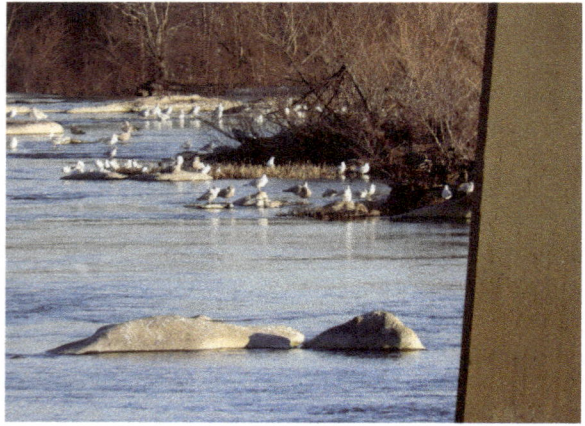

Figure 9-17. Image Taken at Focal Length of 1440mm

Now, look at Figure 9-17. This shot was taken at the same time and place as the wide-angle shot. This time, though, the lens was zoomed in to the full 1440mm extent of the optical zoom, focusing on the center of the previous scene. In this image, you can clearly see groups of birds in the water, which are not visible in the first image. This shot includes a level of detail that is not even hinted at in the first image.

When you use the Coolpix B700's powerful zoom lens at its highest power, you will encounter some issues that you should take into account when deciding whether to rely on the long reach of the lens rather than trying to get closer to your subject.

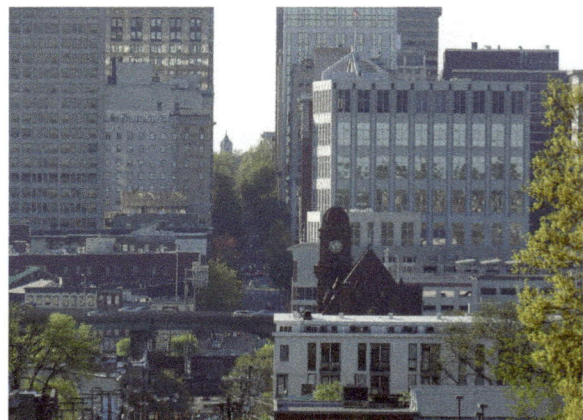

Figure 9-18. Image Illustrating Atmospheric Effects

For one thing, as you may be able to tell from Figure 9-18 (a telephoto view of city buildings), images shot at this power can suffer from the compression of the atmosphere. Shooting through a lot of air can add a foggy aura to an image, especially on a hot or hazy day. In this view, the buildings appear somewhat faded because of the great distance from the lens and the heat of the day.

Figure 9-19, also taken at the full optical zoom range of 1440mm, illustrates another phenomenon—flattening or foreshortening of the objects you are photographing. The result of using such a powerful zoom in this case is that objects appear compressed into a single level, so that it looks as if the people in this image are located right on top of each other, even though they were more widely separated in reality. This same effect is present in Figure 9-18, which shows the various buildings as if they are flattened into a single plane.

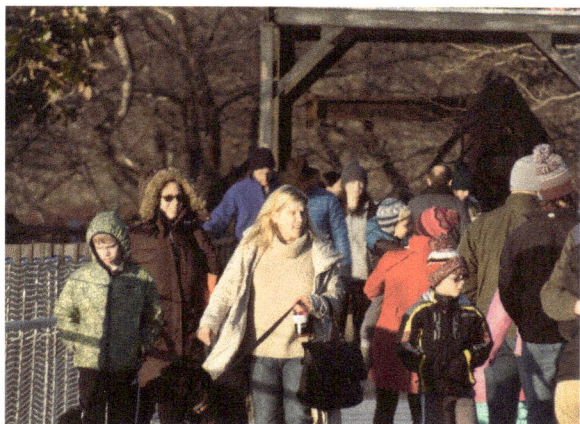

Figure 9-19. Image Illustrating Flattening Effect

However, you very well may find good uses for this effect, which can give a distinctive look to your photos.

Another positive side of the super-long zoom range is its ability to isolate a single subject. If you use the zoom to focus on a particular person in a crowd or on a particular animal in a pack, you can fill the frame with that single subject, thereby limiting extraneous objects and emphasizing the subject you want to concentrate on. Another advantage can be the ability to take a photo of a subject from a distance without disturbing him, as was the case with Figure 9-20.

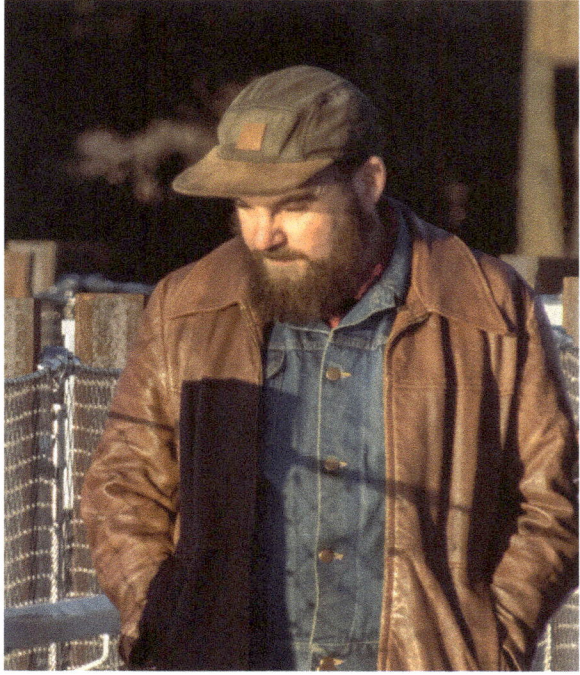

Figure 9-20. Image Illustrating Isolating Effect

When I spotted this man through the lens, I wanted to get a picture of him without getting close enough to disturb him. Using the long zoom range of the B700, I took this shot from a considerable distance, which isolated him from other objects and resulted in an image with the surrounding scenery blurred into an indistinct background.

Another benefit from using the B700's zoom lens in its telephoto range is that, when the lens is zoomed in, it has a very shallow depth of field. As a result, when you take a picture at the long end of the zoom range, particularly when the lens is fairly close to the subject, the background will be blurred to the point of becoming indistinct. This effect, often called "bokeh," as discussed in Chapter 3, can reduce distractions from the background and emphasize your primary subject in the foreground. This was the case with Figure 9-21, in which I took a shot of a mannequin with the lens zoomed in to 400mm, thereby blurring the background of trees and a fence in the distance.

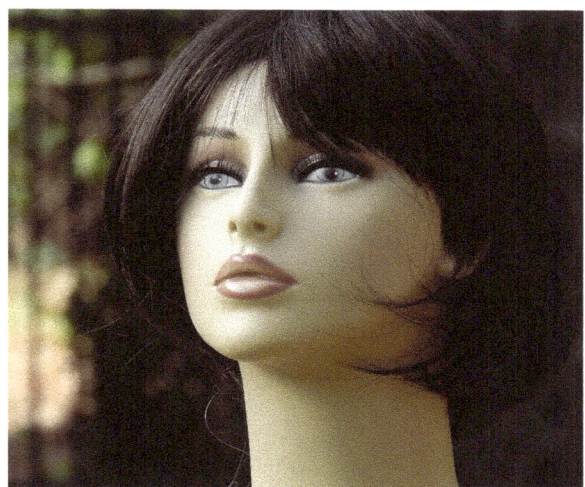

Figure 9-21. Image Illustrating Bokeh Effect

Beyond the specific advantages from the long reach of the zoom lens, perhaps the greatest overall benefit of the superzoom lens on the Coolpix B700 is that it gives you the equivalent of a whole range of focal lengths without the need to carry around a bag filled with heavy lenses. In practical terms, with the B700 you have at your fingertips every focal length that a photographer could reasonably want or need for everyday photography, ranging from the wide-angle 24mm with a strong macro capability to the super 1440mm telephoto.

If you have an opportunity to go on a safari or a cruise around the world and you want to make a complete photographic record of your trip using one lightweight, easy-to-use camera, the B700 can do the job nicely.

There also is a definite problem that comes along with a superzoom capability—image blur caused by camera movement. When the lens is zoomed in to its full 1440mm focal length or anywhere close to that range, any motion of the camera is multiplied because of the magnification of the image. You will notice how jittery the image looks on the display, and it will be hard to keep the picture steady.

There are several steps you can take to reduce the effects of camera movement. First, if possible, use a tripod. It can be inconvenient to do, but using a solid tripod is one of the best ways to ensure high-quality images. If you can't manage a full-blown tripod, use a lightweight travel tripod, a monopod, or any support available, such as a fence post, or just sit on a bench and hold the camera steady on your lap, tilting the LCD display up toward your face to view the image.

Suppose, though, that you are walking through a field in search of wildlife shots and there is no physical support available. There are several things you can do to minimize the effects of camera shake. First, you should make sure the Vibration Reduction feature is turned on through the Setup menu. This system counteracts camera movement quite effectively, up to a point. As is discussed in Chapter 7, I recommend you use the Normal setting for most situations, but consider using the Active setting if you are shooting in a turbulent environment. Also, you may find you can hold the camera steadier if you use the electronic viewfinder, so you can hold the camera against your forehead and look into that window, rather than using the LCD display at some distance from your face.

Next, use the fastest shutter speed you can. If the shutter is open for only a brief instant, there will not be time for camera motion to register on the image. According to one rule of thumb, when handholding a zoom lens you should use a shutter speed no slower than the fraction of a second with the focal length of the lens as the denominator. So, if the lens of the B700 is zoomed all the way in to 1440mm, you would use a shutter speed of 1/1500 second or faster. In the case of the B700, the choices would be 1/1600, 1/2000, or 1/2500, because the speeds faster than that are not available when the lens is zoomed all the way in.

If you want to control the shutter speed, you should use Shutter Priority as your shooting mode, as discussed in Chapter 3. You also could use Manual mode, if you are willing to accept the added task of setting the aperture correctly. Or, if you would like to use Program mode, you can let the camera set the shutter speed and aperture initially, and then use the Flexible Program feature, which lets you turn the command dial to select new combinations of shutter speed and aperture that are equivalent to what the camera selected.

However, you are likely to run into a problem if you use the camera's standard settings and try to set a fast shutter speed. One of the limiting characteristics of the superzoom lens on the B700 is that, as was discussed in Chapter 3, its maximum aperture when zoomed in is quite narrow. When the lens is zoomed all the way out to wide-angle, the maximum (widest open) aperture is f/3.3, which is not exceptionally wide to start with, though it is wide enough for most purposes. But, when the lens is zoomed in, it rapidly loses the ability to use a wide aperture.

When the lens is zoomed all the way in, the maximum aperture is f/6.5. In order to use a shutter speed of 1/1500 second or faster at that rather narrow aperture, there will have to be a good deal of light, unless you change some other settings.

Your best option probably is to increase the ISO sensitivity of the camera, which will mean that the camera's image sensor will require less light to expose the picture, at the risk of increased visual noise in the image. Using the ISO Sensitivity setting in the Shooting menu, you may want to try setting ISO to Auto, in which case the camera will set the value as high as 1600 if conditions warrant. If you want to be sure a high ISO is set, you should use a specific level, such as ISO 800, ISO 1600, or even ISO 3200.

If you prefer not to boost the ISO, which likely will introduce noise into the image, one strategy you can use is to zoom back out until the camera can use a wider aperture, such as, say, f/5.0 or f/4.5. Later on, when editing your photos with software, you can crop them to achieve the same field of view you originally saw with the zoomed-in lens, though with some loss of quality because of the cropping.

Another possible strategy for getting good, clear images with the superzoom lens is to take advantage of the B700's excellent array of continuous-shooting options. With several of these options, the camera will take multiple shots in rapid succession, increasing the likelihood that one or more shots will be usable. You also will experience a decline in image size and quality with some of these settings, though, so you need to consider the balancing factors.

One more note: Don't forget that the Coolpix B700 offers the convenient U slot on the mode dial, for User Settings. If you use the lens zoomed in frequently, you may want to save your preferred settings for those occasions, so you can quickly call them up just by turning the mode dial to the U setting. For example, you may want to set up the camera in Shutter Priority mode, with the lens zoomed all the way in, with a shutter speed of 1/1600 second, with an ISO setting of 1600 and with continuous high-speed shooting enabled.

Next, I recommend that you take advantage of the side zoom control on the B700—the switch on the left side of the camera, below the flash pop-up button. As discussed in Chapter 5, with its default function as an alternative to the zoom lever around the shutter button, this switch can let you hold the camera more firmly in both hands. If you zoom with the left-side switch, you can use your right hand to keep a tight grip on the right side of the camera without having to reach up to use the zoom lever.

Finally, you can use the snap-back zoom button to assist with your long-telephoto shots. You can use that control to quickly pull back from a zoomed-in view, so you can get your bearings and see exactly where your subject is in relation to its surroundings, before quickly zooming back in to take the picture. It can be very difficult to locate your subject when the lens is zoomed all the way in to its 1440mm maximum; use the snap-back zoom to get the wider view quickly when you need it for orientation.

Macro (Close-up) Photography

Macro photography is the art or science of taking photographs when the subject is shown at actual size (1:1 ratio between size of subject and size of image) or magnified (greater than 1:1 ratio). So if you photograph a flower using macro techniques, the image of the flower on the image sensor will be about the same size as the actual flower. You can get wonderful detail in your images using macro photography, and you may discover things about the subject that you had not noticed before taking the photograph.

The Coolpix B700 is quite capable of shooting macro photographs, like the one in Figure 9-22 providing an extreme close-up of a figurine of a knight. This image was shot at a wide-angle setting of 33mm, using the macro autofocus setting, with the camera located a couple of inches (about 5 cm) from the figure.

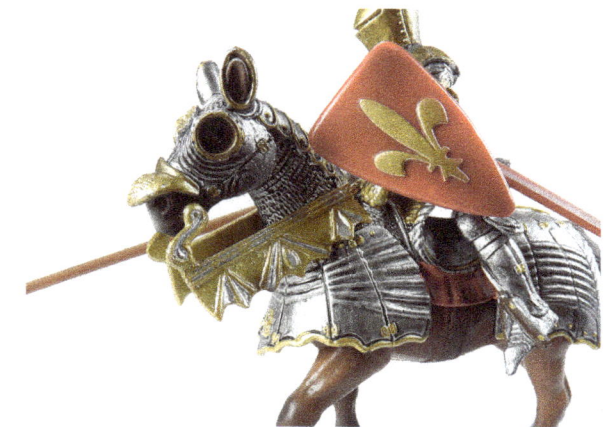

Figure 9-22. Macro Example

As discussed in Chapter 5, to activate macro focus, press the Down button to bring up the focus menu, then use the Up and Down buttons, the command dial, or the multi selector dial to select the flower icon indicating macro autofocus mode, as shown in Figure 9-23.

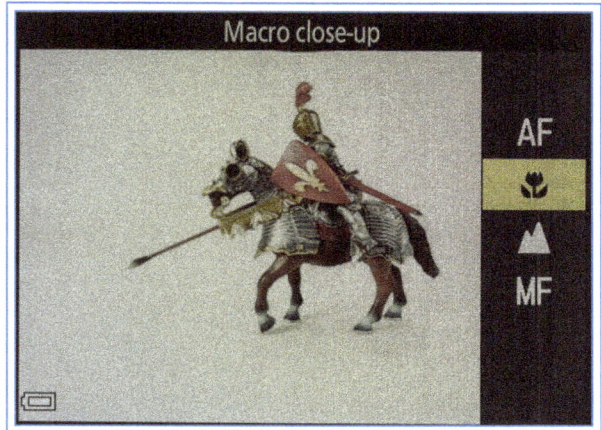

Figure 9-23. Macro Selected on Focus Mode Menu

You can choose macro autofocus in Auto mode or in the Program, Aperture Priority, Shutter Priority, or Manual exposure mode, as well as the following Scene mode settings: Beach, Snow, Pet Portrait, Soft, Selective Color, and Superlapse Movie. In the Close-up and Food modes, macro focus is selected automatically. Macro focus is available for selection with all settings of the Creative shooting mode.

With the autofocus mode set to macro, the Coolpix B700 is able to focus as close as 0.4 inch (one centimeter) when the lens is zoomed out to the wide-angle position or close to that position. The range for which the closest focus is available is indicated on the zoom scale by a small triangle icon, as shown in Figure 9-24.

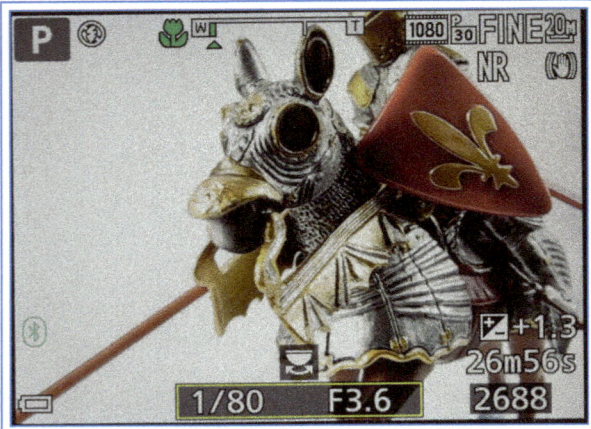

Figure 9-24. Zoom Scale at Macro Focus Point

When the bar of the zoom scale does not extend past the center of that triangle, the lens can focus down to this minimum distance.

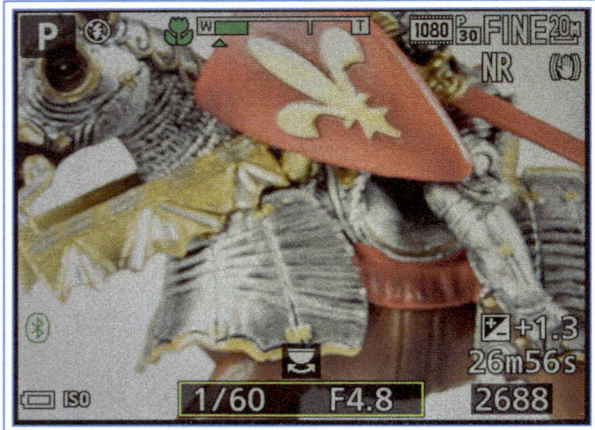

Figure 9-25. Zoom Scale at Maximum Macro Distance

Once the zoom bar goes past the triangle, the lens can focus as close as four inches (10 centimeters), as long as the zoom bar stays green. In Figure 9-25, the zoom bar extends as far to the right as possible before it turns white, indicating the end of the macro focus range.

With normal autofocus, the camera can focus as close as about one foot eight inches (50 cm) at the wide-angle position, and only as close as about six feet seven inches (two meters) at the full telephoto position.

If you don't want to have to fine-tune the zoom position to set macro focusing as close as possible, there are two easier methods. First, you can use the Close-up setting in Scene mode (with the Single Shot option). The camera will adjust for the closest possible macro shooting, setting the focus mode and the zoom position. It also will turn on continuous autofocus and set the AF Area Mode to Manual, so you can adjust the position of the focus frame on the screen. To move the frame, press the OK button and then use the direction buttons or turn the multi selector dial to adjust the frame's position. Press OK to anchor it in place.

Another possibility for close-up focusing is to select the Food setting within Scene mode. This setting is similar to Close-up, except that it adds an adjustment slider so you can fine-tune the hues of the foods (or other items) you are photographing.

You don't have to use the macro autofocus setting or one of the Scene mode settings to take macro shots; if you set the camera to manual focus by pressing the Down button and then selecting MF from the on-screen menu, you can focus on objects very close to the lens. You do, however, lose the benefit of automatic focus, and it can be tricky finding the correct focus manually. However, the camera enlarges the view on the screen when manual focus is in use, and I have had good success using manual focus for macro shots.

When shooting extreme close-ups, you should use a tripod or other solid stand whenever possible, because the depth of field is very shallow and you need to keep the camera steady to take a usable photograph. It's also a good idea to take advantage of the self-timer. If you take the picture using the self-timer, you will not be touching the camera when the shutter is activated, so the chance of camera shake is minimized. Another option is to use the SnapBridge app to trigger the shutter from a distance using a smartphone or tablet, as discussed earlier in this chapter. You should also leave the built-in flash retracted so it can't fire, unless you have a system for diffusing the flash to avoid harsh shadows and glare.

Using Flash

As I discussed earlier, pressing the Up button on the multi selector, the one marked with a lightning bolt, gives you access to the various settings for the built-in flash unit on the Coolpix B700, as shown in Figure 9-26.

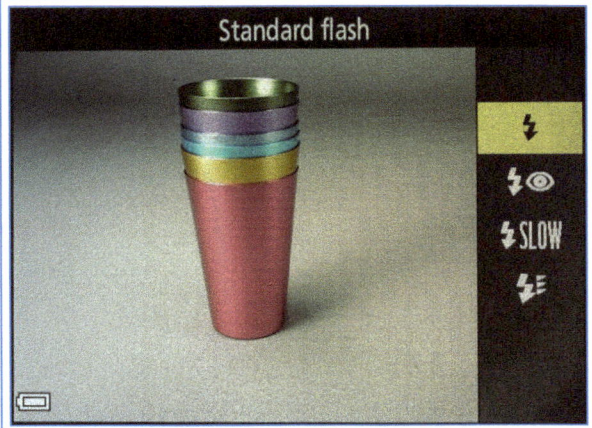

Figure 9-26. Flash Mode Menu

Before I discuss the details of those settings, it's important to recall one basic fact about this camera: The flash cannot fire unless you first pop it up by pressing the flash pop-up button marked by a lightning bolt on the left side of the flash housing, near the top of the camera. If you think there's any chance the flash may be needed, go ahead and press that button to have the flash ready. (If you're shooting movies, though, you should make sure the flash is down out of the way, because it can't be used and might interfere with your shooting.)

The next point to note about the built-in flash on the B700 is that a lot depends on the shooting mode you have set on the mode dial. If that dial is set to Auto, you will have access to all five possible settings for the flash. However, other shooting modes place limits on your flash choices. For example, with the Program and Aperture Priority modes, Auto Flash is not available. With Shutter Priority and Manual exposure, Auto Flash and Slow Sync are not available. With the Night Portrait mode, the flash is set to Auto with Red-eye Reduction—in other words, the flash will fire if necessary, and the camera will use its built-in processing to counteract the red tinge that may result from the red-eye effect of shooting straight into human eyes. You cannot change that flash setting.

In Night Landscape and Landscape modes, the flash is forced off and cannot fire. With various settings available in Scene mode, the behavior of the flash varies according to the particular characteristics of the setting. For example, with the Sports, Easy Panorama, Sunset, Dusk/Dawn, Moon, Bird-watching, and Multiple Exposure Lighten settings, among others, the camera forces the flash off, because flash ordinarily would not be used in the situations those settings are intended for.

Apart from the shooting mode, there are other factors that affect how the B700 uses flash. So, if you have the camera set to Program mode, in which you normally would have four flash modes available, there are some conditions that will disable the flash. For example, you cannot use the flash if you have set focus to the infinity setting, turned on exposure bracketing, or activated any continuous-shooting option other than interval shooting. If you believe the flash should fire but you are unable to turn it on using the flash mode button, check to see if one of the settings mentioned above is in use.

Once you have set the camera to a mode that permits the choice of some or all of the five possible flash settings, such as Auto mode or the Portrait setting of Scene mode, you have to decide whether to choose Auto Flash, Auto with Red-eye Reduction, Fill Flash/Standard Flash, Slow Sync, or Rear-curtain Sync.

Auto Flash is a setting I discussed earlier—the camera's automatic exposure system will fire the flash if it's needed to achieve a good exposure. This setting is available only when the mode dial is set to Auto, certain Scene mode settings, or Creative mode.

Auto with Red-eye Reduction is a special flash mode that uses two techniques to prevent or minimize the effects of "red-eye," the unpleasant reddish glow that can appear in people's eyes when the light from the on-camera flash bounces off their retinas and picks up the red from blood vessels. First, when this mode is selected, the camera lights up the bright reddish lamp on the front of the camera for a second or two before the flash fires, so as to cause the pupils of a person's eyes to narrow. In that way, there should be less chance of light from the flash reaching the retinas.

Second, in order to correct for this effect, if the B700 detects a reddish tint around the eyes in the image, the camera performs internal processing to remove that tint as the image is being saved. In some cases, this processing can affect other areas of the image in unexpected ways, so you should use this setting with care. I personally prefer not to use it; if red-eye appears in an image, it can be corrected with Photoshop or similar software.

The next option, the Fill Flash/Standard setting, forces the flash to fire, whether or not the conditions are dark enough for the camera to fire the flash on its own. This setting is useful when there is enough backlighting that the camera's exposure controls could be fooled into thinking the flash isn't needed. If, in your judgment, the subject will be too dark for that reason, you may want to force the flash to fire. Another such situation could be an outdoor portrait for which you need fill-in flash to highlight your subject's face adequately.

This setting, as indicated above, has different names with different shooting modes. If the camera is set to Auto, Scene, or Creative mode, this flash mode is called Fill Flash. With the Program, Aperture Priority, Shutter Priority, and Manual exposure shooting modes, this flash mode is called Standard Flash. There is no significant difference between Fill Flash and Standard Flash. With either setting, the flash fires whenever the shutter button is pressed, and the exposure metering system takes it into account. If you want to vary the intensity of the flash yourself, you can use the Flash Exposure Compensation item on screen 2 of the Shooting menu.

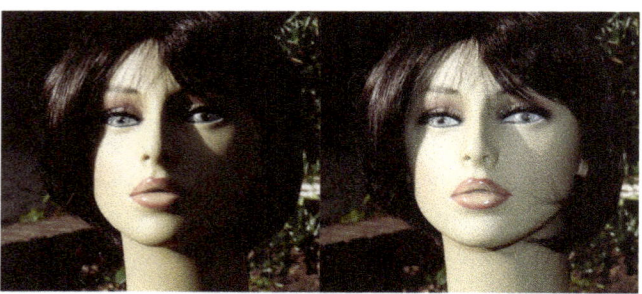

Figure 9-27. Fill Flash Comparison Image

For Figure 9-27, I took two shots of a mannequin's head in an outdoor setting during daylight hours to illustrate the effect of Fill Flash. I placed both shots into this composite image for comparison. In the shot at the left, with no flash, the image was exposed normally, with no special settings. For the shot on the right, I turned on the flash using the Fill Flash setting. As you can see, there is a definite difference. With the shot on the right, the shadows are evened out and the mannequin's face is considerably more visible.

Next, the Slow Sync setting is useful when you are taking a portrait in a dark environment. If you use a normal flash setting such as Auto Flash or Fill Flash, the camera will use a fairly fast shutter speed and let the flash illuminate only the portrait subject. Because the exposure time is short, the surrounding scene and background may be black.

If you use the Slow Sync setting, the camera will attempt to take the picture with a considerably slower shutter speed so that the ambient (natural) lighting will have time to register on the image and illuminate the background also. For example, the two images shown in Figure 9-28 were taken at the same time and in the same conditions.

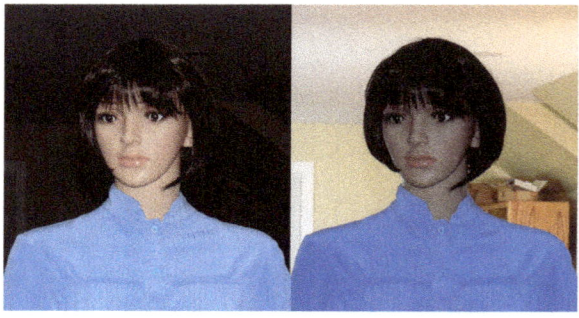

Figure 9-28. Slow Sync Comparison Image

The only difference is that the photo on the left was taken with the shutter speed set at 1/60 second in Standard Flash mode, while the one on the right was taken in Slow Sync mode with a shutter speed of 6/10 second, which allowed the ambient lighting to illuminate the area behind the subject.

The last setting on the flash mode menu is Rear-curtain Sync. This option is one you may not have a lot of use for unless you encounter the particular situation it is designed for. If you don't activate this setting, the camera uses the unnamed default setting, which could be called Front-curtain Sync. In that mode, the flash fires very soon after the shutter opens to expose the image. If you choose the Rear-curtain setting instead, the flash fires later, just before the shutter closes.

The reason for using Rear-curtain Sync is to help you avoid a strange-looking result in some situations. This issue arises, for example, with a relatively long exposure, say one-half second, of a subject with lights, such as a car or motorcycle at night, moving across your field of view. With normal (Front-curtain) sync, the flash will fire early in the process, freezing the vehicle in a clear image. However, as the shutter remains open while the vehicle keeps going, the camera will capture the moving lights in a stream extending in front of, or superimposed over, the vehicle. If, instead, you use Rear-curtain Sync, the initial part of the exposure will capture the lights in a trail that appears behind the vehicle, while the vehicle itself is not frozen by the flash until later in the exposure. With Rear-curtain sync in this particular situation, if the lights in question are taillights that look more natural behind the vehicle, the final image is likely to look more natural than with the Front-curtain (default) setting.

Figure 9-29 is a composite image that illustrates this concept using a flashlight. Both pictures were shot using an exposure of 1/2 second. In both images, I was moving the flashlight from right to left across the scene.

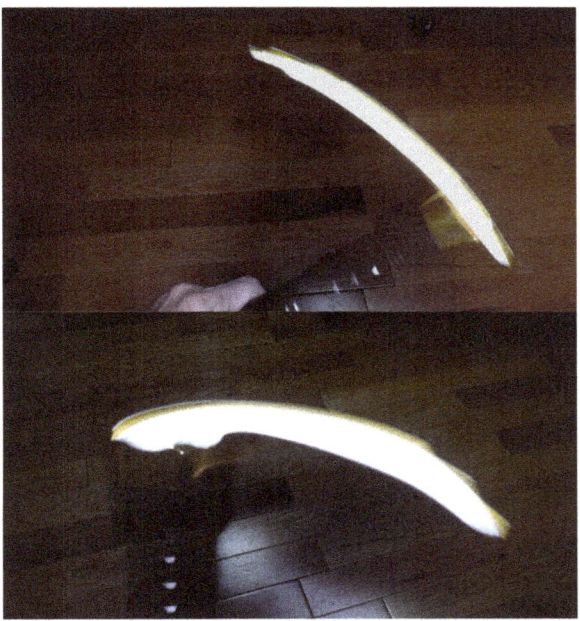

Figure 9-29. Rear-curtain Sync Comparison Image

In the top image, using the normal Front-curtain setting, the flash fired early in the exposure, making an image of the flashlight, while the light beam continued on to the left during the long exposure, to make the trail of light in front of the flashlight. In the bottom image, using Rear-curtain Sync, the flash did not fire until the flashlight had traveled to the left, overtaking the place where the light beam had traced its trail.

A good general rule is not to use Rear-curtain sync unless you have a definite need for it. Using the Rear-curtain setting makes it harder to compose and set up the shot, because you have to anticipate where the main subject will be located when the flash finally fires late in the exposure process.

One other setting you should keep in mind is flash exposure compensation, which is available through the Shooting menu when the camera is set to the more advanced shooting modes. This setting reduces the intensity of the flash, even when the camera is automatically setting the exposure. Just as with normal exposure compensation, when using flash you can adjust this setting if your test shots appear too bright or too dark. Just go into this menu item and set the value to a positive number to brighten the image or to a negative number to darken it, as shown in Figure 9-30.

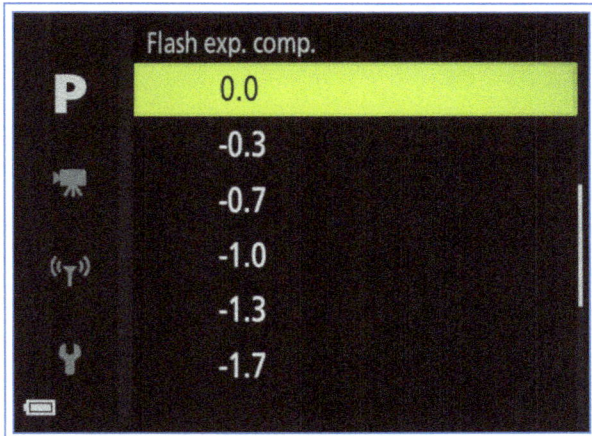

Figure 9-30. Flash Exposure Compensation Adjustment Screen

Be sure to set it to zero when you no longer need the adjustment so it does not affect other shots when you don't need it.

Infrared Photography

Infrared photography involves recording images that are illuminated by infrared light, which is invisible to the human eye because it occupies a place on the spectrum of light waves that is beyond our ability to see. In some circumstances, cameras, unlike our eyes, can record images using this type of light. The resulting photographs can be quite spectacular, producing scenes in which green foliage appears white and blue skies appear eerily dark.

Shooting infrared in the times before digital photography involved selecting a particular infrared film and the appropriate filter to place on the lens. With the rise of digital imaging, you need to find a camera that is capable of "seeing" infrared light. Many cameras nowadays include internal filters that block infrared light. However, some cameras do not, or block it only to a relatively small extent. (You can do a quick test of any digital camera by aiming it at the light-emitting end of an infrared remote control while pressing a button on the remote; if the remote's light shows up in the camera as bright white, the camera can "see" infrared light at least to some extent.)

The Coolpix B700 can take good infrared photographs. In order to unleash this capability, you need to take a few steps. The most important is to get a filter that blocks most visible light, but lets infrared light reach the camera's light sensor. (If you don't, the infrared light will be overwhelmed by the visible light, and you'll get an ordinary, non-infrared picture.)

A good way to make infrared photographs with the B700 is to get an infrared filter. The B700's lens is threaded to accept filters with a screw-in diameter of 52mm. The filter I use most often is the Hoya R72, which is a very dark red and blocks most visible light, letting in mainly infrared light rays in the part of the spectrum that produces interesting images. In Figure 9-31, it is shown on the lens of the B700.

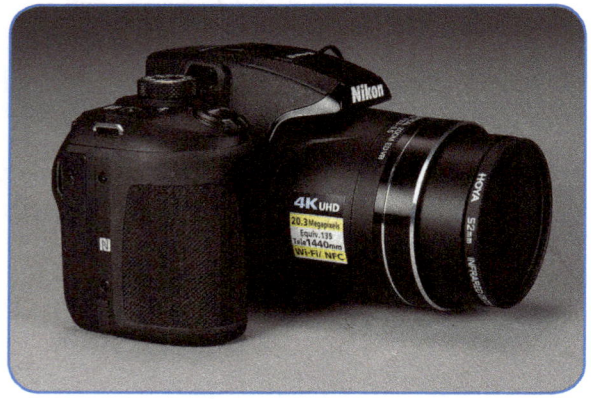

Figure 9-31. Hoya R72 Infrared Filter on B700

The next question is to figure out the settings. For the image shown in Figure 9-32, I set a custom white balance, using green grass as the base. That is, I used the camera's White Balance menu option on the Shooting menu, and, in the screen for setting a Preset Manual white balance, I aimed the camera at the grass and pressed the OK button. The results were essentially what I expect from infrared photography—scenes with grass, leaves, and bushes that look white, and other unusual but pleasing effects.

For exposure, I set the camera to shoot in Manual exposure mode and experimented with shutter speed and aperture until I found good settings at f/4.8 and 2.5 seconds, with ISO set to 100. I set the camera on a tripod and used the two-second self-timer to minimize vibration during the long exposure. You can often get interesting results if you include a good amount of green grass and trees in the image, as well as blue sky and white clouds.

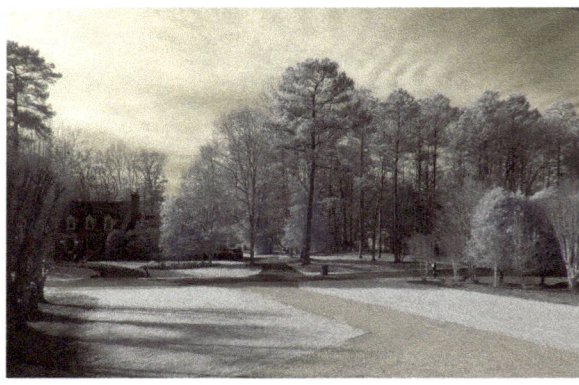

Figure 9-32. Infrared Example

Street Photography

The Coolpix B700 is not the first camera that would come to mind for street photography—that is, shooting candid pictures in public settings, often without the subject's knowledge. In my opinion, cameras that are well-suited for this type of work are small, lightweight, and unobtrusive in appearance, so they can easily be held casually or hidden in the photographer's hand. The B700 is somewhat bulky and not that easily concealed from view. However, it has several good points for this type of photography.

Its 24mm equivalent wide-angle lens is excellent for taking in a broad field of view, for times when you shoot from the hip without framing the image carefully on the screen. In addition, I have come to appreciate the B700's movable LCD screen for street photography, because, if you fold it out so it is parallel to the ground, you can look down at the screen to frame your shots without drawing attention to yourself. With this system, I have found that I can zoom in on a subject a considerable distance away and keep the framing accurate while looking down at the screen. Also, the camera shoots quickly and performs well at high ISO settings, so you can use a relatively fast shutter speed to avoid motion blur. The options for continuous shooting give you a good chance to get a sharp image under difficult circumstances. And, you can silence the camera by turning off the beeps and shutter sounds through the Setup menu.

As far as the best settings for street photography with the B700 are concerned, I will mention some guidelines as a starting point. The answer depends in part on your own personal style of shooting, such as whether you will talk to your subjects and get their agreement to being photographed before you start shooting, or whether you will fire away from across the street with a zoomed-in lens and accept the risk of blurry photos from camera shake at such a long focal length.

Here are a couple of approaches you can start with and modify as you see fit. Some photographers like to shoot in color at the highest quality and image size and then use post-processing software such as Photoshop or Lightroom to convert their images to black-and-white, along with any other effects they are looking for, such as extra grain to achieve a gritty look. (Of course, you don't have to produce your street photography in black-and-white, but that is a common practice.)

I recommend you shoot in Shutter Priority mode at a fairly fast shutter speed, say, 1/100 second or faster, to stop action on the street and to avoid blur from camera movement. You can set ISO to Auto, or possibly use a high ISO setting, in the range of 800 or so, if you don't mind some visual noise. You may want to set the aspect ratio to 16:9 (by selecting an image size of 5184 x 2920 pixels) in order to take in a wide field of view for street scenes. (You could shoot at the maximum image size and then crop down to this size, but then you would not have the benefit of seeing the 16:9 aspect ratio on the screen as you composed your shots.)

Another option is to set image quality and size to their highest settings for JPEG files (Fine for Quality and the largest image size) and set the Picture Control feature on the Shooting menu to the Monochrome option. You can use the standard settings for Monochrome, or you can go beyond the standard screen by pressing the Right button, and tweak the settings a bit. For example, you may want to try boosting contrast by one notch and reducing sharpening the same amount. To get the gritty "street" look, you also could try setting the camera's ISO to 1600 to include some visual grain in the image while boosting sensitivity enough to stop action with a fast shutter speed.

If you don't mind ignoring labels and trying something unconventional, you might consider turning the mode dial to the Creative setting and choosing one of the Noir options, such as Graphite or Carbon. Those settings give you a way to take black-and-white photos without having to fiddle with menu settings.

Also, consider using continuous shooting so you'll get several images to choose from for each shutter press. For the best combination of quality and speed, choose Continuous H, though that setting limits you to five shots at a time. Of course, you can't use continuous shooting with Creative mode; you would have to select Shutter Priority or another advanced mode to use continuous shooting, unless you opt for a Scene mode setting that includes continuous shooting, such as Sports. Also, one drawback to using continuous shooting is that you'll have to wait for the camera to finish recording its sequence of rapid shots before you can start shooting again, so you could miss a photo opportunity while waiting.

Another option is to shoot video footage using the 4K setting (Movie Options set to 2160/30p or 2160/25p), and plan to extract still frames from the final video sequence. However, you will not be able to set a fast shutter speed, so the still frames may have motion blur.

For Figure 9-33, I was shooting scenes from a pedestrian bridge over the river. I used Auto mode and zoomed the lens in to isolate the subjects and blur the background.

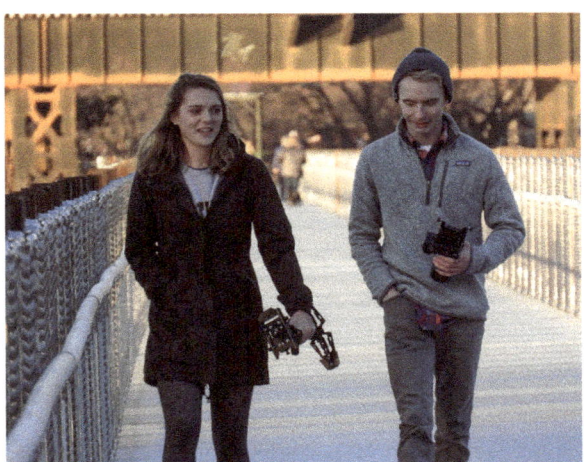

Figure 9-33. Street Photography Example

I generally use normal autofocus for this type of shooting, though some photographers like to use manual focus with the range set for the approximate distance where they expect their subjects to be. If I am shooting at street level, fairly close to my subjects, I often leave the lens zoomed back to its full wide-angle position to maintain a broad depth of field and keep most of the image in focus.

With the B700, you may want to at least experiment with long-range street photography, taking advantage of the superzoom lens. This approach has the advantage of letting you stay at a comfortable distance from your subjects. It has the disadvantage of producing a shallow depth of field, so it is harder to keep the entire scene in focus. Also, you may find that the foreshortening effect of a powerful zoom lens is not the look you are seeking for street photographs. And, you may find it difficult to get really sharp images at a long focal length unless you use a tripod, which limits your options for candid shots. On a bright day, though, or at high ISO settings, you should be able to use a fast enough shutter speed to avoid blur from a shaky camera, even without a tripod.

Connecting to a Television Set

To connect the Coolpix B700 to a television set, you need to purchase an optional HDMI cable and connect the camera to a high-definition or 4K-capable set. Nikon does not offer such a cable, but you can use any generic HDMI cable, as long as one end has a micro-HDMI (type D) male connector, and the other end has a standard HDMI male connector, as shown in Figure 9-34.

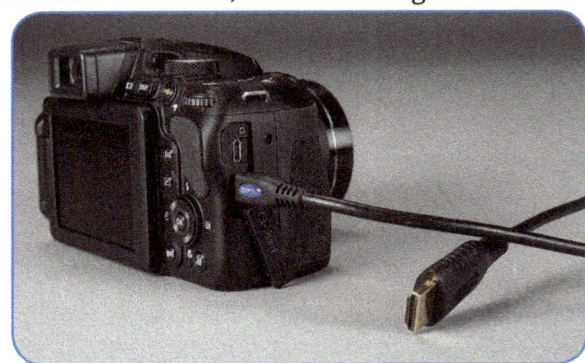

Figure 9-34. HDMI Cable Connected to Camera

Once you have connected the Coolpix B700 to a TV set, the camera operates much the same way it does on its own. Of course, depending on the size and quality of the TV set, you will get a much larger image, possibly better quality (on an HD or 4K set), and certainly better sound for your movies. When the camera is connected to a TV set using an HDMI cable, it will only play back images and videos; it will not go into Shooting mode and cannot record.

Appendix A: Accessories

When people buy a camera, especially a fairly expensive model like the Nikon Coolpix B700, they often ask what accessories they should buy to go with it. I will hit the highlights, sticking mostly with items I have experience with.

Cases

There are many types of camera cases on the market. At least for me, there is no "perfect" case for the B700. The type of case I use with this camera depends on what my purpose is for carrying the camera at a given time. One case that fits the camera well, the Lowepro Rezo 110 AW, is shown in Figure A-1.

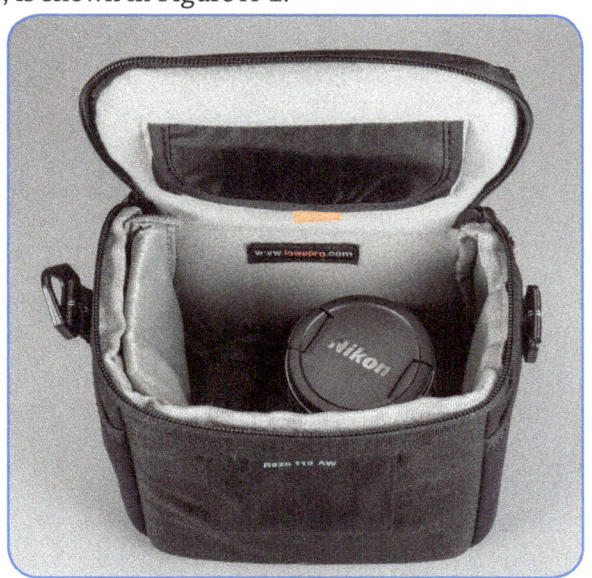

Figure A-1. Lowepro Rezo 110 AW Case

This case is an excellent all-around choice. It accommodates the camera easily with room to spare for extra batteries, filters, and other small items. It has a divider in its main compartment, and also has a built-in microfiber cloth, which is handy for keeping the LCD screen clean and free of smudges. It has a loop for attaching to a belt and a handle on top for easy carrying.

When I'm going on a day trip that's specifically oriented to photography, I often use a case that can hold the camera along with some accessories and a few items such as a water bottle and note pad. One case I have used a good deal is the Lowepro Inverse 100 AW beltpack, shown in Figure A-2 with the B700 in the middle compartment.

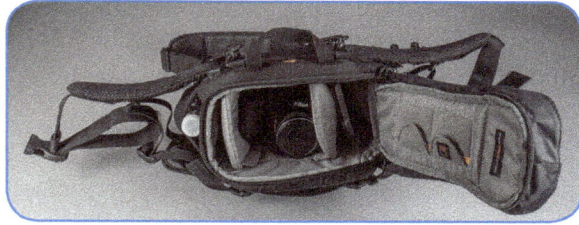

Figure A-2. Lowepro Inverse 100 AW Beltpack - Open

This case can easily hold extra batteries as well as one or two small water bottles and other odds and ends. One feature I especially like about this pack is that it has straps for attaching a tripod to the bottom, as shown in Figure A-3.

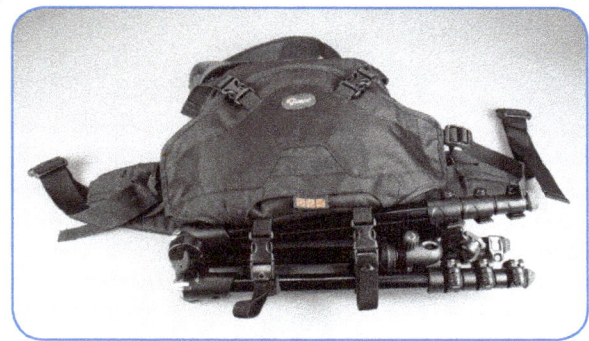

Figure A-3. Lowepro Inverse 100 AW Beltpack - Closed

I also have used the Crumpler 6 Million Dollar Home bag, model number MD6002-X01P60, shown in Figure A-4. This bag is of a generous size, with plenty of room for the B700 camera and several accessories in its subdivided interior compartment and in some small outer pouches. It also has adjustable straps that let you attach a tripod to the back of the case. It has a carrying handle on top as well as a padded shoulder strap. The bright red interior makes it easy to find small, black objects such as batteries inside the case.

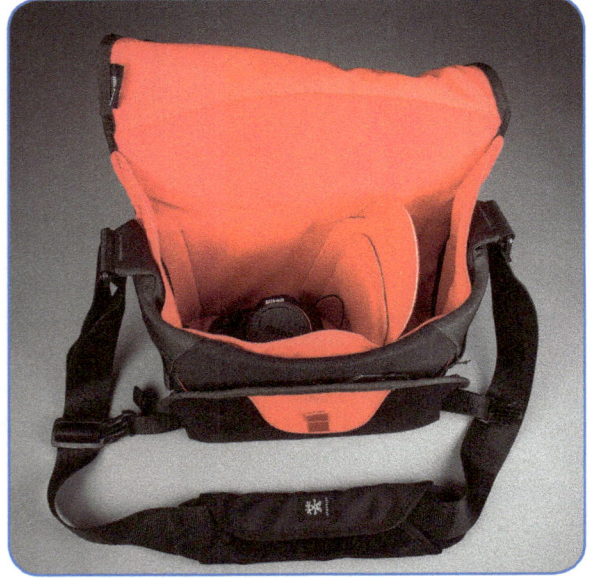

Figure A-4. Crumpler 6 Million Dollar Home Bag

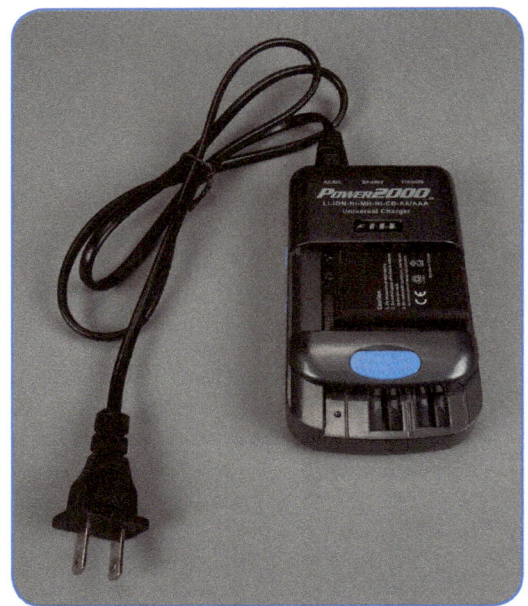

Figure A-5. Generic Charger for B700 Battery

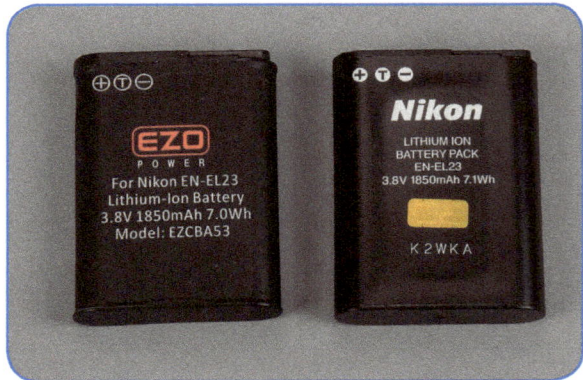

Figure A-6. Generic Battery and Genuine Nikon Battery

Batteries and Chargers

Here's one area where you should go shopping either when you get the camera or soon afterward. I use the camera heavily, and I find it runs through batteries quickly, especially when Wi-Fi or Bluetooth functions are activated. You can't use disposable batteries, so if you're out taking pictures and the battery dies, you're out of luck unless you have a spare battery (or an AC adapter and a place to plug it in; see discussion below). The model number of the official Nikon battery is EN-EL23. You can get a spare Nikon battery for about $45.00 as I write this. It won't do you a great deal of good by itself, because the battery is designed to be charged in the camera, so you can't charge a spare battery outside of the camera.

There is an easy solution to this problem. You can find generic replacement batteries, as well as chargers to charge the batteries outside the camera, inexpensively from Amazon.com and elsewhere. I purchased a package including two generic replacement batteries and a charger for $30.00 on eBay, and the batteries and the charger work well. With this setup, I can have one battery charging while another is in the camera. Figure A-5 shows the charger with a battery in charging position, and Figure A-6 shows a generic battery next to a genuine Nikon battery.

If you prefer to charge the battery inside the camera, the best way to do so is with the charger that comes with the camera, which, in the United States, is model number EH-73P. With that charger, you can operate the camera while its battery is being charged, although you cannot record movies during the charging process. It is possible to charge the battery inside the camera using a generic USB charger, but Nikon advises against doing so, because of possible problems of overheating or damaging the camera. In addition, with other chargers, it does not appear to be possible to operate the camera, at least not with the chargers that I have tried.

AC Adapter

Another alternative for supplying power to the B700 is the AC adapter kit, Nikon model number EH-67A, shown in Figure A-7.

Appendix A: Accessories | 159

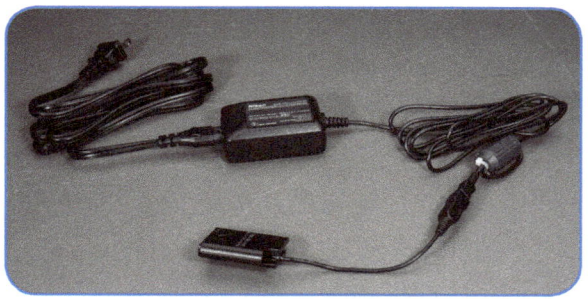

Figure A-7. Nikon AC Adapter, EH-67A

This accessory works well for providing a constant source of power to the camera. It includes a standard-sized AC cord that you plug into a power brick and into an AC outlet. The power brick is attached to a long cable that is then connected to a plastic piece the same size and shape as the camera's battery. You insert that plastic piece into the battery compartment. Before you close the battery compartment door, you have to pull down a small rubber flap that covers an opening where the camera's side meets the battery compartment door. You then place the cable from the AC adapter into that opening, so you can close the battery compartment door fully, as shown in Figure A-8.

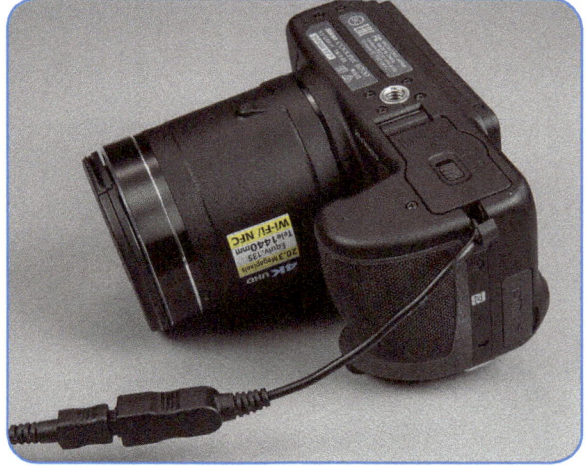

Figure A-8. AC Adapter Cord Going into Camera

Providing power to the camera is all this adapter does. It is not a charger, either for batteries outside of the camera or for batteries while they are installed in the camera. It is strictly a power source for the camera. It is useful if you are using the interval timer or doing extensive work in a studio or laboratory setting, to eliminate the trouble of constantly charging and replacing batteries. It also could be useful if you are recording many images or movie scenes in a setting where you have access to AC power. However, the AC adapter's cables and power brick are bulky, so using the adapter can be inconvenient. If you don't have a real need for this setup, I recommend you invest in one or more extra batteries, and perhaps even an extra battery charger, so you can always have a couple of batteries ready for action. In short, the AC adapter should not be considered a high-priority purchase for most photographers.

External Flash

Clearly, Nikon did not consider the use of external flash units to be a high priority for users of the Coolpix B700, because the camera does not have an accessory flash shoe on top, as many other advanced compact cameras do. You might conclude that this camera does not need a very powerful flash, for a couple of reasons. First, it has a sensor that is capable of taking good pictures in low light, with ISO settings reaching up to 3200.

Second, even apart from the B700's dim-light shooting abilities, for everyday shots not taken at long distances, the built-in flash should suffice. It works automatically with the camera's light-metering system to expose the images well.

The built-in flash is limited by its low power, though. According to Nikon, at the wide-angle focal length, the range of the built-in flash is about 22 feet (7 m), and, at the telephoto setting, about 11 feet (3.4 m), when ISO is set to Auto. This range is not very strong. Also, because the flash is built into the camera, you cannot move it to one side to get the softer look of less direct lighting.

If you will often use the camera to take photos of groups of people in large spaces, or otherwise need additional power or versatility from your flash, you may want to supplement the built-in unit with an external flash unit.

Because of the lack of a hot shoe or any other interface for a flash unit on the B700, your options are limited. However, the camera's built-in flash unit can trigger an external flash if the external unit includes, or is connected to, an optical slave trigger. The optical slave is a device that can "see" the flash burst from the B700 and trigger the external flash instantaneously, adding to the flash output for the exposure that is taking place.

This type of system can be straightforward with some cameras; the built-in flash fires, and the external flash also fires. With the B700, though, there is a complication because the built-in flash fires a "pre-flash" burst. It does this so the camera's metering system can evaluate the light from the flash before the flash fires again for the actual exposure. The problem with this system is that the pre-flash "fools" the optical slave into thinking the flash has fired, and so the external flash is triggered too early, and the additional flash burst is wasted; it fires along with the pre-flash, before the exposure is taken.

The solution to this problem is to use an optical slave that is designed to account for, and ignore, the pre-flash. I have used two units that work very well in this way with the Coolpix B700. The first of these is the Yongnuo YN560 IV.

This flash has a built-in optical slave with several different modes. When I set it to its optical slave mode, using either slave setting S1 or S2, the external flash fired at the right instant, in synchronization with the main flash from the B700. Figure A-9 shows the Yongnuo unit attached to the B700 with a standard flash bracket. This flash is somewhat bulky, but it works well with the Coolpix B700 and is a reliable, solid unit.

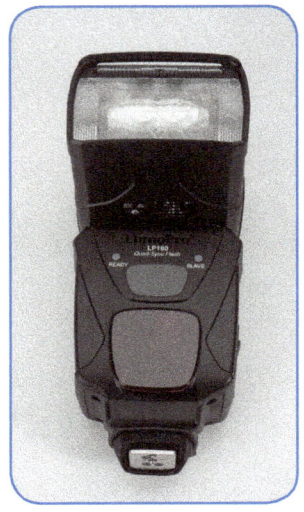

Figure A-10. LumoPro LP180 Flash

With both of these units, I use the B700 in Manual exposure mode, activate Standard Flash or Fill Flash mode, and set the shutter speed and aperture using trial and error to determine the best settings.

When you use an off-camera flash with the B700, you might want to consider using it with a softbox that diffuses the light to avoid the harsh appearance that can result from flash. Figure A-11 shows a Yongnuo flash with a PhotoFlex Lite-Dome XS softbox, which is relatively compact and easy to use.

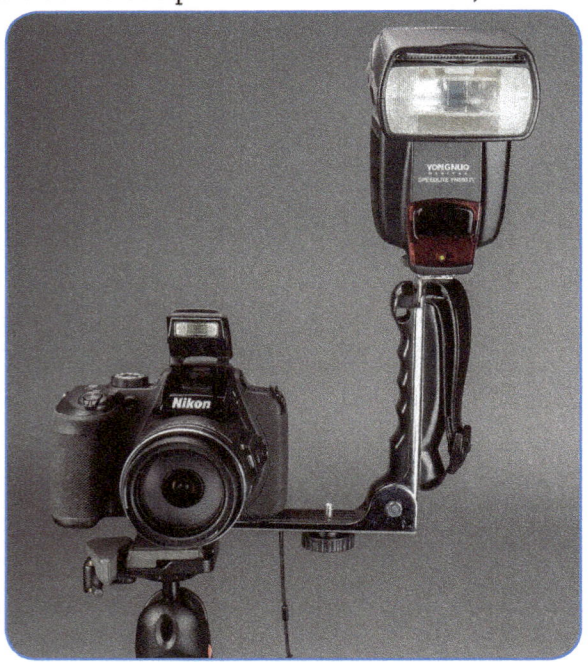

Figure A-9. Yongnuo YN560 IV Flash with B700

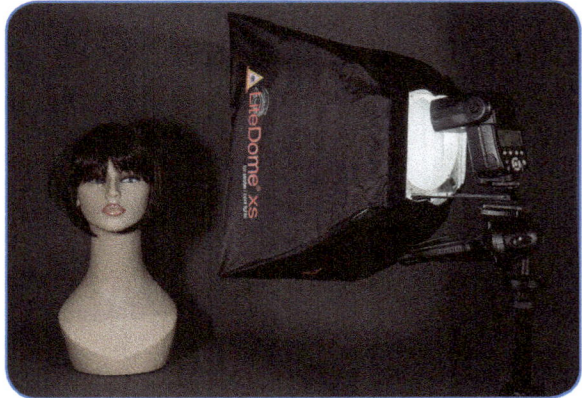

Figure A-11. Softbox with Yongnuo Flash

Another unit that synchronizes with the B700 is the LumoPro LP180, shown in Figure A-10. This powerful unit has several modes, allowing for synchronization with a variety of cameras, including the B700.

There is one other approach to using external lighting that does not involve flash, but is worth considering.

Figure A-12 shows the Coolpix B700 with a portable LED (light emitting diode) fixture attached to it by means of a standard flash bracket. This battery-powered unit has 160 LED lights, and it is continuously dimmable, so you can vary the light from very faint to quite bright. The advantage with this system is that the light stays on, so you don't have to worry about synchronization as you do with flash, and you can see the effects of the light on your subject before you shoot.

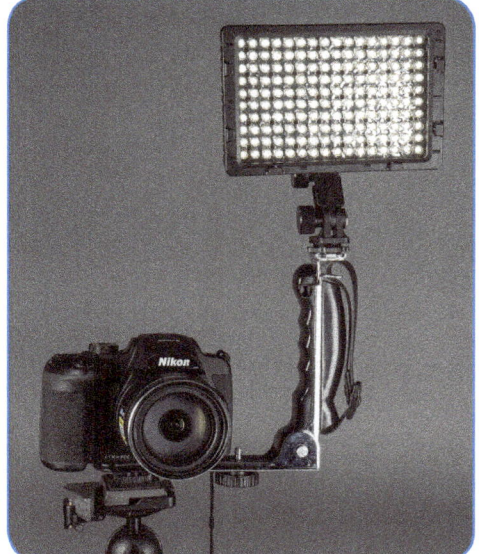

Figure A-12. CN-160 LED Light with B700

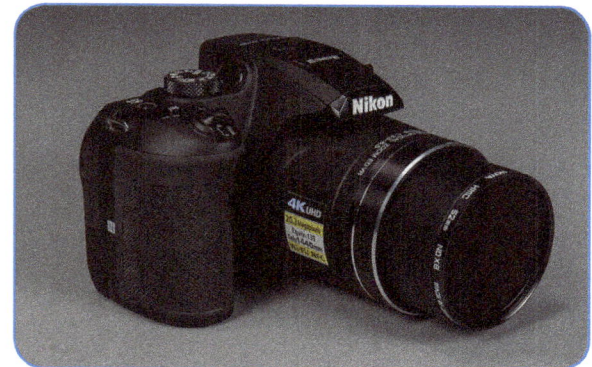

Figure A-13. Neutral Density Filter on Lens of B700

If you want to use heavier attachments, such as wide-angle or telephoto converters, you can use an accessory such as the filter adapter tube shown in Figure A-14.

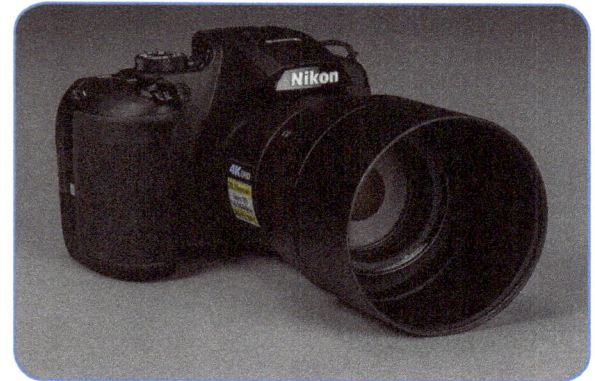

Figure A-14. Kiwifotos Filter Adapter on B700

The light shown here is a model called CN-160, which at this writing is available on Amazon.com for about $30.00. You have to purchase a compatible battery, which adds to the cost; several types of batteries can be used, including some rechargeable models for Sony cameras. There also are larger models available, including one with 216 LEDs.

Filters and Filter Adapters

The Nikon Coolpix B700 is not really designed to use filters or other lens attachments, such as wide-angle or telephoto converters. However, it is possible to use such attachments to some extent. First, the camera's lens has sufficient threads to attach a 52mm filter. Figure A-13 shows a neutral density filter attached to the lens threads. You have to be careful, because these threads are evidently not intended for this purpose, but I have found that a filter will stay attached securely enough for you to take pictures. You can use an infrared filter as discussed in Chapter 9, or you can use a neutral density filter to block some light, to let you use a slow shutter speed to make flowing water look smooth, to give two examples. You also can use various types of polarizing filters, as well as close-up lenses.

This adapter, sold by Kiwifotos, attaches to the outside of the lens using two small screws that provide friction to hold it in place. You can then attach any 72mm accessory, including filters and other accessories. One issue with this item is that, because it is a tube of some length, it causes strong vignetting at wide-angle settings. That is, when the lens is zoomed out, the image will be cut off on the sides, leaving only a circle of the scene in the center. If you zoom the lens in somewhat, the vignetting will disappear, so the adapter is still quite useful at longer focal lengths. If you want to use the adapter at wider focal lengths, you can unscrew the front section and use only the back section of the adapter. In that case, there will be no vignetting, but you cannot zoom the lens to longer focal lengths, because the filter would block the lens from zooming.

I do not recommend using this adapter with the B700 camera, because the camera's own lens provides a wide range of focal lengths, and the image quality will suffer if you use an add-on lens with this adapter. However, it is available if you want to try it. The part number from

Kiwifotos is LA-72P600T. More information is available at kiwifotos.com.

Pole for Extended-reach Shots

Figure A-15 shows the camera attached to a standard pole of the type used by painters or for other purposes, available from hardware stores.

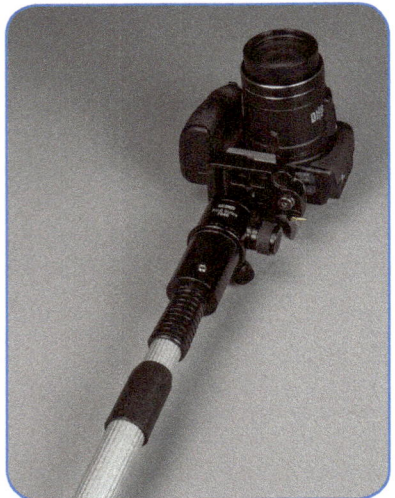

Figure A-15. B700 Attached to Pole

With this setup, you can take shots above crowds of people, or from higher vantage points than would otherwise be possible, for images of real estate properties or anything else that otherwise would be hard to photograph. You can operate the camera using the remote-control feature of the Nikon SnapBridge app, discussed in Chapter 9, or you can use the interval timer function of the Continuous menu option, as discussed in Chapter 4. You also can start a video recording before you raise the camera up above your head. The adapter shown here came from a website called polepixie.com, which has a lot of information about the field called pole aerial photography.

Tripods

If you want to get the best possible results from the Coolpix B700, especially when using its superzoom lens at the longer focal lengths, you should use a tripod whenever possible. Unless I am shooting candid shots when walking around or otherwise am unable to use a tripod, I always attach the camera firmly to a tripod and use the two-second self-timer to trigger my shots.

There are many tripods available at all price ranges. I will not try to discuss the choices in any detail. I will just mention two similar tripods that I have found to be solid and reliable, and sufficiently compact and light to take along on hiking trips without causing a burden. The first is the Manfrotto BeFree carbon fiber version, model number MKBFRC4-BH, shown in Figure A-16.

This tripod, which includes a versatile head, reduces to about 20 inches (50 cm) long just by collapsing the legs, and to about 16 inches (40 cm) if you take the trouble to fold the legs backward. It weighs about 2.5 pounds (1.15 kg).

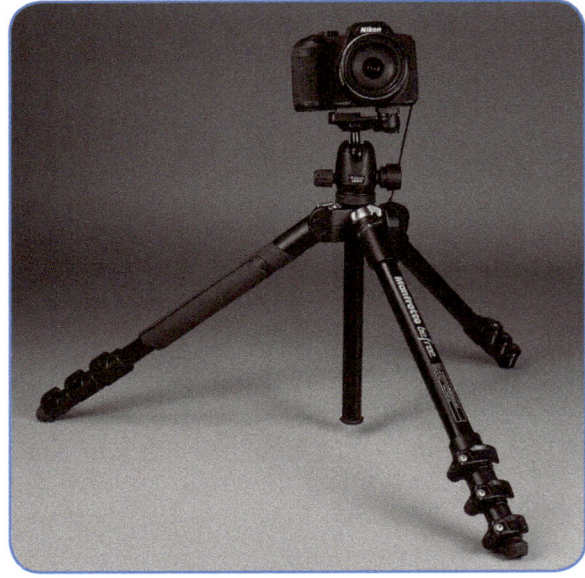

Figure A-16. Manfrotto BeFree Carbon Fiber Tripod

There also is a similar model available in aluminum, which is less expensive and somewhat heavier, model number MKBFRA4-BH, shown in Figure A-17.

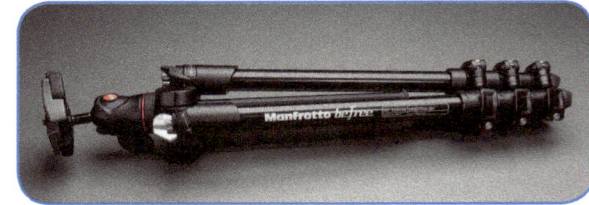

Figure A-17. Manfrotto BeFree Aluminum Tripod

Appendix B: Quick Tips

This section includes tips and facts that might be useful as reminders, especially to those who are new to digital cameras like the Coolpix B700. I have tried to include bits of helpful information that you might not remember from day to day, especially if you don't use the B700 constantly.

Use continuous shooting. The B700 has burst-shooting capabilities that can help you capture images other cameras might not manage. I recommend you consider using continuous shooting as a matter of routine in some situations, unless you are running out of memory storage or battery power, or have a particular reason not to use it. Even with stationary portraits, you may get the perfect expression on your subject's face with the fourth or fifth shot. Go to the Continuous item on the Shooting menu (or press the Function button if it is assigned to this menu item), scroll down the list of options, and turn one of them on. Continuous shooting is not available in the Auto or Creative shooting modes, in most of the Scene modes, or in some other situations, such as when the flash is used.

Take advantage of the User Settings mode. Use this feature to store your most important group of settings. For example, right now I have the U slot set up for my latest settings for street photography: Shooting mode = Program; Image Quality = Fine; Image Size = 5184 x 3888; Picture Control = Monochrome; Metering = Matrix; Continuous = Continuous H; ISO = 800; Sound Settings = Button Sound and Shutter Sound off.

Use the in-camera HDR setting. It's not immediately obvious how to get to the HDR setting on the B700. As a reminder, turn the mode dial to the SCENE position and then press the Menu button. Scroll through the list of scene types to Backlighting, press the OK or Right button, and select On for HDR on the next screen. When you take a picture with this setting, the camera will capture several images and process them to achieve a good balance of light and dark areas in the final image.

Play your movies in iTunes, and on iPods, iPhones, and iPads. Because the B700 records movies in the commonly used .mp4 format, it's easy to play these movies on your computer if you have downloaded Apple's free iTunes software. Open a window on your computer to display the icon for a movie file (Windows Explorer or Macintosh Finder), open iTunes on the same computer, and drag the .mp4 file from the Explorer or Finder window to the Home Videos area of iTunes. You can then play the movie from iTunes. If you want to play it on an iPod, iPhone, iPad, or Apple TV, you will need to take one more step: Select the video in iTunes, then, from the iTunes menu, select File—Convert—Create iPod or iPhone version, or Create iPad or Apple TV version, as appropriate. Then you can sync iTunes with your device, and the movie will play on that device.

Explore the B700's creative potential. The Coolpix B700 has several features for exploring experimental photographic techniques. Some suggestions: Use the Multiple Exposure Lighten setting of Scene mode to take night-time shots with star trails, or trails of lights from automobiles and other vehicles. Use shutter speeds as fast as 1/4000 second to freeze moving motorcycles, track and field runners, skateboarders, and other speedy subjects in mid-motion. Try zooming in or out during a multi-second exposure. Use long exposures (on a tripod) to turn night into day.

Adjust the camera's color settings. The Coolpix B700 has several settings for color-related adjustments: Picture Control, White Balance, the Creative shooting mode, and the Food and Moon settings in Scene mode, which let you adjust a hue slider. Try different values for these settings until you find color and monochrome adjustments that convey what you would like to express with your images. With white balance settings, you can achieve unusual effects by purposely setting a custom white balance while aiming at a colored surface, rather than a white or gray one.

Use a neutral density (ND) filter for some shots. There are times when you need a slow shutter speed, but, in bright light, you can't achieve it, because the aperture can only go as narrow as f/8.2. One solution is to use an ND filter to cut down on the light reaching the sensor, resulting in slower shutter speeds. You might do this to slow down the rush of a waterfall to a smooth, blended look, or to achieve a motion blur in a shot of a passing runner or walker. The B700 accepts filters with a 52mm diameter.

Diffuse your flash or reduce its intensity. If you find the built-in flash produces light that's too harsh for close subjects or other shots, try using translucent plastic pieces from milk jugs, other food containers, or broken ping-pong balls as homemade flash diffusers. Just hold the plastic up between the flash and the subject. Another approach when using flash outdoors is to use flash exposure compensation to reduce the intensity of the flash by about -2/3 EV.

Use Creative mode for recording movies. Don't overlook the fact that you can shoot movies in this mode, with its image-altering settings such as Binary, Bleached, Dream, Sepia, Denim, and others. See Chapters 3 and 8 for details.

Use the self-timer to avoid camera shake. The Coolpix B700 has a self-timer that is easy to use; just press the Left button and choose your setting. This feature is not only for group portraits; you can use it whenever you'll be using a slow shutter speed and you need to avoid camera shake. It can be useful when you're doing macro photography or using the superzoom lens, also, because those are both sensitive to camera motion.

Set zone focusing. If you're doing street photography or are in any other situation when you want to set the camera on manual focus for a general distance, here is a quick way to do so. Set the focus mode to autofocus, then aim the camera at a subject at approximately the distance you want to be able to focus on quickly. Once focus is set, press the Down button, select MF from the focus mode menu to select manual focus, and press the OK button to lock the focus. Now you have locked in the manual focus at your chosen distance, and you're ready to shoot any subject at that distance without the need to re-focus.

Leave good settings in place when you end a shooting session. Several settings on the Coolpix B700 are "sticky"—they will remain set on their current values when the camera is powered off and then on again. Some examples are exposure compensation, ISO, and continuous shooting. It is a good idea to check the camera when you stop shooting to make sure you have not left any settings in place that could cause problems if you have to start shooting again in a hurry. For another example, you might want to leave the Creative mode set to the Somber option, which results in fairly normal shots. Then, if you pick up the camera in a hurry and turn the mode dial to the Creative position, you will get usable images or video footage, rather than scenes shot with Binary, Sepia, or some other unwanted look.

Be aware of the Movie Options setting when recording movies. When you press the Movie button to record a movie, the camera will use whatever setting is currently selected for Movie Options on the Movie menu. If you have this menu item set to one of the HS (High Speed) settings, the resulting movie will be either in slow motion or speeded up, and will have no sound. If you want to be ready to record a good-quality, normal movie, leave Movie Options set to the second setting on the menu screen, which is 1080/30p (or 1080/25p if you have set Frame Rate to 25 fps).

Shoot larger panoramas. When you are shooting a horizontal panorama using the Easy Panorama setting as discussed in Chapter 3, you can increase its height using a simple technique. Select your setting, such as Normal, and then hold the camera vertically, as if you were shooting a tall building, but pan it horizontally. With this approach, the dimensions of the panorama will be 1536 pixels tall by 4800 pixels wide, instead of the normal dimensions of 920 by 4800.

Use the Filter Effects setting on the Playback menu. With this setting, you can add interesting and attractive processing to your recorded images, including Cross Screen, Painting, Fisheye, Miniature Effect, Photo Illustration, and several others. One nice bonus is that you can add these effects even to images that were taken with the Creative mode. In that way, you can use two effects with the same image, such as Sepia with Photo Illustration, or Binary with Fisheye.

Force the camera to use autofocus when using manual focus. When manual focus is selected from the focus mode menu, press the Right button to make the autofocus mechanism operate.

Appendix C: Resources for Further Information

Photography Books

A visit to any large bookstore or an online search will reveal the vast assortment of books about digital photography that is currently available. Rather than trying to compile a long bibliography, I will list a few books that I consulted while writing this guide, which are useful resources for further exploration.

C. George, *Mastering Digital Flash Photography* (Lark Books, 2008)

J. Gulbins & R. Gulbins, *Photographic Multishot Techniques* (Rocky Nook, 2009)

C. Harnischmacher, *Closeup Shooting* (Rocky Nook, 2007)

C. Harnischmacher, *The Wild Side of Photography* (Rocky Nook, 2010)

H. Horenstein, *Digital Photography: A Basic Manual* (Little, Brown 2011)

J. Paduano, *The Art of Infrared Photography* (4th ed., Amherst Media, 1998)

D. Sandidge, *Digital Infrared Photography* Photo Workshop (Wiley, 2009)

Digital Photography Review

http://www.dpreview.com/forums/1007

This is the current web address for the "Nikon Coolpix Talk" forum within the dpreview.com site. Dpreview is one of the most established and authoritative sites for reviews, discussion forums, technical information, and other resources concerning digital cameras.

Reviews of the Coolpix B700

The links below lead to reviews of the Coolpix B700 by several sites.

http://www.cameralabs.com/reviews/Nikon_COOLPIX_B700/

http://cameradecision.com/review/Nikon-Coolpix-B700

http://www.imaging-resource.com/PRODS/nikon-b700/nikon-b700A.HTM

http://www.photographyblog.com/previews/nikon_coolpix_b700_photos/

http://www.steves-digicams.com/camera-reviews/nikon/coolpix-b700/nikon-coolpix-b700-review.html

The Official Nikon Site

The United States arm of the Nikon company provides resources on its web site, including the downloadable version of the user's manual for the Coolpix B700 and other technical information.

http://www.nikonusa.com/en/Nikon-Products/Product/Compact-Digital-Cameras/COOLPIX-B700.html

http://downloadcenter.nikonimglib.com/en/products/333/COOLPIX_B700.html

Photography Information

The site below provides some helpful information about infrared photography with digital cameras.

http://www.wrotniak.net/photo/infrared/

This next site has excellent information and tutorials about many aspects of photography.

http://www.cambridgeincolour.com/

The site listed below has some brief, but useful tips on improving your photographs of birds.

http://www.birds.cornell.edu/AllAboutBirds/bp/tim1

Index

Symbols

4K (UHD) motion picture recording 132
  memory card speed requirement 132

A

AC adapter
  Nikon model EH-67A 158
    using for time-lapse photography 68
Accessory shoe
  lack of for Coolpix B700 1
Active D-Lighting menu option 75–76
  incompatibility with other settings 76
Adobe Camera Raw software 54, 60
Adobe Photoshop Elements software 75
Adobe Photoshop software 39, 75
Adobe Premiere Elements software 68, 137
Adobe Premiere Pro software 137
AEL (autoexposure lock)
  using with motion picture recording 127
  using with Time-lapse Movie setting 45
AF Area Mode menu option 13–14, 71–73
  Face Priority option 71
  Manual (Spot, Normal, or Wide) 14, 37, 72
  Subject Tracking option 72–73
    incompatibility with other settings 73
  Target Finding option 73
AF Assist menu option 116
AF Assist/Red-eye Reduction/Self-timer Lamp 92
  disabling 92, 116
Airplane Mode menu option 144
Aperture
  procedure for setting 27
  relationship to depth of field 25–26
Aperture Priority mode 25–27
Aperture values
  available range of 1, 25, 27, 29
Apple Final Cut software 137
Apple iTunes, iPads, iPhones, etc.
  playing movies from B700 on 163
Aspect ratio 55–56
  comparison images 56
  limitation on selecting with Raw+Fine or Raw+Normal 56
Assign Side Zoom Control menu option 118
Auto Flash setting 151
Autofocus frame, movable 14

Autofocus Mode menu option (movies) 134–135
  using Left button to force autofocus 92, 135
Autofocus Mode menu option (still images) 73
Auto ISO setting 68
  not available in Manual exposure mode 29, 70
Auto mode 22
  incompatibility with other settings 22
Auto Off menu option 119
  pressing button to cancel 81, 85, 86, 119
Auto with Red-eye Reduction flash setting 151

B

Backlighting/HDR setting 38–40
Battery charger
  generic model for charging outside of camera 158
  generic USB charger 3, 158
  Nikon model number EH-73P 158
Battery, Nikon EN-EL23 3, 158
  charging 3–4
    using camera at same time 3, 158
    using USB cable connected to computer 4, 120
  inserting into camera 4
  using generic replacement 158
Beach setting 36
Bird-watching setting 42–43
Blinking aperture or shutter speed value
  meaning of 25
Bluetooth connection to camera 139
Bluetooth menu option 145
Blurred background
  how to achieve 26–27
Bokeh 27

C

Calendar screen in playback mode 96
Cases
  Crumpler 6 Million Dollar Home bag 157–158
  Lowepro Inverse 100 AW beltpack 157
  Lowepro Rezo 110 AW 157
Charge by Computer menu option 120
Choose Key Picture menu option 107–108
Close-up filters 161
Close-up setting 37, 150
CN-160 LED lighting unit 161
Color temperature
  in general 60
Command dial 86–87
  switching functions of with multi selector dial 87–86, 122
  using to control manual focus 87
  using to enlarge and shrink images in playback mode 87
Comment and Copyright Information Screen 97
  not displayed for motion pictures 97
Comments
  adding to images 121
Connecting to and using SnapBridge app 139–143
Connect to Smart Device menu option 139–140, 144

Continuous menu option 64–68
  Continuous H: 60 fps 67
  Continuous H: 120 fps 67
  high-speed shooting 66
  Interval timer setting 67–68
  low-speed shooting 66
  Pre-shooting Cache setting 66
  turning off continuous shooting 66
  unavailability of some settings 65, 70
  using self-timer with 65
Continuous shooting
  advantages of using 65, 163
  in general 65
  playback of continuous shots 20, 65–66, 98–99
    changing key image for 107–108
Contrast
  adjusting 58
Copyright Information menu option 121
Creative Mode 47–49
  list of groups and settings 48
  making adjustments to parameters 47–48
  using to record motion pictures 164
Creative Mode comparison chart 49
Cropping images in camera 96
Cross Screen setting for Filter Effects 103
Custom Picture Control menu option 59–60
  using with Raw format 60
Custom white balance
  how to set 61–62

D

Date and time
  imprinting on images 114
  setting 9–10, 111
Date Stamp menu option 114
  incompatibility with other settings 115
Delete/Trash button 87–88
Deleting images 87–88
  in shooting mode 88
Digital Zoom menu option 116–118
  incompatibility with other settings 118
  limitation when using 2160/30p Movie Options setting 129
  using with motion picture recording 129
Diopter adjustment wheel 7, 85
Direction buttons
  in general 89
Display button 85
Display screens
  playback mode 86, 97–98
  shooting mode 85–86
D-Lighting menu option 101
Dusk/Dawn setting 36
Dynamic Fine Zoom feature 116

E

Easy Panorama setting 40
  scrolling panorama with OK button 41
  shooting taller panoramas 164
Electronic Vibration menu option (movies) 135–136
  incompatibility with other settings 136
Enlarging images in playback mode 96
EVF Auto Toggle menu option 85, 114
EVF Options menu option 112
Exposure bracketing 70–71
  incompatibility with other settings 71
  using to create HDR images 70
Exposure compensation 90
  availability with Auto mode 22, 90
  display of histogram with 16
  incompatibility with other settings 90
  procedure for using 16–17
  using with motion picture recording 127
External flash units
  LumoPro LP180 160
  using with Coolpix B700 159–160
  Yongnuo YN560 IV 160
Eye sensor 85

F

Faces
  setting autofocus to detect 71
File numbering system
  resetting 123
Fill Flash/Standard Flash setting 152
Filter adapter
  Kiwifotos tube model 161
Filter effects adjustment for Monochrome Picture Control setting 59
Filter Effects menu option 102–104
  using to add second effect to image 164
Filters
  attaching to lens of Coolpix B700 161
Fireworks Show setting 38
Firmware Version menu option 124
Fisheye setting for Filter Effects 103
Flash
  diffusing 164
  general procedure for using 17–18, 151
  incompatibility with other settings 17, 71, 151
Flash Exposure Compensation menu option 74, 153
  using to reduce harshness of flash 74
Flashing parts of image in playback mode 98
Flash mode menu 18, 89, 151
Flash pop-up button 17, 83–84, 151
Flexible Program feature 23
  setting which dial controls 122
Focus
  controlling in Auto mode 22
  locking 14, 81
Focus mode
  selecting 13, 90
  with motion picture recording 126–127

Index

Focus mode menu 90
Focus range of lens 150
Food setting 37–38, 150
Format Card menu option 119–120
   incompatibility with other settings 120
Frame Rate menu option 131
Full-time autofocus option 73
Function buttons 82–83, 85
   cannot be used during motion picture recording 130
   changing options assigned to 82–83
   incompatibility with other settings 83
   settings that can be assigned to 82

G

GPS capability
   lack of for Coolpix B700 1
Grid
   displaying in shooting mode 86, 113

H

HDMI cable 93, 156
HDMI port 93
HDR (high dynamic range) photography 39
HDR setting on Coolpix B700 163
   how to use 38–39, 39
Histogram
   playback mode 86, 97–98, 114
      not displayed for motion pictures 97
   shooting mode 86, 113–114
      displaying with exposure compensation button 114
      incompatibility with other settings 113
      relationship to Manual Exposure Preview option 80, 114
HS (high speed) movie options 133–134

I

Icon indicating assignment of command dial 87
Icon indicating assignment of multi selector dial 88
Icon indicating flash ready to fire 18
Icon indicating memory card access in progress 6
Illustrations of controls
   back of camera 7
   bottom of camera 8
   front of camera 8
   left side of camera 7, 83
   right side of camera 8, 84, 93
   top of camera 6, 81
Image Comment menu option 121
Image Quality menu option 53–54
Image Review menu option 20, 95, 112
Image sharpening
   adjusting 58
Image Size menu option 55–56
   relationship to zoom range 117
Image stabilization 115–116
iMovie software 21, 68, 137
Index screens in playback mode 95–96

Infinity focus mode 13, 90
Infrared filter
   Hoya R72 154
Infrared photography 153–154
Interval timer setting 67–68
   location of images 67
ISO setting 68–70
   Auto setting 68
   Fixed Range Auto setting 68
   incompatibility with other settings 70
   Minimum Shutter Speed setting 70
   negative effects of very high setting 68
   procedure for making setting 68
   using to enable use of fast shutter speed 148
   using with Manual exposure mode 70
   value to choose 68

J

JPEG files 53

K

Kelvins
   as used to measure color temperature 60
Key icon for protected images 106

L

Landscape menu option 30–31
Landscape mode 30
Language
   setting 10, 120
LCD screen
   adjusting brightness and color of 112
   characteristics of 1
   switching view by folding in against camera 11, 85
   switching view to and from 85, 114
   tilting and swiveling features of 93–94
Lens cap
   attaching to camera 3
Location Data menu option 122
Locking exposure 81
   when recording motion pictures 90
Locking focus 81

M

Macro autofocus setting 149–150
Macro photography 149–150
Manual exposure mode 27–28
   incompatibility with other settings 29
   procedure for making settings 27–28
   using to create HDR images 27, 40
Manual Exposure Preview menu option 79–80, 114
Manual focus
   adjusting magnification with 15, 92
   general use of 15–16
   reasons for using 15

using autofocus with 16, 90, 164
using for macro photography 150
using Left button to toggle enlargement factor 15
using Peaking with 16
using with motion picture recording 127
Mark for Upload menu option 100
Memory card
    formatting 119–120
    inserting into camera 6
    number of images that can be stored 5
    types and capacities of 5
Menu button 87
    using to crop image in playback mode 96
Menu system
    in general 51
    items that cannot be selected 52–53
    making selections in 52
    names of menus 52, 87
    navigating in 52
Metering menu option 63–64
    incompatibility with Active D-Lighting setting 64
    using with motion picture recording 128
micro-SD card
    using in Coolpix B700 5
Miniature setting for Filter Effects 103
Minimum Shutter Speed menu option 70
Mode dial 82
Monitor button 85
    using to switch between LCD and viewfinder 85
Monitor Options menu option 112
Monitor Settings menu option 86, 112–114
Monochrome Picture Control setting 58
    adjustments to 59
Moon setting 42
Motion picture recording
    effects of shooting mode on 126
    exposure control 127
        autoexposure lock 127
        exposure compensation 127
    focus options 126–127
    general procedure for 18–20
    limitations on length of recording 131
    overview 125
    pausing with OK button 130
    quick guide 125–126
    slow-motion options 132, 133
    speeded-up motion option 134
    still photo settings available 126–129
    still photo settings not available 130
    taking still images during recording 129
Motion pictures
    editing in camera 137–138
        not available when battery charge is low 137
    editing with computer 20, 137
    playback 20–21, 136–137
    saving single frame from 129, 138
Movie button 86

Movie Maker software for Windows 137
Movie menu 130–136
Movie Options menu option 132–134, 164
Multiple Exposure Lighten setting 44–45
Multiple Exposure menu option 76–77
Multi selector dial 88
    switching functions with command dial 88, 122

N

Neck strap
    attaching to camera 3
Network menu
    in general 143–144
Neutral density filter 164
Neutral Picture Control setting 57
NFC active area on camera 93
NFC (near field communication) protocol 93
Night Landscape menu option 32–33
Night Landscape mode 32
Night Portrait mode 31
Nightscape+Light Trails option for Multiple Exposure Lighten setting 44
Nikon Capture NX-D software 60
    using to process Raw files 54
    where to download 3
Nikon Coolpix B700 camera
    advantages of 1
    drawbacks of 1
    items that come in box 3
    new features of 1
Nikon Movie Editor software 21
Nikon Reference Manual for Coolpix B700
    where to download 3
Nikon ViewNX-i software
    where to download 3
No card present error message 4, 111
Noise Reduction Filter menu option 75
Noise Redution Burst setting 31, 37
NTSC video system 131

O

Off-center subject
    setting focus for 14
OK button 88–89

P

Painting setting for Filter Effects 104
PAL video system 131
Panoramas
    shooting 40–41
Party/Indoor setting 35
Peaking feature for manual focus 16, 123
    not available with motion picture recording 127
Pet Portrait Auto Release feature 41, 92
Pet Portrait setting 41
PhotoAcute software 39

Index

PhotoFlex Lite-Dome XS softbox 160
Photo Illustration setting for Filter Effects 104
PhotoMatix Pro software 39, 40
PictBridge printer 108
Picture Control menu option 56–60
   adjustments to settings 58–59
      A setting for Automatic adjustment 58
      incompatibility with other settings 58
   using with motion picture recording 128
Playback button 20, 85
   using to turn on camera 85
Playback menu 100–108
Playback of images and videos
   continuous burst of shots 20, 98–99
      enlarging image 96
      general procedures 20–21, 95
      motion pictures 20–21, 136–137
Pole attachment for pole aerial photography 162
Portrait (Color+B&W) setting for Filter Effects 104–105
Portrait setting 34–35
Power switch 81
Printing images 108–109
Print menu 109
Program mode 23
Program Shift. See Flexible Program feature
Protect menu option 105–106

Q

Quick adjust option for Picture Control 58
Quick Retouch menu option 100–101

R

Raw+Fine setting 54
Raw format 53–54
   ability to change settings in post-processing 53–54
   availability with Auto mode 22
   disadvantages of 54
   incompatibility with other settings
      --Alex White
      White Knight Press 54
   using with Picture Control settings 60
Raw+Normal setting 54
Rear-curtain Sync flash setting 152–153
Reset All menu option 53, 123
Reset File Numbering menu option 123
Reset User Settings menu option 78
Resolution of images 55
Restore Default Settings option on Network menu 145
Rotate Image menu option 106

S

Saturation
   adjusting 58
Save User Settings menu option 77
Scene Auto Selector 34
Scene modes
   incompatibility with other settings 30
   in general 30, 33
SCENE setting on mode dial 33
   list of available scene types 30
Sekonic C-700 color meter 62
Selective Color setting 44
Selective Color setting for Filter Effects 102
Self-timer 91
   using to avoid camera shake 91, 164
   using with continuous shooting options 91
   using with motion picture recording 91, 128
Send While Shooting menu option 143, 144
Sensor
   characteristics of 1
Sequence Display Options menu option 100–99, 107
Setup menu
   in general 110
Sharpness. See Image sharpening
Shooting menu
   in general 51
Shooting modes
   names of 12, 22
Shutter Priority mode 23–24
Shutter release button 81–82
   operating with no memory card in camera 111
Shutter speed
   procedure for setting 24–25, 28
   range of available settings 24, 29, 65
      limitations when using Continuous menu option 65
Shutter Speed Equivalents 25
Side zoom control 84, 118
   assigning to manual focus 84, 118
   using for telephoto shots 84, 118, 149
Single autofocus option 73
Skin Softening menu option 101–102
Slide Show menu option 105
Slot Empty Release Lock menu item 5, 111
Slow Sync flash setting 152
Small Picture menu option 106–107
Smile Timer 91–92
   incompatibility with other settings 91
Snap-back zoom button 84
   using for telephoto shots 84, 149
SnapBridge app 139
   connecting to camera and using features of 139–143
   setting for automatic download of images 141
   summary of options for transferring images and videos 143
   using to control remotely 142–143
   using to download images and videos from camera 142–143
Snow setting 36
Soft Portrait setting for Filter Effects 102
Soft setting 43
Sound Settings menu option 118–119
Sports setting 35
Standard Picture Control setting 57
Star Trails option for Multiple Exposure Lighten setting 45

Startup Zoom Position menu option 79
Step-by-step guides
   basic picture-taking 11–12
   connecting to and using SnapBridge app 139–143
   motion picture recording 18–20
   quick guide to motion picture recording 125–126
Still images
   taking during motion picture recording 129
Street photography 154–156
Subject Tracking autofocus option 72–73
Sunset setting 36
Superlapse Movie setting 46–47
Sync with Smart Device menu option 111

T

Tables
   fractional shutter speed equivalents 25
   limits on shutter speed settings 24
Target Finding autofocus option 73
Television set
   connecting camera to 156
Time-lapse Movie setting 45–46, 68
   compared to interval timer setting 68
Time-lapse photography 45–46, 67–68
Time zone
   changing for travel 111
   setting 111
Time Zone and Date menu option 9, 110–111
Toggle Av/Tv Selection menu option 24, 122
Tone levels in image, checking 98
Toning adjustment for Monochrome Picture Control setting 59
Tripods
   Manfrotto BeFree 162

U

USB port 93
   using to connect camera to printer 108
User Settings mode 49–50, 163
   using to save specific settings 149

V

Vibration Reduction menu option 115
   using with motion picture recording 128
Viewfinder
   adjusting brightness and color of 112
   adjusting for user's vision 7, 85
   switching view by folding LCD screen in against camera 7, 85, 94
   switching view to and from 114
View/Hide Framing Grid menu option 113
View/Hide Histograms menu option 80, 113–114
Vignette setting for Filter Effects 104
Vivid Picture Control setting 57

W

White balance comparison chart 62–63
White Balance menu option 60–63
   Choose Color Temperature setting 62
   making adjustments to preset settings 61
   Preset Manual setting
      using to alter coloring of images or videos 62, 128
      using to set custom white balance 61–62
   using with motion picture recording 128
Wi-Fi menu option 144–145
Wind Noise Reduction menu option 136

Z

Zone focusing 164
Zoom lens
   flattening of subjects into single plane 146
   haze effect from compression of atmosphere 146
   isolating subject through telephoto shot 147
   rule of thumb for choosing shutter speed for telephoto shots 148
   shallow depth of field at telephoto settings 147
   using telephoto power of 145–149
   ways to avoid image blur at telephoto settings 148–149
Zoom lever 82
Zoom Memory menu option 78–79
   not applicable to side zoom control 79
Zoom Microphone menu option 136
Zoom range 116–118, 145
   relationship to Image Size setting 117
Zoom scale
   meaning of white, blue, and yellow bars 117–118

www.ingramcontent.com/pod-product-compliance
Lightning Source LLC
Chambersburg PA
CBHW040540220526
45473CB00016B/2983